Farewell, **Darkness**

FAREWELL
DARKNESS

· · · · · · · · · · · · · · · · · · ·

*A Veteran's Triumph over
Combat Trauma*

RON ZACZEK

NAVAL INSTITUTE PRESS
Annapolis, Maryland

Library of Congress Cataloging-in-Publication Data

Zaczek, Ron, 1947–
    Farewell, darkness: a veteran's triumph over combat trauma/Ron Zaczek.
      p. cm.
    ISBN 1-55750-989-1
    1. Zaczek, Ron, 1947– —Mental health. 2. Post-traumatic stress disorder—Case studies. 3. Vietnamese Conflict, 1961–1975—Personal narratives, American. 4. Vietnamese Conflict, 1961–1975—Psychological aspects. I. Title.
RC552.P67Z33 1994
616.85'21—dc20   94-9188

Printed in the United States of America on acid-free paper ∞
9 8 7 6 5 4 3 2
First printing

In Memory of
Ronald Joseph Phelps
Corporal, United States Marine Corps
Crewchief, Marine Observation Squadron 3
killed in action
over the Co Bi Thanh Tan
Hue–Phu Bai
Thua Thien Province, Republic of Vietnam
16 November 1967

Contents

· · · · · · · · · · · · · ·

Acknowledgments

● ●

Dear Mouse, Saint, Tommy, Mike, Zeke, Moose, Frankie, Smitty, Scotty, CC, Give-A-Fuck, Dutch, Shifty, Jack, Scummer, Rock, Captain Bill, D', and all the flight crews and men of Marine Observation Squadron Three and Marine Observation Squadron One:

Well, guys, I finally wrote that book a lot of you said I should write. It's only twenty-six years late. That's more years than we were alive when we did all this stuff. The book didn't turn out quite the way I thought it would. Remember, Mike, when we used to talk about it in the coffee mess in New River? We thought it would just be a lot of war stories about what we did in Vietnam. Combat action, drinking action, shacking-up action—that kind of thing. Trouble is, I developed Post-Traumatic Stress Disorder along the way, and things didn't turn out quite so simple.

I was in therapy for the first half of the '80s—six years, in fact—and I began writing this pretty early in the process. At first, I intended it for my kids. I thought it would help them understand their old man and why he was sometimes such a pain in the ass. After a while, it occurred to me that other veterans with PTSD could benefit from my experience, that it might shorten their healing process. It was tough to put on paper, and tougher still to organize, since I was living what I was writing about. It wasn't until 1988 (after I finished therapy) that I figured out how to say everything that needed to be said.

The story is woven in two threads. One thread recounts what happened to me—to us—In Country, the stuff that I discovered (during therapy) had contributed to PTSD. Almost everything contributed in some way, but I didn't want this to be a rambling collection of war stories, so I had to be choosy. I decided to write about events that I discovered lay at the root of the symptoms of PTSD, which are described in a useful pamphlet put out by the Disabled American Veterans called *Continuing Readjustment Problems Among Vietnam Veterans*. Not everyone shows all the symptoms—rage, alienation, survival guilt, feelings of isolation, anxiety, flashbacks, night-mares—but I sure did. You remember me, I never did anything by the half-step.

The second thread to the story is the therapeutic process I followed in the Vet Center Outreach program. The sessions at the Vet Center are writ-ten in the exact sequence the therapy proceeded in so that you can see how I worked through everything. I may not have taken the most direct road to recovery, but what the hell—I'm still here. I hope that my experience will lead vets with PTSD and their families to healing, and maybe give every-one a little more insight into what happened to us in Vietnam.

I had to take some liberties in compressing the therapy sessions. I started out with two sessions per week for about a year, then one session per week for several years, then a couple a month, and so on. That was a lot of ses-sions. (Things move a lot faster now, with group therapy as well as indi-vidual sessions. I entered the program before today's clinical processes were fully developed.) Anyway, what appears as a single session's dialogue was frequently a rambling discourse that took place over many sessions. I revisited the subject under discussion many times before a breakthrough in understanding permitted me to move on. I say this in the interest of accu-racy and because I don't want you to get the idea that this is an easy recipe to recovery, six sessions getting your head shrunk and you're out. That would be nice, but those are the breaks of Naval Air—otherwise known as naval air brakes. (Long time since you heard that old saw, huh?)

Some of you will undoubtedly recognize yourselves in this book. My primary "source" was my memory of the thirteen months we spent together In Country, from December 1966 to January 1968, and the letters I wrote to my friend Tom Brooks and to my girl (now wife), Grace. I was in or witnessed all the events I've written about, and have recounted them as accurately as I can. I don't think I've made myself or any of you any better or worse than we actually were. If I misremembered something, I have only an aging brain as an excuse. I visited the Marine Archives in Washington several times, pored through VMO-3's after-action reports and command chronologies, and listened to a number of oral histories recorded by Marine historians while we were In Country. Jack, remember when we pulled that team off the hill? They interviewed you, Dutch, and Captain A. the day after the mission; they interviewed the grunt radio operator the day of the mission, only a couple of hours after we extracted him. Can you believe it? He was all shot up but giving interviews in the hospital at Khe Sanh! The tapes enabled me to accurately reconstruct what happened to us up there. It was very important for me to do that, as that day lay at the bottom of a hell of a lot of nightmares. A lot of the material in the Archives is skimpy, but everything helped me to piece together some of the missions I had difficulty remembering.

I wrestled quite a long time about whether I should identify those of you I crewed with by name. I've learned a hell of a lot about PTSD, and know that nearly half of us who saw heavy combat in Vietnam have some form of it. Before I got it under control, the last thing I wanted my neighbors or employer to know was that I was a "head case," which is how I saw myself back then. I don't mind who knows now because I've learned that neither my role in Vietnam nor my struggle with PTSD was anything to be ashamed of. I regained a great deal of pride, not only for surviving Vietnam, but for surviving homecoming. Still, I know some of you are sensitive about the possibility of having PTSD. I want to respect your privacy, and so I've used fictitious names or the nicknames we used In Country.

I did the best I could in reconstructing actual flight crews, but I couldn't determine exactly who was in the aircraft with me on some missions. As aircrew we shared a common experience, so in these cases I "peopled" the aircraft with the guys I most frequently crewed with. I hope you'll forgive me if you read some section and say, "I wasn't on that mission with you." I tried to represent the squadron as accurately as I remember and did not knowingly create "composite characters."

This book is dedicated to Ron Phelps, who died in November just before we began to come home—the only crewchief of our tour to die In Country. I really miss him; it's important to me to preserve his memory. I hope that anyone who knew him who reads this will understand the honor that I want to do in remembering him.

Dr. Tom Murtaugh was my counselor at the Vet Center and gave me permission to use his name in the book, but this work is entirely mine and represents my understanding of the therapeutic process. I am not, nor ever have been, nor ever will be, a shrink. He knew, of course, that I was writing this while he was counseling me, but always said that this was my story and avoided offering opinions about what I should write or how I should say something. Still, I double-checked some of the shrink stuff—theories on PTSD and the workings of the mind—with him. If there are any errors, they're mine. I literally owe my life to Tom. I came pretty close to checking out before I entered the Vet Center the first time.

Anyone who wants to get in touch with me can write me care of the publisher, the Naval Institute Press. Perhaps we can arrange a squadron reunion; I'd like you all to meet Tom and my wife, Grace.

You remember Grace; I had her picture in the hooch the whole tour. Saint and Mouse, remember how you used to tell me how lucky I was that she stuck by me when everyone else was getting Dear Johnned? She wrote to me nearly every day for the four years we were engaged. We were married in 1970, and she put me through college, so I guess you were right. Grace sort of "pushed" me to the Vet Center when things were getting out of control, and stuck by me all the way. I wouldn't be here without her, either. Wouldn't want to be. I guess I am one lucky son of a bitch. We have two great kids: Chris is eighteen, and Matt is fifteen. Life is really fine. The kids put up with a lot while I was writing this book, and always encouraged me. I hope that by reading this they'll understand how much they mean to me.

Well, I'm going to close now, except for a whopping P.S. thanking a whole bunch of people. I hope you guys feel this book does justice to our experience, and if you were as unlucky as me to get stuck with PTSD, I hope this helps you toward healing. I think about all of you a lot, and wish you well. Of all the things in my life, I am most proud of having served with you, my old friends, of having been a crewchief in VMO-3, and of being a Marine.

Your brother crewchief and gunner,

Zack

P.S. There are many people to thank:

Thanks to the staff of the Office of the Commandant of the Marine Corps and to the staff of the Marine Historical Center. To help me rebuild and work through the past, Tom Murtaugh contacted the Commandant's office, explained that he was counseling a Marine who needed help, and asked, "Did the Marines have detailed records on small-unit actions; could they help?" The answer was crisp "He's a Marine—we take care of our own." The next day, Jack Shulimson, historian at the Marine Historical Center in Washington, called to offer immediate assistance in reconstructing my past. Thanks to Joyce Bonnett of the Historical Center for hosting me, and making available the Center's services to research VMO-3's archives for this book Thanks to the civilian and Marine staff who proved that "once a Marine, always a Marine" is no catchphrase, but a commitment and a matter of pride.

Thanks, Wayne Mutza (240th AHC, U.S. Army), author of *UH-1 HUEY In Action*, *UH-1 HUEY In Color*, and *Bent & Battered Rotors*, for providing details on aircraft configuration, for encouraging me over many years, and for reviewing the final manuscript with the critical eye of a brother crewchief.

Thanks, Bonni Goldberg, poet and teacher, for showing me the building blocks of writing.

Thanks, Harcourt Road Writers, whose constructive criticism, moral support, humor, and encouragement kept me on track and motivated to finish this work. Thanks to Dee (Donut Dolly) Hodges, Kit Norland, Glen (Dr. Richard Gage) Rayson, Sarah Ann (Waiting for Godot) Smith, and Sally Weida. Thanks to member-emeritus Gordon (Saint in Surgical Garb) Livingston. Special thanks to our mentor, Arnold (Skip) Isaacs: journalist, veteran war correspondent, and author of *Without Honor*, a reporter's view of the American and Vietnamese experience.

Thanks, friends Frank Bonini and Patricia Yeatman, who reviewed the manuscript with a layperson's eyes. Thanks to Vet Center Counselor Ernie Baringer, who read sections of the manuscript with a Marine grunt's eyes. Each point of view was essential in making this book comprehensible to a broad audience of veterans and nonveterans.

Thanks, counselors and staff of the Vet Center in Elkton, Maryland: Lon Campbell, Rose Greenleaf, Diane Pizzirusso Hoover, and Sherry Tyler. You made the Center a haven when I needed refuge. You accepted me when I couldn't accept myself. You treated me as a person, not a veteran, until one day I found I was both person and veteran, with pride.

Thanks, Tom Brooks, oldest of friends, for being there when the world turned its back on returning vets, and for the letters you saved.

Thanks, Dick Moskun (VMO-3) and Mike Bisaccia (VMO-6), Huey crewchiefs and old friends, whose encouragement and advice many years ago convinced me that making this a personal story would be the most effective way to tell it, and to help others.

Thanks, Paul Breault (VMO-2), Huey gunner and old friend. You were always a pain in the ass, but that's what I needed.

Thanks, Barbara Rusczkowski, for reviewing the first chapters and for encouraging me during the earliest stage of writing.

Thanks, editor and advisor Esther Giller, president of the Sidran Foundation, for critically reviewing the completed manuscript, for recommending improvements to the narrative, for suggesting publication and marketing ideas, and mostly for giving a budding author self-confidence.

Thanks to the Naval Institute Press; to Shannon Becker, for supporting me during the editorial process, and for giving me a chance; and to my editor, Anthony Chiffolo, for patient, constructive, and most of all, sensitive criticism and editing of the finished work.

Thanks, Jim Astrachan of Hazel & Thomas, Baltimore, for leading this nervous-Nelly author through his first publishing experience.

Then will he strip his sleeve and show his scars,
   And say, These wounds I had on Crispin's day.

—Shakespeare, *King Henry V*

Farewell, **Darkness**
• •

Prologue

• • • • • • • • • • • • • • •

Three times I've lost control.

Only three.

The first was under a crystal sky, in a Huey slick northwest of Khe Sanh. God, it was clear that day! Icy clear and uncommonly fresh over the canopy. It was May '67, I remember.

I lost it only for a little while, just a few dozen seconds as we hurtled down with engine and rotor tachs shuddering around the red line. "Scarface" medevac, Marine Observation Squadron Three. It was just before we broke through the choking smoke darkening the zone, where the flames were closing in on the team. We got them out, the ones that lived. The rest we . . . we. . . . Hill 655, it was. I remember, 655. All these years. A number among so many numbers.

I lost it to fear that time, a hopeless, despairing fear—the kind that washes in with the awareness of certain death. It was the first time I touched it, you know, the Fear they taught us to fear in boot camp, the quintessential Marine fear, to be trapped alone and beyond hope in the zone. But I regained control! I seized it as we burst through the 12 point

1

7, the small arms, the grenades. Our skids crushed the flaming grass, and the fear. I was in control.

I lost it next when we fought the NVA over the valley of the northern A Shau. It was cold that night—God, uncommonly cold. I remember. The cloud mists that swept through the open gunship chilled me to the soul. I don't recall ever being quite that cold. Not in the years before A Shau, never since.

It was the night I shot the Green Beret.

I lost it to guilt that time, the sickening guilt that overwhelms the soul when you make the ultimate mistake. And to shame. I wanted to quit fly-ing—"I shot the Beret!"—but Major Burke convinced me to stay on. The squadron needed crewchiefs, he said, and I needed so badly to be needed. I put it behind me, or thought I did, but I never forgot. I did not forgive. Forgiveness must come from within, and there was nothing in me that remembered how to forgive.

The last time was on a cool noon, in the gray skies above An Hoa. The winter monsoon held back just a little that day—just enough. The air was thick and humid, clotted with the stench of cordite and paddy manure. It was hard to breathe as we dove upon them in the open zones. There were so many zones. It was the day we annihilated the weapons trench, the bunkers. The day we broke the NVA charge and stopped the slaughter with greater slaughter. The day and the hour I learned Phelps was dead.

I lost it to hatred that time, raking my weapon across the red-puddled trenches long after the last fucking slope-headed bastard had fallen. Back and forth, charge and fire. I remember my pilot's face, eyes wide in disbe-lief, watching me raging over the bolt on an empty chamber. The fucking thing would no longer let me kill!

Fear. Shame. Anger and loss.

Control!

Chapter 1

A Warrior I have been.
Now it is all over.
A hard time,
I have.
                                        —Teton Sioux

Vet Center

The Annamese monsoons have long quenched the fires on 655, washed the blood from the A Shau, flooded and filled the trench near An Hoa. And Phelps . . . frozen in youth, forever young, Phelps has lain fourteen years at rest.

How long, the reach of those days.

"You need to get some help."

"I don't need help."

"I think you need to talk to a professional about being so angry all the time."

"You mean a shrink."

"I mean someone who can help."

"You mean a goddamned shrink! You're a nurse and you know enough to say 'professional' because you know shrink will just piss me off. That pisses me off."

By April 1981 the debate is a half-year old. Grace and I have been married nearly eleven years. Good years. Chris is five and Matt just two. Good kids.

I rise from the kitchen table and begin clearing. The meal is only moments finished, but I can abide neither inactivity nor untidiness. I dump the cutlery into the washer with a clatter, which Grace immediately rearranges.

"They won't get clean if you throw them any-which-way."

"Screw clean," I mutter, but I stop to scrape and rinse the dishes before filing them in neat rows.

"If I didn't have such an asshole for a manager, I'd be OK. Do you know what that bastard did today? We had to have our design in for the new system. He insisted on doing it his way even though I told him it wouldn't work. Would he listen? Shit! I just said, 'The hell with you,' and we split the blanket. He got that little kiss-ass to do it. I did it on my own. He didn't have the balls to stop me! They just didn't talk to me anymore. The hell with them. Then guess what! His design didn't work. Know what? He took my stuff, added his name and the kiss-ass's to it, and presented it as work 'from our team.' Team! What the hell do they know about 'team'? Y'know, when I was in Vietnam, I killed people for less reason than these assholes are giving me right now!"

Killing is not difficult, in anger and without. But my rage frightens me. It means I am not entirely in control. I need control. I remember the times I lost it, never for long, but long enough to threaten my survival. Control is essential to survival, and I am trained to survive.

"Ah, hell. What difference does it make, anyway? So we turn out an extra zillion pounds of plastic for some asshole to turn into flyswatter handles. It won't make the merest tremor in the fabric of the universe. I just wish I could find something worthwhile to do."

I cross to our "mail table." A pile, growing for days, has suddenly reached the critical mass for annoyance.

"Do you want any of this shit? Going once! Twice!"

Grace takes the mail and smoothes my hair. "I hate to see you so angry. If you could just talk to someone. If you won't go to the VA, you can go to someone private."

I brush her away, but not far away. My wife, my kids, a few close friends from before the war—they are all I have. Sometimes I go too far and anger them with my anger, my stubbornness. It frightens me when I push away these few who care. I cannot live without them. I would not want to.

"Don't bring up the goddamned VA. Remember when I went down to Baltimore about my legs? That bastard. Treated me like a damn leper." I mimic the VA doctor with a whine, "'We can't find your records. We can't determine that your injury is service related.'

"I don't think he even tried, that arrogant son of a bitch. He made this big deal about being in World War Two, like Vietnam didn't count. Christ. He acted like I was making it all up! Practically called me a liar. I told him if he couldn't treat me like a human being instead of a Vietnam veteran, he could go to hell."

"Then try someone private."

"I am not going to a shrink! What the hell do they know? They'll just take my money and tell me I hated my father and loved my mother."

"Why don't you try the Vet Center in Elkton? Just go in and talk. You keep saying you don't trust anyone. At least you'll be able to talk to other Vietnam veterans. Maybe you can find out something about the dreams. It can't hurt."

"I can handle the dreams. Everybody has dreams."

"You wake up in sweats, choking. You say you can't remember anything. Maybe they can help you remember. You always say you don't have anyone you can talk to about Vietnam. Maybe if you just went in and talked.

"I hate to see you this way."

"Let's not talk about this. I remember enough about Vietnam, and I handle it OK. I don't need to remember any more. The war is over. There's nothing else to understand. I know what you're thinking—you're thinking about that Post-Traumatic bullshit you've heard about. That's a bunch of crap! It's just an excuse the druggies and alcoholics and nonhackers use to blame their screwed-up lives on Vietnam. I was a goddamned combat-Marine crewchief! I saw as much bad shit as anyone and definitely more than any goddamned Saigon Warrior or Da Nang REMF.[1] I turned out OK. I never did drugs or booze, I got myself through college, and I have a well-paying, responsible job, even if it sucks. These sorry peckerwoods who can't get it together should look in their own skivvies for the reason, not Vietnam. They can peddle this pity-me PTSD bullshit to someone else."

I look away to avoid the ache in her eyes. She is a bright person and loves me, and she may be right.

I try to soothe her. "Look, Hon, I know you want what's best, but I just have a lousy job and a lousy boss. If I can just find something else, things will be OK."

---

1. Rear-echelon motherfucker.

I hold her and speak deliberately: "This doesn't have anything to do with Vietnam."

Fourteen years since my war. I have accomplished much: family, degree, career. I function in the community and am acknowledged for my accomplishments. Vietnam is a thing of grainy newsreels, to be viewed less and less often, something the World says should be put behind us, forgotten. Yet it is hard to forget.

A gathering: 1970 or 1980, or times in between.

"Ron, did you see that article about the vet who went berserk? What do you think about that?" someone asks.

I shrug. Why is it important what I think about another Vietnam vet?

"Probably on drugs," notes another, and a third adds, "Yeah. The drug scene was pretty bad over there, wasn't it?"

"I wouldn't know, I was too busy flying." I answer stiffly. "I didn't know there was a drug scene until I got back and read how we were always doped up. In thirteen months no one in my outfit did drugs, and only one guy smoked some grass. He was a crewchief, like me. Flew into his eleventh month and just couldn't take it anymore. You could get a free month of leave anyplace in the world if you extended your tour for six months. He was so desperate to get out of the country, he extended, packed a ditty bag, and blew out the same day. The only trouble is, he went back to the World. The States, you know. It would have been better if he hadn't gone home. Better if he'd just shacked up in Bangkok or Tokyo, or even with a round-eye in Australia.

"He started flying again, but he couldn't readjust. He'd spend the evening with the shakes. He quit flying for a couple of weeks but went stir crazy. I knew how he felt. I'd been grounded once—failed my flight physical. Aircrew couldn't stand to be buttoned up on the deck. So he got back on flight status and smoked a joint each night to calm his nerves. He never smoked during the day or on missions. He was the only one I knew who used anything close to a drug. We lived asshole-to-belly button, and if there were any drugs, I'd certainly have known about it. We worked eighteen- and twenty-hour days. Mostly, we just did our jobs. It wasn't like you think. The Saigon Warriors and other REMFs might have done a lot of drugs, but we were combat aircrew. People depended on us for their lives. This stuff I hear about guys running around stoned all the time—I never saw it. Hell, I couldn't even get enough booze to get juiced more than once. Whatever we had was 3.2; it made you sick before it made you drunk. One time, this buddy of mine, Dutch, from . . ."

"You got him started. You should know not to get him started," the first interrupts, warning the others.

"I forgot," the second answers glumly, "Ron's war stories!"

"You didn't do any drugs at all? Not even a little grass?" a fresh voice asks. He is new to the neighborhood and wants to fit in. Too goddamned bad.

"No," I snap abruptly. I'd been on a roll, talking about what I know best, what I most want to talk about. "Don't get him started"!—fuck them. Fuck all of them. My wife looks in from the kitchen, concerned, eyebrows arranged in moral support. I would leave, but it's my house.

"Then what did you do in your free time?" the new neighbor continues.

"Masturbated." (And fuck you, too.)

The discussion continues around me, an unwilling observer. Talk of Vietnam, once begun, seems to acquire a life of its own.

"It sure seems like a lot of guys came back messed up in the head and stayed messed up, though. My father was in World War Two, and it didn't seem like they had all these problems."

"Yeah, but they were fighting for something, and anyway, anyone with real smarts dodged the draft. I think Vietnam drew mostly coons, spicks, hillbillies, and other losers. I drank a barrel of root beer the night before my physical and raised my sugar level to fail the urine test. I did my thing with the SDS."[2]

"I stuffed myself with peanut-butter sandwiches for weeks. You know how much lard peanut butter puts on?"

"I prefer my way. Peeing sugar was using my head and a lot easier."

New neighbor plays peacekeeper, trying to appease all, including the one who implies I lacked the brains to pee my way out of the service.

"Well, Ron was there, and he turned out OK. It must have been tough, though, being in with all those losers. I guess you were able to put it behind you."

It is my turn to respond, but how? Long years have taught me to conceal the anger I feel, which ties my tongue in these "casual" conversations and makes me feel like a stammering fool. Losers? Can't they see that so heartless a brand scars me as well? How I hate the word. Perhaps they see me as one with them, as able and ready as they to dispassionately and critically analyze the failings of my war, and my failings for fighting in it. Per-

---

2. The Students for a Democratic Society were a militant antiwar group on many college campuses.

haps including me in such conversation is their view of "acceptance."

They are wrong. I do not belong in their World; I will not! Vietnam is central to my being. I am of a kind with the "losers" of whom they speak. The kinships I knew In Country were far stronger than any I've developed since returning to the World. This evening, these people, are transitory. Vietnam endures, its memory as vivid as life. The men I knew were not losers. They were crewchiefs, like me, gunners, and pilots—boys become men, highly trained in a proud service, trusting our lives to one another's dedication and ability. If they were losers, what was I?

But the World is a harsh judge, and truth is undeniable. Vets do run amok. Men with stringy hair and field jackets with the look of the combat zone do cadge handouts in bus stations. More than from any other war? I do not know another war. Vietnam is mine, the only one I had. I know that, somehow, I am held accountable for those of my kind who fail in society. Perhaps it's because we all failed in Vietnam—so the World tells us—and the blame we share reduces us in perpetuity to the least common denominator.

I search for a response; it is so much more complex than anyone imagines. I struggle with the pride I feel for having served, a pride I cannot admit to these who question me. Pride in Vietnam!—they would find it laughable. Yet the failures of my kind shame me; they tarnish the pride. I struggle with the shame, which pride impels me to conceal. I have read of the syndromes, the emotional disorders some claim, and blame. Bullshit! Excuses! Inventions to conceal weak character. I am successful; my success is itself an indictment of their failure. If I can make it, why can't they? The weak shame us all.

Yet I am ashamed of the shame. Do not all who survived that mistaken war deserve more than shame as a legacy, if not for surviving, then for enduring? And who am I to be ashamed of my own kind? How arrogant have I grown to judge more harshly than the World, which understands less well than I, and has no right to judge at all?

The room awaits an answer, these who acknowledge my accomplishment but cannot see nor accept who I am. I frame a reply in sarcasm, an effective weapon and shield; it conceals both pride and shame.

"Well, *if* I had to put up with losers, it was good practice for dealing with you pains in the ass. Let me put it this way. I wouldn't want any of you marrying my sister."

Silence, then quick laughter. The hastily crafted insult closes the subject, and we speak of other things.

Losers! I should have been grateful for the lesser epithet. Once the cries had been "Murderer! Babykiller!" The accusations angered and wounded me, so I forged an armor upon an anvil of bitterness and alienation, quenched it in the guilt of a lost cause to stand between me and the World. The armor made me stronger, so I believed. Sometimes my anger flashed to rage, but I pounded it down, made it one with the armor, turning it inward until it was no longer rage but righteousness, the righteous anger of the wrongly accused. No smith could have worked the metal better.

As the years passed, accusations and epithets faded, but I did not lay down the armor. Its burden had grown familiar. I would not have yielded its comfort had I considered it, and in truth I did not consider it, although I knew it separated me from others. I have read that some primitive societies would isolate and purify warriors returning from battle before accepting them back into the tribe, but we of Vietnam seemed unfit for purification or acceptance. For me, there were only memories and bitterness; the armor kept both alive.

In time I learned to clothe myself in masquerades, to walk chameleon-like and unnoticed through the gentler society about me. They were formidable masques, and expensively earned: the college-educated, three-piece-suit, slide-rule masque of an engineer; or the garden-tractor, two-car, gas-grill good-neighbor masque. Masques of things and accomplishments that made me appear as others. Still, there were times the armor peeked through. I was human, after a fashion, and given to fatigue. But I'd been trained to survive, so I cultivated a coarse exterior and a dry and sarcastic humor, the kind that causes people to remark, "What a character!" when they like it and, "What an asshole!" when they don't. The humor was part me, part masque, and easily maintained. It was my first line of defense, and when it failed there was always the armor. I functioned, and I survived, but in truth no masque or delusion ever obscured my vision of myself.

I was a Vietnam veteran. The words seemed a brand; the experience defined me in spite of family and accomplishment. At times I fancied that all who looked upon me could see the brand beneath the masque. Yet I relished the identity, though the cost was an empty existence. Vietnam had partitioned my life. All that came after suffered in its shadow. No new friendship was so meaningful as those I'd formed In Country. Nothing I'd accomplished in the World matched the value of what I'd done in the war. The knowledge that those days were gone depressed me. Not a day passed that I didn't recall something that I'd done, or seen, or heard, or been in Vietnam. I knew I was a bore, but I needed to talk about the war, the centerpiece of my being. My identity had been forged during the twelve months, twenty-eight days, and two hours of my war, and none of the labels I'd earned thereafter mattered. When asked, "What do you do?" I had to think twice so not to answer, "Marine crewchief—Hueys." When asked, "How old are you," my quick response was, "Nineteen!" Then, laughing, I slowly subtracted 1947 from whatever the year was to compute the answer. Life seemed frozen in the distant year of my war, and in truth, I did not regret it. It was my time, my prime—when I belonged, when I was in control.

I sit in suit and tie. It is office-quiet, the hubbub of electronic noise barely noticeable beyond my manager's door.

I am losing control.

"He has no right to present my work," I say for the third time. "The plant contracted with me to do this job. I did the first phase, not him, and they want me to do the second phase. Not him! I don't care how much he wants to get promoted. I don't care how much you want to get him promoted! Do you know what he's doing today? He's taking my stuff to the plant and presenting it. He doesn't even understand it! I was running this just fine. Everyone was happy, and you know it. You had no right to put him in charge just because he's got twenty years in this place. Let him screw up someone else's stuff. You don't give a damn about the success of this thing, you don't give a damn about the plant, and you don't give a shit about me."

His reasoning is the same as before, but delivered with increasing edge and frustration. "Ron, he's in charge because he's got the contacts to implement this project across the company, not just in one plant. You started it, and you've been successful. You're still key to its success. You know that. He knows that, but if he takes it forward, it means greater success for all of you, doesn't it? You'll all share in that. Ron, you have to start behaving more like a team player."

Images of bursting C-4 blur my vision. The project is mine. The work, mine, and they are taking it away from me, making it their own. You can't trust anyone in this fucking place. Team player! What the hell do these bastards know about playing on a team? We knew what teamwork was, in the Crotch, and it didn't include bending over and taking it up the ass.

"Yeah. It makes me feel real good. He gets to take the bows on the podium, and I get to be the podium. I should trust you? Bullshit! You do what you gotta do, and I'll do what I've gotta do."

My job is in the sewer. Again. I've been working for the company since college in 1973. Though I'm a talented engineer, I'm hostile to anything that looks like authority. From my earliest days, co-workers have cautioned me about having a chip on my shoulder, but I see it as part of being a "character." It keeps people away without pissing them off too much, which suits me just fine.

Corporate culture encourages certain "characters" and tolerates others. I know I've crossed the bounds to "asshole," a condition not tolerated in young engineers, talented or otherwise. Sometimes I invite the conflicts that push the bottled rage to the limits of my control. In the past year, even I have come to suspect that my value as an engineer is fast yielding to my being just too much of a pain in the ass to deal with. But I don't give a damn—or so I've convinced myself. My attitude mimics our perverse translation of the Marine's motto, *Semper Fidelis*: "Hooray for me, and fuck you."

The argument is another in a never-ending series, each built upon the enmity of the last. The mounting tension is unbearable. Me against them. There is one of me and several of them, and I knew when I started that I couldn't win. My arguments are tactically efficient, but strategically and politically off base. I begin a shouting match, then abort it unresolved by retreating to my desk. The subject of the argument can indeed be serious, or insignificant, but my anger discerns no degree. Usually, time dissolves the tension. Hostility cools into the long-term resentment I bear them, whoever *they* might be. Normally, I content myself with images of fragging my enemy, then go home to ruin my family's evening by raving about the assholes at work.

But the images do not content me this time. I return to my office and bury my face in my hands.

God, I hate this place! I hate being pissed off all the time! I hate my life! Is this all I have to look forward to, one fucking battle after another? Why do I have to care so much about what I do? What I do is useless! It's not like when I was in the squadron; that was important. Oh, man . . . back then, with the old guys, Mouse and Phelps and Saint. We were a fucking team!

I'm just so damned alone.

The years of control are dissolving, and over nothing—one argument out of so many. It must be true, what they say about straws and camels. I feel a dizzying numbness, then see myself at the edge of a precipice, so close that I can peer over and see what I have become. Each moment is an ending to all that has come before. I don't know, in this moment, why or how I've come to this end. I am terrified of the endless circle of anger, of feeling alone. So many years have passed, but I recognize the old Fear as I touch it, and then I feel something else.

Suicide.

Some call it taking your own life. Taking means ending, but taking is taking control.

Control!

The notion warms me, enticing. What a boon, to be released from the anger, released from everything! The priests taught that you can't go to heaven if you take your own life, but if you "commit suicide," are you entirely sane? Am I, angrily brooding over perceived and actual insults, entirely sane? Is anyone who has done what I have done, lived what I have lived, sane?

Wouldn't I, wouldn't everyone, be better off if I bit the big one? Checked out? Lifted off to that big hangar in the sky? If I fucking killed myself?

How good it would be to die.

The thought is calming. A direction, any direction, lures me, even if to oblivion. Release is what I seek—from the rage and the cold depressions, the loneliness and fear of loneliness I carry in spite of my family. *They'll miss me, but I'm no use to them. No use to myself.*

How would I do it? *There's the bridge at Susquehanna. I'll be sure to drown; they were never able to teach me to swim in boot camp. Boy, did the DI kick my ass.*

Dummy, the drop will kill me, not drowning.

The 12-gauge in my closet! Fast, but messy. I just wallpapered. Besides, I saw a picture of someone who ate a 12-gauge and lived. It would be just my luck to fuck up my own suicide. If I use the car, they can collect the insurance. Eighty into an abutment, no seat belts. I can make it look like losing control. Losing control! Get it? Ha!

Lots of choices. I would need only one. It is a thing I can control.

I'm not afraid of death, not since the Hill. Not since. . . . I was afraid before then, but not since. I should have died there, you know. I was prepared. But I didn't, I made it out, and I've never been afraid of dying since. Someday I might be afraid of it again, though. I won't like that. I was pretty afraid going in . . . going into the Hill. It frightens me, to be afraid of death again. I'll probably be afraid someday. Maybe when I get old. It's better to do it when you're not afraid. It's an easy thing to do. Easy . . .

Yes, it will be a good thing to die. Yes. It will be good.

A distant fear grows, exploding inside, shattering the false calm of oblivion. Nauseating, choking.

Oh, God! What's happening to me? What am I thinking? I must be insane. I am trained to survive. I am a survivor, goddammit! I am a fucking Marine!

OK OK OK. Get a grip. No suicide. See, God, I'm doing OK. I'm the crewchief here. I'm in control. I just have to come in from this zone. Like the old days, remem-

ber? Control! "Crewchief, mark your target!" I can't handle dying. "Taking fire four o'clock!" I can't handle living. Lord, don't do this to me. I've been good, like I said. Remember, I promised to be good. I have been fucking good! I never killed anybody who didn't need it. Never. We lost a crew today. Kelsey, Carter. Hochmuth was on board—the general. Who else? "Taking fire, taking fire." We lost . . . Phelps! Hey, guys, let's have a hymn! "Lord, guard and guide the men who fly through the great spaces in the sky. . . ." Fuck the hymn.

God, I need help.

An image stirs within the fear, taking form as it moves slowly forward. Help . . . I have heard something of help, but where? My mind churns, something "help for Veterans." My own kind, not like these *enemies* beyond my door. Something Vet . . . Vietnam Vet . . . Vet Center! Grace wants me to go. It must be a good thing if she wants me to go. What's the number? Oh, Christ, I can't remember how to find a lousy number!

I call Grace. "Things . . . things really suck, hon. Those bastards— they're really getting to me. I think I'm . . . I'm really losing control. I just don't. . . . God!"

"Do you want me to come get you?" she asks quickly. I hear the fear and concern in her voice, and do not like it, do not like being responsible for it. How useless I am, that I bring such distress upon those I love.

"No, no. I'll be OK. I need . . . I just need, uh, that place for Vietnam veterans. I think I need that place for veterans, like you said."

She reads a number, and I write it down. Again she offers to come for me, but the fear in her voice shames me, makes me want to be stronger, or at least show strength. "No," I say quietly, "this is something I have to do myself. I'll be all right."

My eyes ache, my temples throb as I dial the number, not knowing what to expect, hoping only that help will be at the other end.

Rose answers, the Center's secretary. Her voice is calm and clear, the kind of voice that reassures.

"Vet Center, may we help you?" My mind races in circles. Help, God, I need help. I try to answer, but my throat runs dry. Pain and shame fight against the boiling need and prevent me from speaking.

"Is anyone there? Is anyone on the line?"

I grow afraid that she'll hang up. "My name is . . . Ron. I need help." It takes so much, just to say my own name, and I give it unwillingly.

"Are you in an emergency?" A steady, controlled voice. "We can get someone to you right away."

A medevac! Like the old days, like the zone, but I translate her question to mean, "Are you going to kill yourself or someone else?"

"No," I whisper, beginning to feel more than a little foolish, "it's not an emergency." The pain begins to melt.

The voice is friendly, but urgent. "We can get someone to you, or you can come in now, or later today. There will be someone here to help."

Again, the word I need most to hear.

I check my watch. Noon. I need some time to . . . I don't know. I just need time. "I'll be there. I'll be there at, uh, three." I try to concentrate as she gives me directions and makes me repeat them until I've understood.

I replace the phone with a shaking hand. A wave of relief washes over me, then shame for the weakness I've betrayed. I feel committed and nervous, afraid, lost and embarrassed all at once, but I have a goal, three o'clock and whatever it will bring. And most of all, I no longer want to die.

The Veterans' Outreach Center lives in Elkton Commercial Plaza in Elkton, Maryland.

Elkton is a small town, an old town. John Smith named it "Head of Elk" after the river, a tributary of the Chesapeake a few miles away. Town history records the visits of the great and the near-great. Even the not-too-great and the never-great have left their mark. Elkton is a friendly place and welcomes all. George Washington passed through on his trips from Mount Vernon to destinations north. He even slept there a night or two. And Lafayette with his army, and again on his Grand Tour in the 1820s. Grant slept there as president, and William Howard Taft delivered a speech from the wooden porch of the Howard Hotel.

Elkton's heyday came during the middle years of this century. It was the marriage capital of the East Coast. No waiting period. Sixteen thousand couples got married there each year. Traffic dropped off when the state introduced a cooling-off period before tying the knot, then died altogether. Still, you can buy flowers in the front of O'Marie's and get hitched in the back chapel under the valentine arch. Folks picking crabs on the Howard's porch will grin rock salt and Old Bay smiles as you leave, and raise frosted mugs all finger-smudged with crab mustard to your new life and health. It's a friendly town.

My job brought me to northeastern Maryland, a neutral reason for choosing a home but one that proved fortunate. The Davises are next-door grandparents to Chris and Matt. I've lost count of the nights Red and Wilma ran over to help with a fever, or babysit one while we rushed the other to the emergency room. Frank and I swap stories about our kids every spring at the Agway. I like to buy seed and shrubbery from someone I trust. Charlie's kids go to school with mine. He drives his state police unit to the playground when he's off duty to play catch with the kids. The whole school said prayers when my mom was in the hospital, and the Klings and Brophys drove to Baltimore to be with me at her wake, and my dad's too a couple of years later. The Klings spend Christmas Eve with us, and we spend Christmas night with them. We're always invited to the Fanueles' family affairs, and the Robertsons', and they come to ours. Matter of fact, there are more friends and neighbors at communions, confirmations, and graduations than family.

I should be content, at ease and at rest. Yet in the time I've made my home in this town that welcomes all, among friends and neighbors who welcome me, I feel a stranger and, more, an intruder. On the cathedralled streets of tulip and oak, where so many have forged bonds to last a lifetime, I live, still, the isolation of the zone.

Those of my kind know the chill of *the zone*, a thing foreign to those reared in the shelter of the World. The zone is the centerpiece of my centerpiece, as it were.

The zone was a place, become an abstraction, the embodiment of old terror. It lives within the dark memories of those who learned its meaning In Country, and is the place I visit when my mind loses its way, or finds its way. Its sounds and smells take form so easily from shadows that are never shadows.

There were many zones: drop zones, landing zones, and others, all alphabetized and neatly catalogued. I learned them as I learned all the lexicon of Vietnam. There were loading zones and medevac zones, places we scouted and probed by fire from the tenuous safety of our gunships before risking troops and crews. There was even the combat zone, a term so generic it had lost its meaning. It was simply all around us; it was where we lived. Some zones were crowded, some not, but any zone was lonely when it became the one where you could die.

Three o'clock approaches. I study the Center from the safety and cover of my car. The building is unimpressive, a door framed by two curtained windows in a blue-collar shopping center, tightly sandwiched between a deli and an obscure office; it could conceal a walk-in laundry or the local Democratic headquarters. I am parked in a pot-holed lot, close to the Valu Food, reading a paper as if waiting for someone inside. The CIA approach makes me feel foolish, but I have not survived through carelessness.

The hours of operation are posted on fly-specked pasteboard next to a hand-drawn sign that reads, "Help Without Hassles." A lighted "VET CENTER" sign almost as wide as the office hangs above the door, conspicuous not only for its size but for the fact that it has not yet been punctured by rocks. I watch the building suspiciously. I did not give Rose my full name, nor a return number, so I can escape quickly. It looks government after a fashion, but not quite regular government. Nothing like the VA. I remember that bitter experience: I'd paused at the inscription in the hospital's lobby, "To care for him who shall have borne the battle and for his widow and his orphan." Lincoln's words were empty, in my experience, and I flipped them the finger as I left. I do not know the Center's affiliation. I assume it isn't with the VA, else I'd not have come.

I cannot see the staff inside, but several men have come and gone. Vets, I assume. A few working stiffs in jeans. One wearing an old Army field jacket. Some wear ties loose at the neck. Most seem older than me, a little balder, a little thicker in the waist, but then I've always looked younger than my age. I loosen my own tie, not so much for comfort as to fit in. Everyone appears "normal." There are no crazy, frantic eyes, although to me they all have the look of the zone. I mark myself in the rearview mirror. The look is there; it is in the eyes, you see. It never leaves the eyes. I scan the terrain and fix on the deli. The customers seem oblivious to the presence of war next door; a fat, balding man shuffles back and forth, slicing bologna, and a couple of long-haired teens play the arcade.

Quiet zone. I take a breath, put myself on automatic, and enter the Center. I don't think I can walk in if I have to concentrate on it. I step immediately into a tiny waiting room. There is a vestige of military about the place, but it is the slovenly military we cultivated to such perfection In Country. American Legion magazines, pamphlets on PTSD, and the remnants of the week's *Whig* litter several vinyl chairs. A scattering of posters are thumbtacked to the walls: Agent Orange warnings, MIA remembrances, competitions for Vietnam-memorial designs, roughly drawn pencil sketches of tired men in action that recall the days in the zone. The postings seem to have hung for ages, destined to remain until the paper rots through. It reminds me of the hooch and the squadron's line shack—the grungy familiarity of places that men inhabit rather than call home.

The secretary rises from the paperwork cluttering her desk. "Hi, come on in. I'm Rose. Are you Ron?"

She knows my name! Does the tension show that much? She smiles. An inviting, disarming, nonjudgmental smile, the kind meant to set nervous people at ease, yet it seems genuine, not practiced. She seems soft, yet as solid as her voice.

"Yeah. I, uh, called." I motion vaguely to my car as if driving and calling were somehow linked. What am I supposed to say? *Hi there, I'm the guy who wigged out this morning. I was just thinking about blowing myself away and called to see if you had any better alternatives.* Accurate, but how can I say that? Worms crawl in my stomach. I try not to appear as agitated as I feel.

I notice the paperwork on her desk and stiffen. I point to it, accusingly.

"Do I have to fill anything out?"

She laughs, "Oh, no. We don't do that sort of thing," and motions to the sign. ". . . Without Hassles," I read again. It sounds good, but the place still looks government, and I don't like the yellow, government-looking memo flimsies on Rose's desk. If this is the VA, I am history.

"Uh, yeah. You know, I think I left my car unlocked."

I move to the door, but Rose intercepts me with her voice.

"You're going to be seeing Tom in a little bit. He's the team leader for the Center. He's running a little late, so you can go back with Diane here and get started right now. I can take care of your car," she insists.

I turn in surprise to find a woman in her mid-twenties standing behind me. I'd not heard her approach. I mutter, "Hi," then extend a weak hand to meet hers. Her gender, youth, and gentle face are not what I expect in any case, much less in a room that evokes the hooch. She takes my hand firmly, briefly. It is a good handshake, good as any, better than most.

"Would you like some coffee?" Diane asks, and I nod.

She fixes cups for us both and adds the chemicals I request. I sip it gratefully as she leads me down the Center's single corridor to an office near the end.

It is a warehouse of a room, dimly lighted and windowless, rivaling the waiting area with its clutter. A desk and several bookcases crammed chockablock with family pictures, novels of Vietnam, and treatises on Post-Traumatic Stress gives it the semblance of an office. I cock my head sideways and scan several titles. Santoli's *Everything We Had, Edition III of the Diagnostic and Statistical Manual.* There is no apparent order to the place and barely space to draft a letter on the desk. Memorabilia inscribed "Tom Murtaugh, Ph.D." and "World's Greatest Dad" are arranged in a rough arc centered on the empty spot where the owner works.

Two gray swivel chairs occupy the space in front of the desk and bookcases. Diane brings in a third. I choose the one away from the entrance; it has long been my preference to sit with my back to the wall. Diane closes the door and takes a seat a yard away. The third chair forms a tight equilateral triangle, and I feel both claustrophobic and relieved by the closeness in the room and the clutter. I can hear the deli's arcade games ringing dimly through the wall, but the Center is otherwise quiet. I sip my coffee and study Diane, puzzled. She is attractive, slender, a dark-eyed brunette with a Mediterranean complexion, dressed demurely and comfortably in pants and blouse. I'd anticipated being greeted by someone attired in anything from a three-piece suit to battle fatigues, not by a woman who was a preteen when I fought my war.

She reads the confusion in my stare. "I'm an intern. I work with the counselors. Tom is the team leader, and is on his way. He'll be here any minute. You look like you need to talk."

I squirm in the chair, the vinyl sticking to my flesh.

She is correct. I am burning to speak, but where do I begin? I can't even remember why I came. I am so confused. I do not know what to reveal. What does she expect me to say? What am I allowed to talk about in this place? I don't like being here. It shames me; I feel so weak. *God, what a loser I am. I can't even keep my own life in order. I can't even stay in control.*

"Uh, I really don't know why I'm here. I mean, I don't know what this place is about. I just heard it was a place for veterans to go to get help. Things got a little shaky for me this morning. I've been having these, uh, problems."

I rattle on, talking to the floor. "I don't know that there's anything anyone can do to help. What do you want me to tell you, anyway? This place is about Vietnam, isn't it? I was there in '67. Marine helicopters. Crewchief. Do you want me to talk about that? What do you want me to talk about?"

I raise my hands to my cheeks, force them roughly up my temples, grind the palms in my eyes, run them farther into my scalp, stretching the skin taut, then down across my ears and neck, gripping the nape in a vice of knitted fingers. I have to get in control. I have to get in control!

Diane reassures me. "This place is for helping veterans, in any way we can, so you qualify. You're in the right place. Sometimes it has to do with Vietnam, sometimes not. The counselors work with you to sort things out. What's important now is what brought you here. Tell me what happened this morning."

I begin to laugh, a nervous, breathless, snuffling sound like someone going around the bend. I speak—to her, to the corner, to the books jammed in the case. My head twitches this way and that, spasmodic, without focus.

"I lost it this morning. That's what happened. Over nothing, a lousy argument with my boss. I've had a zillion arguments, and I never lost it before. But I got it back again, you see, or else I wouldn't be here. I mean, if I really lost it all the way then I wouldn't be here, because I had to find it again in order to get here, didn't I? See? So maybe I didn't lose it after all like I thought I did. No, no. I think I did lose it, or something close to it. You know."

I can't stop babbling. *Why doesn't she make me stop babbling? This isn't me talking. This is some fucking fool, and I can't get his mouth to connect to his mind. I can't say what I need to get out. I don't know what it is that I need to get out.*

Diane watches closely, as if every utterance is wrought with meaning. "What is it that you think you lost?"

I concentrate, knitting my brows so hard the flesh begins to hurt, rubbing the furrows with rigid fingers, forcing my breath to come in measured drafts. I concentrate on control.

"Control. I think I was losing control. I came that close." I illustrate my proximity to meltdown with a narrow space between thumb and forefinger, then return my hands to my lap, rubbing them, popping knuckles, grinding fist into palm.

"You don't have any idea how tightly I'm wrapped. I feel so much anger. I got really angry this morning. It scares me. I don't know what I'll do if I ever get really, really angry and lose control."

I snuffle again. "You know, I lost it on this one mission. I killed a lot of people." I study her closely, testing for a reaction. "It was understandable, though; they were the enemy, and my friend, uh, my friend. . . . It wasn't like I killed civilians, or anything, but still, I really, really lost it You have no idea. I killed a lot that day. I did. I don't ever want to lose it again."

This conversation is so bizarre. I am in Elkton, in a warehouse of a room a wall away from a butcher slicing bologna, speaking in spasms about the men I have killed. The line of dead stretches before me.

"Staying in control is important," Diane rejoins, calmly, understanding, as if she can see the line of dead through my eyes, but chooses to look past them, to me.

The answer comes quickly. It seems the only thing that makes sense. "Lady, when you're in the zone, staying in control is staying alive."

The door opens; a large man stands breathless. He is a very large man, and Diane slides her chair to allow him to enter. He moves to me immediately, extending his hand. The wind of his passage flows over me, cooling. I half rise and place my hand limply in his firm grip.

"Hi, I'm Tom Murtaugh."

He is of medium height, wide, thick, his voice a flowing tenor in the body of a bass. Mephisto brows arch densely over eyes sparkling in intense concentration, as if he is entering the scene of an accident and is quickly sizing up the situation. A salt-and-pepper beard sprouts Freudian from his chin, but his chipmunk cheeks are clean-shaven and rosy. Large, white teeth smile warmly between generous red lips—a fleshy, muscular smile that is born in the eyes and feeds the rest of his face through fanning crow's-feet. He reminds me of Dickens's Ghost of Christmas Present. He wears a hunting shirt of red and black plaid and corduroy jeans, all extra-large. A tan suede Stetson, saddle-brimmed with a silver Mexican band, is in his hand. He cocks his head, still smiling, as if distracted by some fleeting thought, and pauses with arm extended before dropping the hat deliberately on a bookcase. A paper ruffles to the floor, noticed and dismissed with the same smile. He crosses between Diane and me and settles into the third chair, its springs protesting, leans forward, and waits for Diane to take her seat. The room seems smaller now, and is again quiet.

"Go on. Don't let me interrupt anything," he says, as if trying to melt into the background. It doesn't work. I stare wide-eyed. He is the strangest shrink I have ever seen, though I am no expert.

What the hell is going on? But I feel assured by his appearance that this cannot be the VA.

"What?" I ask, confused.

"You were speaking when I came in. You said that staying alive meant staying in control. Just go ahead."

"I don't know why I'm here. I'm so embarrassed. I just didn't know where else to go," I continue plaintively. "I mean, this place is for veterans that have real problems, right? Drugs. Alcohol. Can't hold a job. I've read about this Post-Traumatic stuff, but I have a good family and a good job, except my boss eats shit and sometimes I want to blow him away, and I never do drugs or booze. I . . . I don't have the right to be acting this way."

I have never felt so impotent, so craven. I have been afraid before, and though afraid have walked the places where dwell fears too terrible to imagine. Some say that is the measure of bravery. I have been that brave. What am I doing here, begging help I don't understand, seeking shelter from . . . what? How low have I been brought, to crawl to this shabby room, whining, "Pity me, pity me"? What can they possibly do for me; can smiles and encouraging words serve as anodyne for shredding pain? *What happened to my guts?*

"I don't know what's happening to me. Is it the war? Jesus Christ, it's been so many years. I hear about this shit, Post-Traumatic Stress Disorder. What the hell does that mean? Is the Center for you only if it was the war? What if it's not the war? Do I have to leave? I don't know where else to go. I think I better leave. I don't trust anyone. Except my wife. You people are nice, but she's the only one. She said I needed help. She wanted me to come here. But, Jesus Christ, look at me. I have a fucking coat and a fucking tie and I make a fucking lot of money. What right do I have to be here?

Christ, I know she's right. I guess that means a shrink. I can't go to a shrink. Do I need a shrink? Am I mentally ill? I thought about killing myself today. Do you know that? I did, for the first time in my life I did. I sat there and figured out life insurance, and what I had to do to get double indemnity. I'm very analytical, you know. I work with computers. I'm a real good engineer. I was a crewchief in the Crotch. We're real good at figuring things out. Does that mean I'm insane? I felt so alone. I mean, I have my wife and kids, but if it weren't for them, there'd be no one. I don't know what I'd do if they left. Jesus Christ, it scares me. I'm so fucking scared. I just don't know what to do. I don't know where to go."

Tears well. God, I don't want to cry. The weakness I show shames me. I haven't cried in years, not years. I can't let myself cry, couldn't let myself cry, not even then. When was the last time I cried? Not even when Phelps . . . not even . . . I can't cry in front of these strangers. But the tears won't stop. Hold my breath, swallow. *Make them stop!*

I stare bleakly into my hands.

"I don't belong anywhere anymore. It's so lonely, me against everyone. I just don't know where to go."

I chance a glance at Tom and Diane, covertly, my eyes scarcely leaving the damp hands wringing in my lap. I search for the revulsion I feel for myself but see understanding in their eyes, not pity.

Tom leans forward and takes both my hands. It is a gesture of intimacy that would shock and embarrass me in another place, but here it seems a lifeline.

"All right. You're in the right place, and we can help you here. You're in the right place."

The quiet voice seems distant, barely heard against the rush of noise filling my ears, but it gives me direction. His hands grip more firmly, growing stronger as he speaks, as if trying to convey his strength and understanding to me.

"You belong here."

It seems not the words he chooses but how he says them. His voice and touch are an anchor. How long has it been since I felt the touch of a stranger? Not since the zone, and even then it was my hand that did the holding, cradling the body of a wounded grunt on some nightmare flight to safety. "Two more minutes—hold on. Just two more. You can make it," I used to whisper. Sometimes they made it, but sometimes I lied. Oh, God! I remember the times I lied!

His hands are blurred in a haze of burning tears, but his grip holds fast, like an anchor firmly set, and the hysteria slowly fades. If I'm not completely in control, at least the hysteria is gone.

"How do you feel?" he asks.

"Ha," I laugh aloud, shaking my head. I feel like shit. What a stupid fucking question. I pull my hands from his and rub at my eyes. A box of tissues appears. I seize a handful, blot my face, blow my nose, and drop them in the basket Diane extends. It seems to be the normal order of things.

How do they think I feel? My eyes burn, my throat aches from tension and bile. My clothes are sodden with sweat, crushed where I'd clenched them in my hands. Christ, I can smell myself! The sweat holds an oily stench of fear. My head aches—God, it throbs—and I will vomit if I even think in that direction. How do I feel?

"Uhhhh, fine!" I blurt, and blink my eyes wide, trying to look convincing. I don't know what else to say. I am coming here for help. I am not about to tell them to shove their ridiculous fucking question.

"Right," Tom says, his voice starting high, then driving lower in friendly sarcasm. An eyebrow moves up, then another, not quite so far. The eyes squint, then fix upon me. Diane turns her head and swallows a smile.

"OK. Just relax and listen for a while. You're in the right place. People from all backgrounds come in here with exactly the same problems you have, feeling the same way you're feeling right now. You're not alone. I know you think this is a personal failure. It isn't. I know it's hard for you to say these things to us, but that means you've already taken the first step to getting help and regaining control. You've accomplished something just by coming in here. Bankers, bricklayers, and people who are down and out have all been here for the same reason. Some have drinking problems, some are on drugs, some are not. They're all veterans, the same as you. We can help you to work through this, and we will help you if you let us!"

Tom pauses; I draw confidence from his words, but one phrase more than all the others has my attention: "Not alone." The sense of shame, of weakness diminishes as I consider it.

"How. . . ?" I wonder, but do not complete the question. There are so many things I need to know.

Tom continues slowly, to allow each word to penetrate my anxiety to know. "PTSD robs you of the ability to choose. Traumatic events, like war, create hot buttons that get pushed by things which may have no apparent connection to the war. We try to find those buttons in therapy and disconnect or redirect them. If you have PTSD—and we don't know that yet— the most we can hope to do is to give you back the ability to make choices. What you choose to do will be up to you."

The ability to choose. What is this? I am a most apt student of myself;

it is how I survive. My choices are my own, and buttons—yes, they are a minefield, but I've mapped it well and know the course. What can he discover, that I've not probed and evaluated within myself many times?

"Look, maybe I do need help. I'm here, right? I just don't understand something. I've thought about Vietnam for years. There's nothing that I did there that I haven't rolled over and over in my mind. I know I have problems, but I'm an intelligent, logical guy. How can you solve what I can't solve inside my own head? I mean, honestly, what are you going to be able to do in there that I haven't tried?"

"Are you saying you're going to resist?" Tom challenges. It is flat remark, without rebuke or offense, but a challenge nonetheless.

"Jesus Christ, no!" I nearly shout. Does he think I would fight him, or that I do not truly want the help they offer? For a fleeting moment I'm afraid they'll show me the door, tell me to go off and think about things and come back when I'm fucking ready to cooperate. Oh Christ, what if they kick me out? What's left, but the loneliness and anger I can no longer endure?

"Jesus Christ, I want to settle this. I just don't understand how. I didn't mean I wouldn't try. I just meant I don't understand. I need to do this. I want to do this!"

"Then you want to go on?"

"What?"

"You choose to go on?" he asks quickly.

Tom studies me; his small smile marks the moment of guile. I return his gaze, nodding slowly. The stratagem was so artful that I did not perceive, nor resent the manipulation. Perhaps there is more to understanding than remembering. I begin to see that options do lie before me, and perhaps it does require someone to rub my nose in them. I have been what I have been for so long. Perhaps there is another way. Choices, choosing, choose. I like the idea. In choice there is control.

I will cooperate with these strangers, for a while. Cooperate only. Trust I will reserve.

"Do you want to go on?" he asks again. I breathe a simple "yes" in a way that seems less an answer, more of an expression of need—and a commitment. Tom studies me, as if weighing my commitment.

"OK, then let's begin. Tell us about yourself."

"Where should I start?"

"Anywhere."

"Uh, I don't know where to start." If I knew where to start I wouldn't be here. "Do you want to know about my childhood, like 'I was born at a very early age,' my parents, the war, my job, or what? You have to tell me where to start."

"Where do you want to start?"

Another choice. "OK, I want to talk about the war. This is a Vet Center, right? Let's start with vet things."

"Then tell us about the war."

I begin, and clearly; I choose to do it.

"I think about the war every day. Every . . . fucking . . . day," I pronounce in even tones, each word a separate entity to emphasize the weight of the statement.

"Every day I remember some story about the war. Everything that happens around me reminds me of the war. I relate everything back to that time. I want to talk about it, but no one wants to listen."

"What do you think about?"

The question hangs. There are so many things of which to speak. Does it make a difference? Tom and Diane sit patiently. There are so many memories, but I want to give them an important one. I want to give them the most important one. He hadn't asked for that, but it is what I seek to give—the definitive memory of Vietnam. It is something I've never done—categorize my memories.

"I think about a friend of mine who died," I say plainly. The answer is suddenly clear. How could it have eluded me for even an instant? It is, after all, my essential memory of Vietnam. "I think about him every day." I feel strangely calm. His memory is a solace to me, something I keep inside that binds me to that time where he still lives, where I belong.

"Not a day has passed since I came back that I haven't thought of him. His name was Phelps. Ron Phelps. Ronald Joseph, just like me. He was just like me. I have to remember him, you see. If I don't, it'll mean he died for nothing, and I can't let that happen. I couldn't live with that thought. If I forget him, it'll be like he never was. I mean, nobody else can remember him the same way. Not his parents, no one. He died out in the zone, and I was the last one he spoke to before he lifted off."

I feel the tears coming but do not care. I've held them in for too long. I grasp at the old images, binding my eyes with my hands to see them more clearly. I have to see them. I have to remember.

I begin our story, in a voice distant and small.

"We became friends in Pendleton. We were so much alike; both only children, both Catholic. And straight! God, we were so straight. Not like real-Marines at all. One time we went on liberty in San Diego, out with the 'big guys.' They went off to get laid, and we just chickened out. We went to the goddamned zoo! Some Marines! He had a girl, and I had a girl, and we didn't mess around. The truth is, I was a virgin and didn't even know how to go about it. I think Ronny Joe was too, but he wouldn't admit it. That's what we called each other, you know? Ronny Joe. I hated it. He hated it. Our mothers called us 'Ronny'; one of us started calling the other 'Ronny,' and the other added 'Joe' just to piss him off. We did it all the time."

Tom and Diane sit quietly. I haven't spoken of Phelps for a long, long time, and never in this way, though it is how I most warmly remember him. It feels good to speak the tale aloud.

"He was a good looking guy, very lean, blondish hair a little lighter than mine. Sharp featured, not like me, with almost Oriental eyes. Sometimes he'd pull them back into even more of a slant and imitate Mr. Moto or Charlie Chan. 'Ah so.' What a jerk. He was a couple of inches taller than me, which I guess made him average height. I could kick his ass, though. We had some hand-to-hand training before we shipped out, and I guess I remembered the shit they taught us in boot camp better than he did. I pinned his ass and kept hollering for him to say 'Uncle.' He was a stubborn bastard, and I just kept twisting his arm, and twisting, and twisting."

I smile at those days, so long ago. We were so young. Phelps and me, tussling like kids on the hand-to-hand mat in Margarita Area, on our way to war.

"Did he ever say, 'Uncle'?" asks Tom.

"Phelps? Nah. Not him. That bastard would have lost his fucking arm before he gave me the satisfaction," I laugh and shake my head.

"I went In Country with an advance party of VMO-3. We disassembled four Hueys in El Toro. C-130s ferried us and the aircraft to the Philippines. We rebuilt everything and boarded a helicopter carrier—the *Iwo Jima*—for an air-amphibious operation in the Mekong. Ron went over on ship with the main squadron and arrived In Country a couple of weeks after me, so that put me ahead of him to rotate home. I enjoyed rubbing it in. 'Boot, FNG,' I used to call him.[3] It was only two weeks, but any advantage one of us had over the other was always exploited.

"We lived in the same hooch. He was at one end, and me at the other. We were the only ones who finagled private cubicles; the other guys shared by twos. We'd have had more room if we shared, and we got along real well, but both of us liked our privacy.

---

3. FNG stood for fucking new guy.

"We read a lot. I introduced him to Tolkien, and even haiku. That was, uh, that was just before he, uh. . . ." The image shimmers, and I shake my head hard, blinking my eyes, forcing myself to focus on the memory. "He liked the haiku. I'd never heard of it until my girl started sending me these thin little books of Japanese poetry. I shared them with him and some of the other guys. There was one that we especially liked. I still remember it. 'At dawn my castle / Was stormed / By a flight of ducks / Quacking in a mist.'

"Neat stuff, huh? It reminded us of the crap the Crotch dished out every-day, and the ships we flew in were UH-1Es—Hueys. Get it? Huey duck? Monsoon mist? Poetry! Some 'Marines' we were. Some 'Marines.'

"You know, he went to the Philippines for Escape and Evasion School to learn how to be a real-Marine—the kind of training where they dump you in the jungle and you develop a taste for grubs. I had no interest in being a real-Marine. I liked being fucking–air wing just fine, and so did he. He hated E & E school. Brought me back a present, though. Well, I guess he really didn't bring it back for me. I kept bugging him for a present, like kids do, and he gave it to me to shut me up. It was a Filipino butterfly knife. You know how a butterfly knife works?" I direct the question to Diane, who shakes her head.

"The handle is in two halves, not like a switchblade, and there are no springs. You open it with wrist action. Each half pivots on the blade's shank so that the handle folds over the blade and conceals it when the knife is closed. Opening a butterfly is much more dramatic than using a switch-blade. You hold the handle by one of the halves, flip it forward, then back and catch the free half on the return pass. The handle and blade flutter like a butterfly. You can conceal the folded knife in the palm of your hand, and if you're good, you produce five inches of stiletto in a fraction of a second, like magic. I was good," I added smugly.

"Do you still carry it?" Tom asks.

"Oh, no. Not any more. I used to carry it, when I first came back. When I was in college during the peace demonstrations, some asshole tried to stop me from getting to class in the engineering building. I flashed it out and put the point to his throat. He almost shit his pants!" I laugh, then fall quickly silent. *Too much. I've said too much.*

Tom waits for me to continue, but I sit, watching. *OK. I just told you how dangerous I can be. What are you going to do about it?*

"You know you can't stop there," he says after a few moments.

"Stop where?" I ask. *That fixes you for that footwork a while ago. Fuck with my mind!*

"It's against the rules to say something like that to a shrink and stop."

"So what do you want to know?"

He answers with a smile, "Just go on."

"Well, I was at the University of Maryland. I always liked carrying the knife; it's sort of a talisman. The peace creeps were blockading the entrances to the math and engineering buildings, and I needed to get to class. I guess they had a right to their views, but I had a right to the education I was paying for. The main entrances were blocked, so I went to another that was out of sight under a breezeway. This ape told me I couldn't come in, and started spouting his socio-repressive-political bullshit. He was a big asshole."

It was quiet in the breezeway, no one but me and the fat, long-haired creep. The bastard. I wonder if he knew how close he came. If he had twitched, I'd . . .

"Tell us what you see."

"Huh?" I answer, startled. The breezeway is so close. I can almost feel the space around me, the anger.

"You and the other student. Outside the engineering building. What do you see?"

"I didn't plan it. He was spouting his bullshit. The next thing I remember is pulling him down by the collar, and I had the point pressed into his throat. Didn't draw any blood, just close to it. He was big, and his breath stank, and the stubble on his fat fucking neck felt like sandpaper on my knuckles. I don't even remember how the knife got in my hand. I carried it in my right trouser pocket, and it just appeared when he tried to stop me."

I laugh, nervously, remembering the rage I felt, like in a dream, and like today.

"Christ, I was mad. I said, 'Get out of my way, you asshole. I'll kill you.' It was more like a growl than human speech. I wonder if he understood a word I said. My vision went red. Literally red. Funny, you hear about seeing red. It really happens. It was like I was looking at him through a film of blood. I kept him bent over with the point at his throat, then I moved him away from the door, pushed him on his ass, and went in."

I lean back in my seat, feeling the rush of adrenalin, just as in that moment near the halls of academe. I had come so close.

"What did he do?"

"Huh?" I answer, tiredly.

"What did he do?" Tom repeats.

"I think he went looking for toilet paper. I would have."

"And you?"

"I was really scared. When I got inside, every muscle in my body tensed. It was excruciating. My back locked up from my neck to my ass; my head was pounding. I couldn't breathe. I think I was a little crazy."

The rage and fear are as one to me. I cannot tell them how truly frightened I was. The rage seemed overpowering. If he had resisted, I'd have killed him. It was as frighteningly simple as that.

"You said you stopped carrying it. Why?"

"I started getting worried that I might lose it. I would never want to lose that knife. I keep it in my dresser. I see it twice a day, morning and evening."

"Didja bring me a present? Didja, didja?" I'd badgered Phelps.

"Get out of here."

"Let's see what you got. Hey, a butterfly knife. Nice, and shiny. I want it."

"That's mine."

"Jeez. You got two!"

"You can have the one with the aluminum handle."

"No. I want the pretty one."

"Take it or leave it."

"Oh, OK!"

"I wouldn't want to lose that knife."

"What's a talisman?" Tom asks suddenly.

"I, uh, think its something that has special meaning. A memento," I dodge.

"I think you know exactly what a talisman is. People don't walk around using the word 'talisman' without knowing what it means. What's a talisman?"

"I give up, you tell me."

"I think a talisman is something that has power over good and evil."

"Is that what a talisman is?" I answer lightly.

"You know exactly what it is."

"I think I don't wanna talk about the knife anymore."

I watch Tom and Diane watching me, thinking that I do not want to be pushed.

"OK. Let's get back to Phelps," he answers, but leaves me feeling we will cross this path again.

I continue eagerly. It feels good to walk the familiar paths of the past when we had faces and lives together. We had little, but the memories are soft silver, untarnished by all that came after.

"Phelps was the only one in the hooch who cared enough about the Vietnamese to learn something of their language. I never gave a damn; few of us did. I guess I'm ashamed to admit that, after a while, I didn't think they were even human, but he would tell me what good people they were. That's kind of how he was. He used to sit with Babysan—she was the youngest of the two mamasans who cleaned our hooch—and teach her English. Phelps and I were probably the only two who weren't trying to crawl into her skivvies. Not that Babysan let anyone in; she was a pretty good kid. Sometimes I think he quit flying gunships because he just didn't want to kill anyone like her anymore. I don't blame him. Phelps was no coward, and it happened to me just before the end, so I understand. The killing—it got to you after a while. But maybe he'd still be here if he had stuck it out in the guns."

"What do you mean, 'stuck it out'?" Tom asks.

"He began his tour flying in gunships. I started in slicks, but we switched about halfway through.[4] I remember going to Gunny in the line shack. I wouldn't tell him why I needed out of the slicks, why I had to have a gunship. I just said I'd done my time and I wanted to rack up more missions in the guns. It was a very 'Marine' thing to do, you see. I couldn't tell Gunny that I couldn't stand the screams anymore. The wounded . . . I couldn't. Oh man, I could take the blood, but the screams. . . ."

You can get used to blood. But the echoes of the screams never go away. Someone asked Helen Keller which sense she most deeply missed. She said that blindness separates you from things, but deafness separates you from people. God! if only I could shut out the screams. Their pain is still fresh.

"Phelps . . . we used to talk," I continue, letting the sound of my voice drown the echoes. "After six months, he just couldn't hack the killing. He still wanted to fly—hell, we were crewchiefs! He switched to slicks the

4. Unarmed UH-1Es, or "slicks," were used for observation, chauffeuring VIPs, resupply, and medical evacuation (medevacs). VMO-3's slicks were frequently crewed by a pilot (officer) and an enlisted crewchief. There were insufficient flying officers to place a copilot in every slick, so the crewchief often flew in the copilot's position. The pilots taught the crewchiefs to fly and land the aircraft in case the pilot was disabled. A slick carried a single M-60 machine gun, stowed in flight under the crew's seat, for use in emergencies; the aircraft was otherwise unarmed—hence the term "slick." VMO-3's gunships carried four M-60s and two 7- or 19-shot, 2.75-inch folding-fin rocket pods fired by the pilot; two M-60s in a chin turret controlled by the copilot; and one M-60 for each crewchief and gunner to fire from inside the aircraft. The typical ordnance load was 4,000 rounds of 7.62-mm ball and tracer, and fourteen to thirty-eight 10- or 17-pound High-Explosive Anti-Tank (HEAT), flechette, or white phosphorous (Willy Pete) rockets.

same time I got a gunship. Gunships covered slicks on combat missions, you know. Funny, how we went from him covering me on missions to me covering him. I don't think it ever occurred to either of us just to quit flying. We could have quit anytime, but we didn't.[5] Most guys kept flying until they couldn't fly anymore. Mouse flew 'til he was hit, and Saint, and the Jew, and. . . . I guess quitting wasn't the 'Marine' thing to do, either."

"You were better Marines than you thought," Diane notes, softly.

"Yeah, yeah. Maybe." My eyes mist; it feels good to recall the pride.

"You had some good men in your outfit," Tom adds.

I clear my throat, straighten myself in my chair and hold my head erect. For an instant, the soft flesh of middling age grows sleek and firm, as it was before, and for the first time since entering the room I meet their eyes with a confident gaze.

"We were all good men."

The old pride fills me. But it is a moment only. Pride seems the most fleeting of memories, yielding always to the pain. The pain cuts as fresh now as it did so long ago, drowning the pride. I see Phelps, clearly, painfully, on that last day.

"I brought him this lunch, see. He was on standby and couldn't leave his aircraft, so I brought it back from the mess hall. It was a meat loaf sandwich and a couple of hard-boiled eggs. The meat loaf was wrapped in waxed paper. We had a new mess sergeant, finally one that wasn't a thief. If you told him you liked his cooking, he'd even give you extra. Phelps was in his slick, VT-4. I can't remember the name of his ship. We named all our aircraft. Maybe it was named after his girl. I can't remember her name anymore. I remember her picture. It sat on the shelf in his cubicle. She was a pretty girl. High-school picture—you know the kind, where you look like your head is dislocated relative to your shoulders. She was pretty. I think her name was. . . . I was in VT-11. You know, we were two of the last ones still flying from the original crews. Most of the rest were gone. None killed; a bunch wounded, unfit for combat, couldn't fly.

"Mouse took it bad in the arm—M-16, shot by our own guys. The bastard was wearing my flightsuit, too. It only nicked the sleeve, but -16 rounds were bad shit—took a hunk cut of his triceps, like pulling a spoon

---

5. Flying (at least for enlisted flight crew) was voluntary. Any crewchief or gunner could "turn in his flightskins," which referred equally to his flight equipment and flight pay, anytime. In practice, most flight crew continued flying until they were wounded and disabled so as to be unfit for combat; however, it was common practice for long-time crewchiefs and gunners to stand down from the flight line during their last month In Country.

through pudding. Bled all over the suit, the bastard. It was my best one. He never even replaced it. What a shit! I had to walk around in his blood stains.

"Tommy's face split open when the CO crashed into a .50-caliber nest. It got to him; he never flew again. What a shame. Tommy was the best. He trained me to be a crewchief. What a way he had with the broads! Used to say he had 'fuck-me' eyes. Dutch got hurt bad in Tommy's ship—same mission; so did the CO and Captain Brucie.

"Tiny went quadriplegic when his gunner's belt severed on impact. The goddamn aircraft broke apart, he was thrown clear, and the tailcone landed on him. Really bad luck. Tiny was so skinny, one foot either way and the 'cone would have missed him. Brucie's legs were broken on the same mission, but he recovered.

"Then there was Patrick. I think his mind went after he saved Chandler's life, when that piece of heavy caliber took off Chandler's hand near Dong Ha Mountain. Y'know, that fucking gunship rolled all the way down the hill? Tossed its blades, of course. Cabin stayed intact, but it was a wonder any of them got out. Poor fucking Patrick. I saw him eating shoe polish once, but they wouldn't send him home. That was when Lieutenant Stevens got it—12.7 right in the flak vest. We had good vests Damn, they were heavy. Forty pounds, I think, some kind of ceramic. The round exploded; he never knew what hit him.

"Patrick and his crewchief used the tail rotor tie down as a tourniquet on Chandler's wrist. They had to carry him back up the hill to get to the relief ship. Steep slope, all under fire. They had balls. And Stevens— Stevens held that fucking gun all the way. Head-on fire, point blank. Damn, he had guts. Can't remember the crewchief's name. Langley? Longley? I remember he was interviewed by a Marine historian. The historian asked him if he was still going to fly. He said, 'I was born to fly.' We called him 'Born to Fly' after that. What a joker. I think his name was. . . . Jesus, it must have been hard leaving Stevens's body in the ship, even though they went back for him later. We tried never to leave anyone in the zone. I think it got to Patrick. Shit.

"The Jew's ear bone got blown out by the shockwave of a 12.7. If his head had been this much to the left . . . hell. Frankie transferred, and Vest, and Give-A-Fuck, and Sergeant Adams. Saint got ripped in the thigh with heavy caliber—clipped a femoral, I think—almost bled to death. When he came back all he did was walk around showing off his scar. Jerk!"

Some memories shine like gems of many facets, polished soft and warm; others, cruel, hard, and chilled as ice. I see them—us—as we were in the beginning. Tommy, Saint, and me posing in aircrew leathers at El Toro, hours from beginning our great adventure. I see us at the end—the end of all things, it seems. I miss them. I miss them all.

"Were you wounded?" Tom asks.

"Me? No. Phelps neither. We each had about three hundred missions when he. . . . Can you believe it? Three hundred missions without a scratch. Phelps and me. We thought we'd make it. We really thought . . .

"You know, we had a month left to fly! Phelps and me. One lousy motherfucking goddamned month. I was heading south to do gunship support off the *Tripoli* for a few days. He was going to pick up General Hochmuth[6] to head north on a VIP run. A lousy VIP run. I was busting his balls about crewing VIP ships. The sandwich had mustard on it! He liked mustard. On that coarse white bread they baked in the mess hall. Wasn't bad bread, really. Sometimes, all I'd eat was gravy and bread. I couldn't take the pot roast anymore. You know what they called it? Phu Bai steak. And at Marble Mountain, they called it Marble Mountain steak, and at Dong Ha. . . . He ate it in the aircraft while we waited for our pilots, and I smoked his last cigarette. Maryanne. I think her name was Maryanne. Can't be certain. We always used to joke about that—taking the last cigarette. I'd say, 'Give me a butt,' and he'd say, 'You wouldn't take a fella's last cigarette, would you?' And I'd say, 'You bet your fucking ass.' Or he'd do it. Didn't matter. Depended on who had the last butt. Didn't matter. We did it all the time.

"We always shared, you know. Always. I lit his last weed, and he mooched it back, and we shared it out on the mat where we weren't supposed to be smoking. It was a Philip Morris Commander. Filterless. I hated filterless. The tobacco sticks to your tongue, and you spend all your time spitting it out. Disgusting. Philip Morris Commanders. All we had. Twenty cents MPC in the PX for a whole carton.[7] Can you believe it? Still, they should have paid us to smoke them. Don't remember who held the match. I think it was him. Yes, it was him. But he used a lighter. C-rat matches were lousy. Had the Marine emblem on it. Zippo, just like mine. I'm certain it was Maryanne. It was overcast; monsoon season but no mist. Neither one of us smoked back in Pendleton. I don't know what we talked

---

6. Commanding General 1st Marine Division.

7. MPC was Military Pay Currency or Military Payment Certificate paper notes in denominations of five, ten, twenty-five, and fifty cents and one, five, and ten dollars. PX was the Post Exchange.

about. I should, though. I really should. I'll have to think on it some more. He used to talk about being a teacher. He'd have made a good teacher. He liked teaching Babysan."

My voice is thin and cracking. I summon the spirit from a hidden place to drive it.

"Then . . . the pilots came, his and mine, see? You know what he did? Jesus Christ! He flipped the cigarette butt out behind the extinguisher. What an idiot! I mean, a real fucking idiot! It had an inch left. Made me mad. 'You tossed my butt, you asshole,' I said. He said it was his butt. My butt. His butt. What a shithead. He was a real shithead. I don't know why he threw that goddamned butt out! Phelps could be an asshole, sometimes. You know, he . . . he. . . . We had ashtrays in the aircraft. Hell, you could smoke in flight, just not on the deck. He could have pulled off the fire with his flight glove and put it in his pocket. Or my pocket. Didn't matter. Perfectly good butt. I mean, I didn't have any cigarettes with me, and he had to go and throw out our only one, and someone would just have to police the butt, anyway, and . . . and . . ."

Tom interrupts softly, "You're angry with him, aren't you?"

"He shouldn't have thrown that butt away! It was dumb. Plain fucking dumb!"

"No, you're angry *with him*. For leaving you."

"He didn't leave, dammit. You can say what it was. I'm no fucking kid. You know. I know. We all know. He, he d . . . d . . . d. . . . Don't call it 'leaving.' Don't you . . . don't you call it that! People don't leave, not when fucking helicopters blow up. They d . . . they. . . ." *God, he was my friend! Damn him! Damn him for never coming back!*

I see him, through the tears—the last image, forever the last.

"He waved as his ship taxied across the mat. You know, he flipped me the finger! That fucker! Then he folded his arms and looked away on purpose, with his nose stuck up in the air so he wouldn't see me flip it back. I flipped it back, but it didn't count if he didn't see it. I got the finger for the last time. That fucker. That crazy fucker. I never got to say . . . I never got . . . never. . . ."

I have no breath to continue. Oh, it hurts—the emptiness so deep. Tom and Diane watch without question. I must finish the story. It is his story, and I am its keeper.

"We lifted right after him, and taxied to the arming area. He continued onto the runway. Must have gotten straight clearance to go pick up the general. I was occupied arming the guns, so I didn't see him take off. I wish I had. It would have been nice if I. . . .

"We flew south. And he, he went and picked up the general and headed north, and then he went out into the zone."

I take a last breath, close my eyes, lock my voice, my soul in a vice of iron. Control. I must control, this above all other things. I must speak the final words quickly. It is easier that way. I say them quickly and maintain control.

"And then he died.

"He . . . died."

I pound one fist against the other.

"He fucking died!"

I stare blankly at Tom and Diane.

"I never saw him again, not even his. . . . Mouse packed up his stuff, and they just took him away. When I got back, his place was empty. It was like he'd never been."

I point to my head, where the memory is imprisoned. "All that's left of him is locked up in here."

How is it that I should feel this way, after all these years, as if it happened yesterday, as if it were happening at this instant? All these years.

"There's a verse in 'Eternal Father,' the Navy Hymn, for aircrew, did you know that? It was engraved on a bronze plaque in the vestibule of the chapel in Pendleton where we went to Mass. 'Lord, guard and guide the men who fly, Through the great spaces in the sky. Be with them always in the air, In darkening storm or sunlight fair, O hear us when we lift our prayer, For those in peril in the air.'

"Ron and me, we liked that verse. You know, we read that prayer every Sunday. Every Sunday, we read that goddamned prayer, every fucking Sunday! It's not right, you know. It's not right! Three hundred missions. Oh, Jesus, we almost made it. We almost made it . . . h . . . ho. . . ."

Each word is a blade that tears itself through my throat. I taste the blood of each one.

"We almost made it home!"

Suddenly, I am in the thatched-roof church the Vietnamese built for us In Country. I hear the Hymn, sung dirgelike by the squadron, the men and crews of VMO-3. "Eeeeeee-terrrrrn-al Faaaather, strong to save, Whose arm hath bound the ressssstless waaaaave."

It makes me feel so dead.

"I stood next to Mouse at the service. Mouse recovered the bodies. I kept asking him, 'Mouse, what did he look like?' and Mouse just kept saying, 'He looked fine, Zack, like he was asleep.' I kept asking Mouse the same question, and Mouse kept giving the same answer. He was really patient. 'He looked fine.'"

The tale is done. Once again, I have fulfilled my duty, to remember.

"You've kept him alive all these years," Tom says slowly.

I nod through tears. I am so tired, without even the energy to sob. "I have to. I'm the one who remembers. Out there," I point to the World with accusing finger, "no one cares. Just me. I'm the one who remembers."

Again Tom grasps my hands. The human contact is foreign. Not since the zone have I felt the touch of a stranger who cared.

"We want you to come here and let us help you work through this."

I feel the caring, and sense the hope he tenders. I want so badly to come in from the zone.

"We know how to work through this, if you're willing, and we know what's on the other side. It'll take time, but we can do it, and you won't be alone. I can tell you what to expect. It won't be easy, but you're a survivor. You can do it."

"Me?" I laugh nervously, "Some survivor." I motion to my disheveled clothes, wipe the wetness from my face. "Do I look like I survived?"

"Yes," he says with conviction. "You were a Marine crewchief. You flew an entire tour in combat. You didn't accomplish that through weakness. You are a survivor."

A final grasp, strong, and Tom leans back.

"We'll work with you," he continues, "and talk about the things that happened. We'll talk about Phelps again. It will be painful for you to visit these thoughts, but we'll work through it all together." Tom speaks slowly, emphasizing each phrase. I concentrate as hard as I can, sensing the weight in his words.

"When we're done, everything that's troubling you will be in one of two places. There will be no in-betweens. We'll work with you until then, and as long as you need help, we'll be here for you. That's a personal commitment."

His voice carries a steadiness and an urgency that, despite the years of mistrust, I believe. I want to believe. I need to.

Tom continues, "Some of the things that affect you will go away. Some dreams and intrusive thoughts that you have now simply won't come again. Right now, you're struggling to control a lot of things. You won't have to control as much. You won't need to."

I shift nervously in my seat, suddenly afraid. Vietnam is all I have, all I have had for so long. I cannot lose it. Can they take away my memories? What will I be, if not a Vietnam veteran? It is what I am.

"Are you saying that I won't remember things anymore?" I think of Phelps, my duty to remember. I cannot not remember. It is unthinkable, not to remember.

"You won't lose anything. You won't forget anything. You're a veteran, and always will be. No one can take that away from you. What'll happen is that when you remember certain things, you'll remember them as they happened, not as if they are happening. You'll remember them with sadness, or anger, but you won't be angry. The memories won't drive you. They won't control you anymore, and you won't have to try to control them."

His claim seems both promise and threat, at once too much to hope for and something to dread. I am what I have been and know no other way to be, in spite of the pain. I am proud to be a veteran. The pride is linked to the pain; it has always been this way. If I will not surrender that pride, how can I be delivered of the pain? Still. . . . I must think on this.

"OK," I answer warily, "you said that I wouldn't have to control some things. What about the rest?"

"There are other things that'll never go away. This was a war. Your friends were wounded and killed; your life was altered by that. No one can change what happened to you. Some experiences from the war will always affect you. They'll always try to drive you to behave in certain ways. We'll help you learn ways to control the things that'll never leave, so that they'll no longer control you."

"I need you to be more specific," I interrupt, wanting to understand.

Tom scans the ceiling, then watches me steadily, stroking his beard. I sense caution in him, rather than evasiveness, as if my demand for specifics is premature. I return his gaze, unflinching. They seem to offer much, in the Center, but they ask much as well. Trust, resolve, faith—things I have not granted in years. I will understand what lies ahead.

"Trusting people is difficult," Tom begins. "You spoke earlier about only trusting your wife, that you can't trust anyone else. In Vietnam, trust was black or white because being alive or dead is pretty much black or white, but if you think about it, you might see that there are degrees of trust."

"How can you trust someone in degrees? You either trust them or you don't," I counter.

"Look at the different ways you relate to other people. Some people you share confidences with, some you share only a cup of coffee and the weather, some you just nod to when you pass them in the hall, and some you don't turn your back on. Do you place the same degree of trust in someone you share an idle moment with, as you place in a close friend? Do you need to? It's not that you *mis*trust this person, it's just that the relationship is different, less established or demanding than one you'd develop with a close friend, and you relate to them on a different scale."

"No," I mutter grudgingly, "but I'd like to think I could. I'd like to think I could have the same kind of relationships I had In Country. Are you telling me that's impossible?"

"No, but those relationships were special, and those were special circumstances. Only in something like a war, where everyone depends on everyone else to survive, can you have a large body of people with such deep relationships Usually, people only develop close friendships with a select few. The war changed the scope of everything. Even your Marine training taught you to depend on 'the team' more than most people rely upon others in civilian society.

"It's possible you'll always have a tendency to mistrust people. But, you don't have to mistrust everyone simply because you don't share the kind of relationship you had in the war. You just trust people to the level of the relationship that you do have, and recognize the value in having the relationship. Granting levels of trust is something you can learn to control. Make sense?" Tom finishes.

"Yes," I answer evenly. No, I think suspiciously. In the war I gave all of myself; I held nothing back. I was crewchief! The word itself defined trust, dependability, dedication. I cannot fathom doing things by halves or quarters. How do I grant one-eighth trust? I watch Tom watching me. I think he can read my mind, that he sees this is much for me to swallow. Nevertheless, his premise is defensible, and logic impels me to consider it. Besides, what the hell do I have to lose? Trust is mine to give. It is something that I control.

"Remember a little while ago I said you had lost the ability to choose. When you regain control you'll be able to choose how you react to the feelings that remain. You'll control their influence on your life. No one can ever make you do anything you don't want to do, but we'll help you regain control, and through control you regain the ability to choose. Then, if you never trust anyone again, it'll be because you want it that way, not because you can't. For better or worse, the choices you make will be up to you."

Tom brings the session to an end. "This is enough for today, I want you to go home and get some rest. You'll come back on Monday, OK? If you need to talk before then, you can call or you can just come in. There are two other counselors, and Diane is almost always here. You can always get hold of someone if you need help."

We walk to Rose's desk and set a time for Monday's meeting.

Another vet waits in the anteroom and nods to Tom. "Two minutes, and I'll be right with you," Tom says quickly.

"Take your time. I've got Rose, so who needs you?" the vet responds agreeably.

Our eyes brush; I see patience, sadness, understanding, and take in the full meaning of this place. I am not alone.

Tom and Diane walk me to the door. It is an awkward moment. I think but for these people and this place, I may have ended my life. I cannot say that it would have been so, only that I had never come closer. "Thanks," I say. The word seems at once too little, and as much as I can give. I recall a nameless grunt who visited my riddled gunship after we'd pulled him from the zone. He touched the jagged holes that framed my battle station, looked at me with feeling eyes, and said, "Thanks." I'd shrugged and answered, "That's OK." Now I know how he felt.

With a final handshake in the doorway, Tom says, "Take care." I smile to Diane and Rose.

I leave the Center and walk slowly to my car. Along the way I remember to raise my head and match my stride to a silent cadence as once I was trained. It is not the way I entered the Center, but it is how I choose to leave. The evening's breeze is cool against my cheek, still damp with Phelps's memory, but the memory and cadence warm me.

Tom waves from the doorway. I gesture a last good-bye and drive off. As I glance in the mirror I see him watching after me, like a friend who waits for your return.

So much to think about. Choices, belonging, hope, control—all attributes of survival.

Maybe I did survive. Perhaps I will.

Chapter 2

To thee old Cause!
   Thou peerless, passionate, good Cause,
Thou stern, remorseless, sweet idea,
   Deathless throughout the ages, races, lands,
After a strange sad war, great war for thee,
   (I think all war through time was really fought, and
      ever will be really fought for thee,)
These chants for thee, the eternal march of thee.

—Walt Whitman, *"Leaves of Grass"*

"Why did you go to Vietnam?" Tom asks, then watches through steady, patient eyes.

The days since we first met have passed quietly. I've maintained a tenuous sort of control, enough to sustain me to this first return to the Center. I feel good, the best I've felt in a long, long time. Still, I sense it is impermanent, and I think I've begun to know how a recovering alcoholic must feel, taking things one day at a time. I do not know what to expect, but I've prepared myself to begin Treatment, whatever that is, and I've vowed to cooperate.

But "Why did I go to Vietnam?"? I am disappointed. I want to ask, "Can't you think of anything more original?" but that would not be cooperating.

"That's an old question. People have been asking me for years. It's an *old* question," I answer, hoping he takes the hint.

"Why did you go?" he repeats, insisting.

I know the answer. At least, I know an answer; I've had much time to ponder it.

"Look. What difference does it make? I tried to explain for years. No one wanted to listen. No one cared. It's too late now. It doesn't make any difference anymore. Who fucking cares?"

The steady, knowing gaze, questioning, challenging.

I know who it is that cares.

"Tell me," Tom demands softly, "Why did you go to Vietnam?"

That's how the questions began, with the Question. It rose from the ashes of the beginning, as the promise of glory faded and the long hurt was born. The Question lives still, unremitting in its demand to be answered. But it was not a question at first. An accusation it was, and my inquisitors cast it down in challenge like a brutal gauntlet, demanding justification, not explanation. Never explanation.

The war grew old. It died, at last, of its excesses. So many lives devastated, the living no less than the lost. Yet the Question survived, feeding on the carrion anguish of a generation. In time the war passed into remembrance—for some, perhaps, but not for me. The Question lived, and with it the hurt, but it was never a question.

A judgment, a damning judgment: "Why did you go to Vietnam?"

I subtract the year of my war from today to compute time's passage. Fourteen years. It shocks me. How have the questions—the Question—kept alive?

Once, accusation met justification in passionate battle. Now the answer comes more slowly, thoughtfully, framed with the greater care of middle age. And my eyes reflect a sadness earned at an age that deserved better. When I study them in the mirror, it seems but moments have passed, not years, since the sadness was burned into them. I have but to close them and I am nineteen again! as if there is nothing worth remembering between then and now.

After fourteen years, you would hope the Question would have been forgotten. You would think the sadness would have yielded long ago to happier times in better places, that other things of more worth or less pain would have taken the place of the old memories. But it is all so important, the past. Nothing has faded.

"You want to know why I went to Vietnam," I respond with an evenness I do not feel.

"Yes."

"And who do you want to answer that question?"

"You."

"Which me?" I parry.

"What?" Tom asks, confused.

"Which me?" I demand, "Which me, damn it. Which me? Pick one!"

In the beginning it was simple. The World made it so, to deal with us on its terms. I was a pariah when the Question was new. A babykiller. Bad soldier, bad war. I hadn't planned it that way, but that's how it played. Nothing could be simpler. I was a hit on the small screen, a late-show villain—the scarred veteran atop an office tower with his M-16. It was my fault, of course. Being there as a country, being there as a Marine, being there as me. One cannot be a pariah without taking things personally. I was a pariah from 1967 on into the late seventies. I call it my Pariah Period. It is, perhaps, the most bitter of my memories of the things of Vietnam.

It began at mail call, Christmas 1967.

VMO-3's mail room is a CONEX box just behind the flight line. Steel and green, the ubiquitous CONEX stores all things of value In Country, from ammo to icy corpsesickles in green body bags. We swarm to it each day and jostle each other to steal a glimpse into the clerk's bag. It's the closest you can get to home. Mail, any mail, lights up a day, and any day turns black when there is none.

The clerk enjoys his duty, and calls the names slow enough to satisfy his lust for power but fast enough to avoid getting his ass kicked.

"Shitslinger!"

"Yo." Charlie Shits is married, and an envelope heady with perfume passes from nostril to nostril on the way to his reaching hand.

"Hey, Charlie, I think she wore that between her tits for a week. Holy Christ!"

"D-cup. Look at how sharp the crease is. Has to be a D-cup. All that pressure . . . ooh!'

Charlie buries a grin in the letter, inhaling deeply.

"Moose. Package, Moose." A chorus of "Dibs!" runs around the CONEX. Moose grabs the package and smiles a greedy Semper Fi smile—"Hooray for me, and fuck you"—then hammerlocks the package under one arm. "No dibs, you assholes."

The expectant atmosphere turns carnival for those rewarded with precious words, a brownie, a mold-encrusted salami. Disappointment consumes the rest, who linger when everyone else has left, hoping the clerk will find something in his sack that isn't there.

There is nothing for me, but there is a letter addressed, "To Any

Marine." I snatch the precious thing before anyone can lay hands on it. Mouse and Shifty fight me, but I am fast and nasty, and run off like a thief. Word from the World, even anonymously, is better than none.

I sit on a pile of rags in a corner of the hangar where the wind wash of turning aircraft cannot reach. The card is red and white, with sleighs cavorting in a Currier and Ives world. "Merry Christmas." I open it slowly, needing the moments to last.

> Dear Babykiller,
>
> How can you live with yourself, knowing you have slaughtered innocent children?

I read it once, twice, looking for the punch line. No return address, no name to hate. "Babykiller. . . ." The sting of it won't go away, not even after I burn the thing and scatter its ashes. It won't go away.

In three weeks I left Vietnam, and for the year and a half that followed I lived secure in the company of my fellow pariahs in VMO-1, a Huey training unit in North Carolina. The Marine Air Facility near Lejeune was a bleak limbo between Vietnam and the World, a place to endure until discharge. Job One was to cram enough information into the heads of fresh-faced air wing boots so they could go to Vietnam and become FNGs and, if they were lucky, live long enough to return and abuse others as we abused them. Job Two was to fuck off on Job One, usually in the line shack.

Marine line shacks are ramshackle affairs built to store bodies and equipment between flights. In Vietnam there had been little time between missions, but what there was we'd spent in the shack, playing cards or honing a blade, and planning our future in the World.

"What're you going to do when you get back to the World?"

"I am going to get laid."

"I'm going back to school. Gonna eat the apple and fuck the Corps; gonna get an education. I won't even talk to any of you bastards. I'll see you on the street and just look the other way. No—I won't even see you on the street, 'cause I'll have so much money I won't ever have to go to the same scumbag places you assholes will be. Hah!"

". . . and laid, and laid, and laid, . . ."

"I got a job working for my old man."

". . . and laid."

"What's the first thing you're going to eat when you get back?"

"Steak."

"Lobster. I'm going to eat all the lobster I can stuff in. Then I'll stick a finger down my throat 'til I puke, and make room for more."

"I am still going to get laid."

In North Carolina we passed a great deal of time in the shack. We lounged on torn Naugahyde, burning cigarette holes into the arms of the big stuffed chairs, and regaled each other with the old stories to remind ourselves of who we had been.

"Remember Mouse?'

"Yeah, fuckin' Mouse."

"Fuckin' Mouse got hit wearin' my fuckin' flightsuit, y'know?"

"Yeah, his pilot sent him out into the zone to capture a zip. Can you believe it? Got popped by our own grunts, coming out of the trees. Fuckin' Mouse."

"Musta took a pound a meat off his arm."

"Nah, half a pound.'

"Nah, 'bout a teaspoon. M-16 rounds leave big holes, though."

"The asshole didn't even clean my flightsuit when he gave it back."

"Yeah. I got stuck with that pig gunship of his, too. What a pig. Leaked oil. Shit!"

"What did he call that pig? Frito? Dildo? Mungo?"

"Frodo. Frodo was a hobbit, kind of a cross between an elf, a dwarf, and an orangutan."

"Yeah, Frodo. Fuckin' Mouse."

"Remember the night we ran blood to Dong Ha?"

It was mindless and safe in the line shack, yet it was the security of a cage. Beyond the base the World we had yearned for was foreign and unfriendly. In an airport or bus station the satin ribbon of yellow-green-red pinned to my chest drew the epithets the World applied to my kind.

"Hey, look! Here's one of the warmongers. Here's one of the killers. What did you do over there, killer? Why did you go?"

"Look, I was fighting for my Country. They killed my buddies. They . . ."

"You were killing innocent people. It was their country.

"Stop The War. Stop The War."

"I was . . ."

"Stop The War. Stop The War. Stop The War."

"Babykiller!"

In the shack we also told these new stories.

"I went home this weekend, and one of them fuckin' long-haired freaks braced me in the bus station."

"Fuckin' freaks. What'd you do?"

"What d'ya think? I kicked his fuckin' ass! Ha!"

"I heard of a grunt who walked into a bar, and the peace-creeps ganged up on him. He pulled a frag and greased them all."

"That's the way. Grease the motherfuckers. What we need is some dungaree liberty. Show 'em how we did things In Country. Kick some ass!"

"Fuckin-A!"

The world is very clear when you're a pariah. There is only "us" and "them." I am an us. Pariahs always get to be the us's. It's one of the few perks of being a pariah. Another is self-righteousness. Pariahs revel in self-righteousness—it's what we have instead of pride. From it I forged my armor, strengthening it with bitterness each time they pressed the Question beyond academic query into insult and accusation. Us and them. It was a simple equation.

A Christmas party. The conviviality of the crowd sets me apart, though I want so desperately to belong. I stand ill at ease at my wife's elbow, feeling safe only in her proximity. She is invited; I am there because I must be. She is popular, I am "husband," and because she is popular someone rapidly whisks her away. I do not resent it, but it is discomfiting; I do not like to be alone, without someone to watch my back. Soon she is speaking animatedly to strangers across the room. Her eyes seek me out every now and then, and I nod wanly: "I'm OK." She puts up with a lot and has a right to a good evening, but the sweep hand of my watch ticks the time off too slowly.

Someone who is her co-worker or acquaintance or friend approaches.

"You're Grace's husband."

"I'm Gracie's husband."

"I hear you were a marine."

It is 1970, and no honor to be a Marine. I am tuned to the inflection of his words. The emphasis is on all but "marine," which he says quickly, in lower case, like a minor invective.

"Yes."

"'Nam?"

"Vietnam. Yes."

"Bad situation."

A fencing observation. The open-ended statement is a test to feel me out; am I pro- or anti-war? My response is a terse "Yes," which admits nothing.

"Why did you go?"

The tension builds, and my defenses rise. Does he feel it, or is it just me who defines the roles? I am the stag, baited by the dogs.

He answers himself quickly, "Oh, I know, you were drafted," extending to me the option of an excuse.

"No," I bridle.

Why don't I take the "out"? I could avoid so much if I did. "Yeah," I could mutter ruefully, "I lost my exemption." I could shuffle and bad-mouth the military, and bring the interrogation to a halt. It would be the easy way, but I rebel at the thought. Can I not reject that which is so integral, or have I come to need the confrontation?

"I volunteered for the Marines after high school," I add, spurning the lie and the excuse.

"Oh, uh, OK." He wonders why anyone would be so dumb as to volunteer for the Marines, but he doesn't say it. People don't tell former Marines that they're idiots for enlisting. It's unhealthy. Being a Marine has few advantages. Being a former Marine has many, not the least the reputation of being an animal.

"You didn't see combat, did you?" he asks. I sense that combat makes him uncomfortable. Why? Does the specter of war threaten him so ominously from my shoulder?

"Yes, I did. I flew in helicopters." I force a quiet answer. Anyone who watches the late show knows that real combat types are always soft-spoken.

"Choppers, huh? Wow."

"Helicopters," I correct him, bristling even more. "We called them helicopters, airplanes, 'planes, ships, pigs, fucking pigs, birds, slicks, gunships, or aircraft. Not choppers, whirlybirds, helos, whoppy-copters, floppy-copters, or any bullshit like that. And the Marines in '67 called the place Vietnam, Down-south, or In Country. I never heard 'Nam' until I got back to the World and the *civilians* were using it."

"Were you wounded, or shot down?" he asks, ignoring my sneer.

"No," I answer, and sense his disappointment. Would he have asked to see the scar, or is he simply, clinically, curious about pain? I have no scars, but of pain I know a great deal, and my knowledge is neither simple nor clinical. I know what fear and pain sound like when they're locked behind clenched teeth. I know the muffled grunts of those who fight for control. I know the screams of those who do not. I know the pleading whimper of those beyond the need to control, and I know the rattle and the hiss, the scream and the silence of life as it passes out from the lips of someone too young to comprehend the end.

Yet I share his disappointment, and once again feel the guilt of escaping without injury, none the less real for being undeserved. Does it mean I didn't do my part? Or hasn't one really been in combat unless they've been wounded? Is it my guilt that I see mirrored in his eyes, or is he so callous? One of these days I'll lie and say I was wounded, just to see what happens differently.

"You were lucky," he offers. Again the disappointment—mine, his, both, I cannot tell. No matter; I have never felt especially lucky. I remember the missions when someone took my place in the path of a round. I'd spend hours, days, flying in the same aircraft, in the same zones, only to be relieved to go to chow or R & R so that another crewchief went up in my place, to be shot down in my stead. Is that how you define "luck"? Someone else gets it instead of you? I remember how it was at the end. Three hundred and ninety-three missions without a wound. It seems as if I lived through it all without having survived anything. If I have survived, I should feel good, shouldn't I?

"You were lucky." My feelings are too complex to frame a response, and the guilt too keen to display. It is locked inside. I answer slowly, with words I do not feel, "Yes, I was lucky," and leave it there. He wouldn't understand. Hell, I don't understand.

I study him, awaiting the next question. I know what it will be, and soon he knows as well, but propriety briefly ties his tongue. Some things are difficult to demand of another human, even a Vietnam veteran. A spectrum of emotions play upon his face: fascination, embarrassment, eagerness. Empathy, sympathy, and respect are absent. Curiosity wins.

"Did you ever . . . did you actually kill anyone?" The words come in a rush, all blurred together, but "kill" always stands out.

"Yes," I answer simply and add nothing.

"How many? A lot?" He demands a number, any number, as if killing many differs from killing once.

I answer with a stare as malevolent as I can muster.

"I never counted. At least one. How many did you want?"

He is shocked. How can I make such a response? Don't I take this seriously? This is Death we are talking about. Only an animal would answer thus.

Yes, I take it seriously. Quite seriously, for my experience is neither poor nor antiseptic. I know what Death looks like and feels like and smells like when you meet it inside a very small place—the flight deck of a gunship, the loneliness of the zone. Do you know how much blood there is in a human body when it erupts suddenly, and the rose mist of it freckles your face, lingers metallic on the lips, the tongue, until the wind stream's screaming blast crusts it upon your flesh, then scabs it off like flakes of dun snow? It is more than you ever imagined. More than you think possible. So much that it never, ever drains from its corner in the back of my mind where the memories pool. The blood glistens in that corner, and the sight and the smell and the feel are as real now as then. I take it too seriously to let it show, to let anyone get close enough to penetrate the simplicity of the equation that is the shield against the things that I have seen, the things that I have done. The thing that I became, to survive.

How many did you have to kill for it to be enough? How many before you stop remembering them one by one? Something they never tell you, when you learn to kill the enemy, is that the enemy is only the enemy during the killing. I don't remember how many I killed, but I remember each time. If I count the times, and the men and the women within the times, and blur them all together by sterile addition, would it make the question easier to answer? The memory easier to bear? Would the scar in that corner of my mind be smaller, heal more easily, glisten less wetly? Tell me, someone, what's the number so that I can stop remembering each time, and remember one time, and then maybe it won't hurt so much to remember.

Then the final question.

"Did you . . . you mean . . . you actually liked killing people?"

I used to argue the premise, that killing does not equate with wanting to kill. I'd plead my case with logic, explaining that few men in war actually like killing other men. They do it because that's what war is all about. I'd try to explain that 90 percent of the time there's no opportunity to ponder the sociopolitical ramifications of killing the enemy, not if you want to survive. Killing machines survive. Philosophers die pondering, and sadists grow sloppy. Combat tolerates neither extreme.

I killed because it was my job, and I did it before some motherfucker did his job first or better. I killed without emotion, without feeling, without cursing or anger or tears, or cracking a smile; without breaking stride, missing a heartbeat, or taking a deep breath. I did it without noticing much more about the fucker than that he was emphatically dead. Then I moved on, to blow away the next poor bastard who crossed my path.

Five percent of the time, I was just too damned scared to feel anything, sometimes too scared to think. That's when those reflexes drilled into me during boot camp took over, preserving my life when the rational mind succumbed to terror. They carried me through the moment, until something inside told me I was going to survive, or woke me to the greater terror that others would die unless I quelled my fear and regained control.

But then there were other times, mercifully few, when someone wearing my face killed another man and reveled in the killing, and wanted it to go on and on forever. This couldn't have been me! I had once worn black and white and carried the precious Eucharist to the altar, and my nearest urge to kill had been frying ants with a magnifying glass before Mom took it away. No. This was someone with my face who'd gazed back from a shaving mirror on the morning of a day that began like all others but became a memory never to be forgotten. Whose thin control over passion failed for too many moments. Who for once *felt* during the killing. And what he felt was a craving for revenge. For the lives of his friends, but also for himself, because he realized that he had become a kind of *thing* that could kill another man 90 percent of the time and feel nothing! Feeling nothing frightened him so. To lose the ability to feel angered him, so he struck back! In hopeless, helpless rage, and in striking cut the scars even deeper.

But it was not me.

"Did you like killing?" I used to argue, justify, explain, until I realized that anyone who asked that question had already made up their mind.

"Yeah, it was a real trip. Killing, man! It just made you feel like you had The Power!" I feign a far-off look, as if gazing in wistful longing for a time long past, and say, "Man, it's been years since I've blown anyone away. God, I miss it."

Us and them—a simple equation, but it served.

"That was when I was a pariah," I say to Tom.

The memory churns, but I bury the bitterness with a flippant tongue.

"I was good at being a pariah, you know. We learned to be the best at whatever we did in the Marines. Pariahs, Babykillers, fucking butchering animals, whatever. Hell, my father did I-don't-know-how-many years in World War Two, blowing the shit out of Japs, and never made it to Babykiller. None of those bozos from the Good War did. I guess they just didn't have what it took. Of course, my generation was better educated— my parents grew up in the Depression and never learned how to be properly sadistic. I remember Sister Agatha telling me in the eighth grade, 'Now, Ronny, when you grow up be certain to join the Marines and become a Babykiller. Ronny, you know you can do it if you try.'"

"All right," Tom interrupts, gently, "it hurts."

"Me? Nah, not me!" I shout. "It doesn't fucking bother me! I was a fucking Marine Babykiller. It was all I ever wanted to be. When fucking Kennedy said, 'Ask not. . . ,' I said, 'Hell yes, I want to serve my country and be a fucking Babykiller!' I wouldn't have enlisted unless the Sarge said 'Zaczek, you can kill kids.' I . . . I . . ."

The tears come. I summon the old anger to contain them. It has hardened with the years, and grown brittle, but it is the only way I know. To whom should I direct the anger? The adversaries of my youth are memories! The war is remembrance to them, and I am nothing! It is only the epithet "Babykiller!" that echoes through the years—the memory of its pain, the pain of remembering. How it hurts. Is it possible to be angry with a memory, or can you only be bitter? When anger hardens, so that it can no longer distinguish memory from reality, what is it but bitterness? If anger was my shield, what is bitterness but a prison, and what does bitterness do but consume? Us, them. The old equation is false in its simplicity. It does not protect, and serves no longer.

But it is the only way I know.

Nonperson oozed in as the war died, or after the war died. It's hard to tell when; nonpeople are pretty quiet.

The closing act of the war was messy and unfulfilling. Bad war, fucked-up soldiers, all-around disaster. The show had run too long; everyone was tired and just wanted it to end. Even I wanted it to end, no matter what the ending. I watched with a sense of numbness as North Vietnamese armor overran the places I'd fought for and lived in, lacking the energy for either anger or sorrow. It seemed nonreal.

Nonperson lasted three or four years. Lonely, it was. No more "us." Just me. All the way down to me. Nonpeople don't belong to anything, not even to one another. My Nonperson Period, I call it.

The Question didn't come so often, but when it did it was posed academically, not with the emotional assault of the Pariah days. There was a touch of the supercilious abstract, the philosophical, the "Let's be adults and analyze why you went to Vietnam while I went to college or work or Canada, don't you agree?" manner with which subjects are discussed over cocktails or beer or Perrier. The war was a distant matter for objective discourse. Only in retrospect do I realize that I thought the war was behind me as well! I wanted to get on with my life. I wanted to deny the war's continuing influence. I wanted to belong.

So we'd analyze why I went, objectively of course. I was (a) suckered by my leaders into volunteering, (b) too fucking dense to avoid the draft, (c) a Babykiller, or (d) all of the above. "Why did you go to Vietnam?" they asked, but "You were a fool" is what they meant, and it was no question. The judgment was clear, and objective. I tried to see it that way, sometimes quietly and sometimes not. Nonpeople aren't very good with words, and my arguments were weak against the logic.

I graduated to Victim when the Iranian hostages came home. It wasn't an especially enviable position, but it was a step up from nonperson: good soldier, bad war, and none of it was my fault! Damn! The hostages were the best thing that ever happened to Vietnam veterans. I was grateful for every one of their four hundred days of captivity. (Sorry, hostages, but things were tough all over.) The hostages' homecoming got the country wondering about things like "parades" and "welcome home," and how there were a hell of a lot of nonpeople around who'd had neither. More importantly, it got us nonpeople wondering as well, enough that we started to become an "us" again. Things were looking up! I was a victim from 1981 until November 1982. My Victim Period, I call it, but I didn't enjoy it as much as I should have because too many things were starting to hurt. The Question became a statement, a sad acknowledgment: "You went to Vietnam. . . ," and the incomplete sentence tapered off into pity.

I couldn't handle being a victim, and neither could the nation. It just didn't work. Although media coverage for victims runs high, no one loves victims. They spoil too fast, like fish. They make everyone around them feel uncomfortable because the nonvictims suspect they might have had a role in creating the victims. Or at least in keeping them victims. That makes people feel guilty, and no one can stand that for very long. One way out is to have the victims apologize, accept pity, and then vanish, but I wouldn't apologize or act the role properly because I didn't feel like a victim. I certainly wouldn't disappear; by now there were too many of us breaking out of our cocoons, and we wouldn't shut up.

Four hundred and ninety-three feet and four inches of black granite from Bangalore, India, changed everything. The Wall was dedicated in 1982 and ushered in a new role. Tragic Figure. It plays very well. It's still playing. They make movies about me, and I don't even have to shoot people from office towers.

Tragic Figure is right up there with Noble Savage. Good soldier—hell, damned good soldier, a fucking hero!; lousy war, but y'know, we had to fight those Commie bastards somewhere, after all; and by the way the nation didn't treat you guys so hot, and maybe, well, we're kind of, uh, sorry.

Tragic Figure plays so well that aging preppies who escaped or avoided the ordeal tell me how much luckier I am than they because I "got" to experience Vietnam. They speak ruefully at parties, mumbling in semiembarrassed tones about missing the draft and missing a meaningful experience in their lives. I never know how to respond. Do they seek commiseration or absolution? I have neither to grant, not even if I could.

Sometimes I get to operate outside the pale of society and be an Antihero, the flip side of Tragic Figure. Rambo. Chuck Norris. The romance makes a good story. Deep down, most people would love to be an anti-hero for a day just to get the rush of doing great and evil deeds outside the pale. Anti-hero is sort of a victim who won't say, "I'm sorry." An anti-hero says, "Fuck you, World!" and the World cheers. Everyone loves an anti-hero, and no one ever feels guilty next to him, so they're willing to keep him around longer.

I enjoy Tragic Figure, especially the Anti-hero side, and I hope it plays for a long, long time. Now people ask, "Why did you go to Vietnam?" with genuine interest, and they mean well in their own way. It's finally a question, with no hint of accusation, no judgment, and damned little pity. (The pity's not all bad, you know. It gives you an edge. It hasn't been fashionable to publicly castigate Vietnam veterans since late Pariah/early Nonperson, or at least since we became a protected species under Equal Employment. Vietnam works well on resumés.)

Those who ask the question do so with a genuine fascination. I am living history, still the bug in a museum bottle but no longer loathsome. So many people seem to think that the historic sadness I carry is worth brushing against, even if too frightening to touch. Mine are the eyes that saw Kennedy and Khe Sanh! It's an updated version of "Tell me what the war was like, Daddy," and they want me to inform and educate, not justify.

But I can't stop trying. I don't know how to stop trying.

Pariah. Nonperson. Victim. Anti-hero. Tragic Figure. At least no one calls me Babykiller anymore.

But I remember when they did, and when I brood on how I have been perceived through the years, a shrill, silent scream erupts: *What do you want me to be? I've been here all these years. I am I. Why do you make me another thing, subject to your whims, subject to your inability to deal with me as I am?*

Then I curse my shallowness and wonder why I so readily yield my definition of self to society's perceptions. Why is my definition of self so tied to the dead war that will not die? Will I ever find my self again? If I do, will I ever be accepted for myself? I wanted to be accepted, at least in the beginning. When I wasn't, I convinced myself I didn't need acceptance, and then I would have spurned it, had it been offered, for it never came on my terms. I know that. But if the hand of acceptance is extended now, will I know how to take it? Will I even recognize it? Has it been extended? Us and them! Am I too blind, still working through the old equation, to see?

"You know, you never answered the question," Tom says.

"What question?" I ask, tiredly.

"Why did you go to Vietnam?"

"What the hell have we been talking about?"

"Everything else—what happened to you, how much it hurt, how you survived. Perceptions—all things that have to be said, that you need to say."

"People have been asking me why I went to Vietnam for years, especially after I tell them I volunteered. They could never understand. I tried to tell them. I tried."

I shift in my seat, feeling nervous and exposed. The anger I wrapped myself with is gone. Only the hurting remains. I can't explain. I can't even justify. I can only hurt.

"They didn't let you explain. No one wanted to hear," says Tom.

"But why? Why wouldn't they give me a chance?" I plead.

"Why do you think?"

"You know, what everyone says. The war ran on too long, too many casualties, too much shit on TV. Walter Cronkite said it sucked. My Lai and all that. You know."

"But why do you think?"

The answer is hard, and it takes long moments to respond despite the years I struggled to admit it.

"I think, because we lost."

We lost. There is no explanation for losing, for either a good soldier or a bad war. No justification. There are only excuses.

We lost. That's simple enough to say, a statement of fact: "Won the battles, lost the war." How many years passed before I could say it so simply. Five years before I could admit it to myself in silence. Eight before I could mouth the words, even in pain, and as I said them I could meet no one's eyes. Ten years before I could say the words and know I had begun to believe them myself. Twelve before I could say them and stifle the pain and hide the shame.

"We lost." See, I can say it again, in my best good-soldier, bad-war veteran victims' tragic bravado. "We tried, but we lost. We lost, we lost, we lost." Want to see how many times I can say it?

I can shout it too.

"We lost!"

See how well I handle it? Ask me in a bar why we lost. It won't even cost you a drink. I'll tell you that the politicians did it to us, that we had a "no-win" policy, that we shouldn't have been there but as long as we were, we should have, and could have, if we, if they, if someone, if . . . if . . . if!

"Do you feel like you lost?" Tom asks.

"We lost." I speak to an empty corner, as barren as I feel. "Everybody knows it. It was just a big waste. All those guys; just a waste. Everyone knows it. Most everyone. Some vets say we didn't lose, that the ARVN lost it for us. What the fuck difference does it make? Saigon isn't Saigon anymore. It's Ho Chi Minh City. Wars are for turf. We lost."

"Did you lose?"

"Me?"

"Yes."

"Me, personally?"

"You."

"Hell, I don't know. Yes. No. After a while, after I'd learned just how bullshit the war was, I knew that the only winning was to get my friends out alive, get the grunts out alive—hell, get myself out alive! Sometimes I let myself remember that there are a few grunt bastards walking around out there who wouldn't be if not for me. If that's winning, I guess I won something. Doesn't usually feel like it, though.

"Ah hell, I must have lost! Know why? No parade. It's as simple as that, isn't it? Winners get parades; we got to sneak in the back door, coming home in dribs and drabs. That says it all, doesn't it?"

Tom purses his lips, joins his palms across his belly as if in prayer, and taps his forefingers, studying me. "You don't strike me as someone who did it for the parade."

I mimic his pose, studying him in return. "It would have been a nice touch, yes?"

"Did you really need it?" he asks seriously.

"No." I drop the pose. "That's something the vets from World War Two and Korea just won't understand about us. They think we're whining because no one welcomed us home. Sure, we'd have enjoyed a real parade, but 'parade' is just a metaphor. I don't remember ever thinking about a real parade until people pointed out that we didn't get one. I don't think a parade is what we really wanted—at least not what I wanted."

"What did you want?"

"I think . . . just for someone to say, 'Zack, you done good.' I mean, we lost the War, but we all fought our own little wars. A few of us killed innocent people; most of us just did our lousy job the best we could with what we had—My Lais didn't happen every day. I saved a lot of lives, at least as many as I took. That deserves something, doesn't it?"

I stare away, recalling the sad history of those days, the emptiness and rancor of return. We deserved something for our pain. Was it so much to ask?

"I guess all I really wanted was for someone to say 'Zack, you done good.' That would have been parade enough."

"Would you do it again?" Tom asks, softly.

I breathe slowly, a cleansing breath. A sense of calm grows within me. There is, at least, one thing of which I am certain.

"Yes," I say quietly, "yes."

Tom leans close; almost whispering, he asks again, "Why did you go to Vietnam?"

I pause, remembering, rebuilding, reliving why. The years seem to fall away as I touch the shining memory of why, and answer as softly as he asks.

"It seemed like a good idea, at the time."

There is so much baggage that obscures the early years of Vietnam. It is easy, or perhaps convenient, to forget that in the middling '60s, graduating from high school to become a jarhead, doggie, airdale, or squid was as honorable as joining the Peace Corps—not to mention that it could get you laid. ("Peggy Sue, I may not come back. Let me take your love with me.") Of all the answers to the old questions, "I volunteered" is the most difficult for people to understand.

It was all Lenin's fault.

The national obsession of the '50s and '60s was Communism. Communism was new on TV. TV was new itself, and everything seemed so threatening on the small screen. The encroaching Red Menace, casting godless tentacles over the Free World, was terrifying, even in black and white. My leaders were afraid of the Communists; so were my teachers and parents. They spoke of the Red Menace often, and I listened to them and worried. I was nine when Red Army tanks crushed Hungary on my family's ten-inch Capehart. Nothing seemed to stop them; what if the Reds decided to come for me as well?

They seemed to be everywhere. Joe McCarthy was looking for Communists and finding them under our beds. I didn't know what he stood for, but he was gross and always needed a shave, and made people sweat in the hearings on TV. His witch-hunts spread suspicion and fear throughout the land, and I saw my elders believe him—or at least not disbelieve him, which was the same thing.

The Reds lived up to the yellow press. We'd upgraded to a Zenith console when an ugly man with a wart on his nose, the leader of the entire godless horde, beat his godless shoe on a god-fearing table in New York and threatened to bury us. Behind Khrushchev loomed the H-bomb. I didn't know what an H-bomb was, but I'd devoured photographs of Hiroshima and Nagasaki and figured it had to be worse. It was seven letters up from "A," and these were modern times, not moldy history, nor some distant enemy who could be stopped by the oceans. The threat from Moscow could become a reality over Omaha in less time than it took for Jimmy Durante to finish saying good night to Mrs. Calabash.

Still, annihilation hadn't entered our lexicon. America was invincible—hadn't lost a war yet. If we could just figure out a way to survive the first blast, we'd kick godless ass when the Commies came ashore.

To fight, you had to survive, and to survive, you needed a fallout shelter. Construction was based on three factors: how close you planned to be to ground zero, which way you expected the wind to be blowing for the two weeks following the blast, and how much you could afford. There were plans to survive any one, two, or all three forms of radiation. Alpha particles had little penetrating power and could be stopped by a few inches of dirt and positive thinking. Beta particles were slightly hotter, but a couple inches of brick could handle them, no sweat. Gamma rays were the problem. They could penetrate several feet of concrete and damned near any thickness of money the average private citizen could afford to pile around himself. If alpha, beta, or gamma radiation didn't get you, fallout could. It took an order of magnitude more money for the air-filtration systems you needed to keep out the ash.

When I worried about fallout, I worried about half-life and radiation poisoning. The half-life of the nuclear material scattered from a cheap bomb was two weeks; from a "top-of-the-line" model, eternity divided by two. In between, most half-lives were still several thousand years longer than the maximum amount of time you could survive on Spam in a home shelter.

Radiation poisoning burned your skin, turned you into a leper (best case) or mutant (worst case), and made all your hair fall out before it killed you with massive hemorrhaging. It made you sterile, too, but I didn't know what "sterile" meant back then. I thought sterile meant especially clean. I figured they'd use a cheap bomb on Baltimore. What was worth hitting in Baltimore? I'd lived there all my life—ten years—and knew there wasn't shit in Baltimore, but I knew they'd use an expensive one on Washington, right down the road. It was a well-know fact that the Commies were sloppy bastards, and near misses were near certainties.

My family could survive alpha and low-wattage beta, but not gamma or fallout as we were quite limited in our ability to afford anything that had a half-life much longer than my dad's vacation. His earnings were definitely in the low-half-life range—two weeks tops. I didn't know why they said two weeks and not thirteen days or even fifteen, and I was analytical enough to be suspicious of round numbers. Suppose they meant fifteen days and I came out early?

I worked earnestly to protect us against the threats we could afford to survive. Our basement was halfway beneath ground level and had windows that needed to be sealed. My father wouldn't let me deepen the basement, so I piled bricks and wood beneath the windows to await the Conelrad alert heralding the attack. I kept a radio in the food locker, and tuned it back and forth from 640 to 1240, searching for the Conelrad signal. There were times I couldn't find it, and this worried me too. It popped up on the Zenith in the living room all the time, but we had only the one TV. What if I missed it? If I spent too much time in the living room I'd get blasted by those lousy alpha particles that even we could afford to survive. How would I know when to dash from window to window to raise our shields? And why couldn't I have been born into a family that could afford gamma protection?

Years passed. The timbers rotted. Green slime covered the bricks, and so many soft bugs crawled beneath them that I'd have preferred a dose of beta to picking one up. About the only thing that remained intact was my paranoia. It was almost disappointing, after all that work and worry, that I didn't get to try it out.

Thus passed the '50s. They weren't boring—far from it. The sheer quantity of fads and teen idols was overwhelming. It was heroes that we lacked. This isn't to say heroes didn't exist. They just belonged to the older generation. When my parents looked at Dwight Eisenhower, I'm sure they saw the old D-Day commander exhorting his paratroopers on to victory. I saw a tired old man who spent the bulk of his presidential day chasing a small white ball while I fretted about gamma rays and fallout. He was not a hero to me; he was a legend. Heroes convey a sense of immediacy; they establish a standard or ideal to live up to. Legends are what heroes become as they pass into history. Sometimes you can work on a legend enough to turn it into a hero, but it's hard to do that with someone who's dead or close to it.

In 1961 John Fitzgerald Kennedy captured the imagination of a generation searching for ideals. A bona fide hero from my parents' era, his vitality and appeal to youth seemed to make him more a part of my generation than his own. We named him Blessed during his campaign, canonized him before his death, and deified him after it. Kennedy inspired us, appealed to us, influenced us. "Get in shape," he exhorted, and the nation's youth pulled on PF flyers and clogged the highways. He convinced us that we could make a difference; more than that, that we had a duty to make a difference. He convinced us that we could do more than survive, that we could overcome, and that we could stand tall in the face of adversity. Clutching our books on the No. 15 trolley, we shivered schoolboy goose bumps, chattering excitedly about Cuba and Castro and kicking Commie ass as JFK stood against the same warty gnome whose scuff marks marred that New York table.

Truly, those were glory days. Many of my kind heeded the siren call, too young to see the dangers of blind idealism. Patriotism, youth, energy, and intellectualism were in vogue. We thought we had something new, something our parents had lost. Such is the privilege of youth. We were the guardians of democracy. Kennedy set so much in motion: "My fellow Americans, ask not what your country can do for you—ask what you can do for your country." Often quoted, always inspiring, this was the keynote speech of my generation. We were the brightest and the best, who would serve the Best and the Brightest. That speech marked the true beginning of "my time."

Yet, it was the second phrase, seldom quoted and unfairly lost in the shadow of its predecessor, that raised the ideal to its loftiest: "My fellow citizens of the world, ask not what America will do for you, but what together we can do for the freedom of man."

The Freedom of Man!

I never recall his words without feeling the thrill and commitment to "do something," to "make a difference," that I felt while watching him take office. I cannot hear the strains of *Camelot* without thinking of him—and myself, as I was. It was like being at the birth of something great, and appreciating the good fortune of your presence to the depths of your soul.

John Kennedy was my first hero. And my last. Vietnam soured my taste for heroes, who are, after all, only men. Now I yearn for legends, who are inviolate. So many years have passed since I first heard his words. The new generations know him not and regard him only as legend, and must search for their own heroes. In spite of all that followed, I am sad that they cannot know him, even only as I knew him.

When I need to blame someone for my going to Vietnam, I blame John Kennedy. Lyndon Johnson sent me, and Communism was the cause, but Kennedy was the reason I volunteered. The need to blame is less strong with the years, but it comes. Sometimes I wish I could hate him, but I cannot.

For what should I hate him? For inspiring me to join a losing battle? Did he know that we would lose? So many, now, claim to have known it was a fool's errand from the start. Surely he knew as well. Perhaps I should revile him because I took inspiration from his ideals and, in the fragile naivete of my youth, wove them into my being. Perhaps I should hate him because I was suckered, because his ideas failed, his ideals proved false. Or perhaps it was me, perhaps I failed in the execution. Was it me and my kind? Were we unworthy?

Does it matter where the blame lies? I still cling to those ideals. Sometimes I feel a fool (and the coarser judgment of history might agree), admitting that they still move me. But they do, for they are warm even if only in the folds of memory, and the disillusionment I feel is cold. So very cold.

With all the sadness, how much sadder would it be to have passed one's youth without ideals. Without a cause.

Oh, the Freedom of Man.

"Why did you go to Vietnam?"

It was Camelot . . . yes, Camelot.

I have thought what I should do, had I the power to relive the past. It is good that I cannot, for the gift would be wasted upon me. There are few decisions that I would have made differently, given the world as it was. Given as I was. The long, painful journey, pariah to hero. They are all me. I would be less than what I am were I to have done differently.

There is sadness, but no regret. Truly, it seemed a good idea at the time.

Chapter 3

In the clearing stands a boxer,
 and a fighter by his trade,
And he carries the reminders
 Of ev'ry glove that laid him down
Or cut him 'til he cried out,
 In his anger and his shame,
"I am leaving, I am leaving,"
 But the fighter still remains.

—"The Boxer," words and music
by Paul Simon

Crewchief

A wall map of Vietnam hangs above the coffee pot. After six months of coming to the Center, I still find it fascinating, and I study it before each session. It is the genuine article, not some remnant from a geography text. It warns of firing zones and areas of probable enemy contact. It is odd to see the whole country displayed. When I lived inside the map my view was confined to a few grid squares. There was much of Vietnam that I never saw, although as a crewchief I saw more than most.

I follow the roads and rivers with my finger. There is Phu Bai, our home base; Khe Sanh; and Dong Ha, where we maintained support crews. My eye lingers on the contour lines that circumscribe my memories. There is Mutter's Ridge, the Street Without Joy, and the Razorback And Helicopter Valley, with its litter of -46s, -34s, and Huey gunships rusting on the poisoned and defoliated floor.

A strange thing it is, what war does to a name, altering it forever. Antietam was a sleepy creek before its landscape turned red. Tarawa, a polyglot of syllables before the seawall and the long pier. To speak these names is to recall cordite, flashes, echoing screams. I gaze upon the terrible, romantic names that consumed my youth. No longer need I sit enthralled in schoolboy wonder, mesmerized by the names of my fathers' battlefields. These are my own.

Rose, as always, makes me welcome with my first cup. I wander the hall, studying pictures of fire bases, reading postings for job openings in the VA. Tom had told me there were more than one hundred Centers, and grimly noted that business was picking up. In the early weeks, coming to the Center, I craved only the help the window sign freely offered. As I began to regain a semblance of control, curiosity got the better of me, and I asked how the place was operated. "We're the VA," Tom had explained under pressure, "but we don't advertise." By then he'd demonstrated the truth in the sign; "Help Without Hassles" is the code of the Center, and the vaunted bureaucracy of the VA affects me not. I am at ease in this place.

I am less comfortable with the veterans who wait with me, shuffling magazines and chain-smoking in the anteroom. There are times I desire conversation, and times the last thing I need to hear is a voice that echoes the zone. Like me, I know these men have days of well-being and days of disturbance. I wish neither to intrude on their private counsel, nor to be intruded upon. By unspoken agreement, we test the mood for conversation.

"Coffee sucks, huh?"

"Yeah, about as bad as it was In Country."

"In Country" was variably pronounced Koon-treh and Cunt-tree; it was also referred to as Down South, and sometimes even Vietnam. In VMO-3, we said "In Country" or "Vietnam." However spoken, naming the place opens the subject.

"When were you there?" I ask.

"'66 and '67."

"Same for me, and early '68. Where?"

"Gio Linh, Camp Carroll. You?"

"Phu Bai, Dong Ha, Khe Sanh mostly. Some Marble Mountain. Little bit in the Mekong, not much."

"Who were you with?"

"VMO-3," I see the questioning look. Air-wing designations were always confusing. "Huey gunships. Crewchief," I add. The vet nods familiarly. "Gunship crewchief" is as defining as "grunt."

"You?" I ask.

"Grunt." The vet mentions his specialty and names some designation like 3/26 or 1/16. Grunt unit designations always confused me. I nod familiarly.

"Follow the Orioles?" he asks.

"Nope."

The vet nods agreeably, says, "I need more coffee," and buries himself in the map. The conversation is over.

Tom appears with a greeting, pours a cup for himself, another for me, and leads me down the hall. We take our accustomed positions.

"How ya doing?" he asks.

"OK. Been out on the boat?"

"Got it in the water and sat in it."

"Is that all?"

"Are you supposed to do something with boats other than scrape paint at the dock?"

I am at ease, and welcome the small talk with which we begin the sessions. The pleasantries are a nonthreatening way to ease into a subject. Sometimes he introduces a topic, sometimes I do, but I feel no need to break new ground today. I am content, and have been for some months—on vacation from PTSD, as it were. We've resurrected many memories over the months but delved none too deeply into any one. The angst of the early sessions has been vented. Not resolved. I am comfortable and wish to remain so. I "maintain an even strain," as we said in the Crotch. The Marines taught me not to fuck up a good thing.

"How about telling me about a day In Country," he begins.

"What day?"

"Any day."

"Any day?"

"A typical day."

My mood changes quickly. A day, he asks. I shift in my seat, shake my head, using my entire body to display the instinctive reluctance I feel.

"Oh, man, you don't want a day," I warn, bestowing my best querulous glance.

"Why not?"

"Because a day is deadening. Fucking dead. And long. We'll be here forever."

The aspect of describing a day—a whole day—seems enormous to me. I cannot, I do not wish to wrestle with it. Even the tiniest pieces of those days consume and fatigue me when I summon them from the archives. To relive a complete day—it is daunting.

"Look," I continue, "all I did was eat, shit, fly, and fix fucking aircraft. Over and over again, like a goddamned machine. I didn't get to sleep enough so that it counted. I used to jerk off, but I stopped after I couldn't get it up anymore. You don't want to hear about a day. I'll give you a piece of a day. How about that? Pick a mission, I'll tell you about some nice traumatic mission. Isn't that what we're supposed to be talking about, anyway? I mean, is this PTSD-101, or what?"

"A day," he persists, "a whole day. Pick one."

I remember few days from my time In Country. While no two were exactly alike, it was rare that any two were very different. It seems only the worst of days are preserved whole in my memory, for they are the easiest to recall. As for good days—well, I've grown pessimistic since Vietnam, and good days require too much effort to remember. The rest, those endless passages between good and bad, lie row on row in ill-marked graves, their sad tombstones crumbling into earth, easy to ignore in their numbing sameness.

Tom is insistent, as ever, so I try to recall some time that merits telling. Time In Country! I laugh at the thought. "Doing time," we used to call it, like being in prison. What was time that I should recall one day of that faceless many? There was only "now" and "when." There was no past. Time done was time forgotten. Why should I recall a past?

"I can't remember a whole day—a real one. Christ, there were 393 of them. They're all blurred," I dodge.

"Then put one together. You've got all the pieces in there somewhere," he said, tapping his head.

"And just what the hell good will that do?"

"Humor me."

"This is going to take some time," I warn again. It is my last shot, but I will do as he asks. Nothing in the Center is without purpose.

Tom responds by leaning back in his chair, shifting to a comfortable position.

"OK, wet or dry?" I sigh, thinking how much this session is going to suck.

"What?" he crinkles his eyes.

I explain, patiently. "I only remember things by the aircraft I flew in, and whether I was wet or dry. I crewed slicks from January through May, then gunships until December. Rainy season began in October, so if I was in a wet gunship, it was October, November, or December. If I remember being dry and in a slick, it was April or May. Wet or dry?"

"Dry."

"Gunship or slick?"

"You choose."

"Gunship?"

"Fine. A day," he urges once again.

I settle into place, hands clasped behind my head, and close my eyes to conjure from the old images, a past.

OH-FOUR-SIXTEEN HOURS, THE HOOCH, PHU BAI, THUA THIEN
PROVINCE, REPUBLIC OF VIETNAM

The night watch touches me lightly, chanting over and over, "It's oh-four hundred . . . it's oh-four hundred. . . ."

"Fuck you, asshole," I mumble, but I am a light sleeper; aircrew with 229 days left In Country wake easily.

"You up, Zack?"

"Fuck you," I mumble again, horizontal. The watch stands in a ghostly halo of moon glow, blurred by the mosquito netting that drapes the cot. I cannot see his face. I do not want to see his face. The beam of his flashlight washes the coarse floor, uninviting. I close my eyes tightly against it.

"C'mon, put your feet on the deck."

"Fuck you. Who're you?"

"Crisp."

Crisp is a Remington raider—admin clerk. Medium height, skinny. Sculpted hair, pencil-thin mustache, even In Country. Sort of an anorexic Wayne Newton. Good ol' Crispy Critter—I like him.

"Fuck you, Crisp."

He stands away; it is the wise thing to do. Dragging flight crew into consciousness isn't the safest job.

"Feet on the deck, Zack," he insists.

It takes forever to swing upright. I pull my legs from the sleeping bag. The warmth trapped inside washes briefly against my face, a heady perfume of sweat and farts. I push back the netting, the nylon slippery against my arm. It falls itchy-cool across my naked back. I rub back and forth, letting it caress my shoulders. God! No wonder broads like stockings. I sit on the edge of the cot, shoulders hunched together, wrists tucked inside my crotch, rocking. Oh God, I am so fucking tired! Skivvies are no protection in the predawn dampness, and I tuck my arms closer, locking my wrists between my legs, sapping the warmth from my nuts.

The watch stands over me. "You awake, Zack, you awake?"

"Yeah, yeah," I rock. "Fuckin' yeah. Go away."

Crisp moves on to wake another luckless bastard, chanting softly, without feeling, "It's oh-four hundred . . . it's oh-four hundred. . . ."

I steal a few moments, slouched on the edge of the cot. *Fuck*, I think. "Fuck," I bark, to no one, to anyone. Weariness coats my mind and I can't focus thoughts, but "fuck" sums it all up just fine. It is the first word of the morning, every fucking morning, and there is none finer to express how I feel.

"Fuck."

Down the hooch other crewchiefs and gunners stretch, scratch, curse their way to consciousness. A few cigarettes glow. Flashlights gleam miserably, their batteries as sapped of strength as their owners. An ache grows beneath my thighs, and I look down. The air mattress is deflated, again, and the wood frame is hard.

"Fuck."

My left hand gropes for the Winstons atop the rocket crate doing final duty as a desk. My right snakes into my skivvies, ready to scratch. I expect my nuts to itch. They always itch when the short hairs lift off the dampness and grit of the sleeping bag. I like them to itch; it is the only thing they can do anymore. I scratch eagerly. "Ah!" The first and best feeling of the day. The morning's coolness infiltrates my skivvies; shrinking balls hurriedly retreat into the heat of my body. I clutch them; the chill of my hand shrinks them farther. The cold dissipates slowly within that narrow, happily focused reverie of warmth. I roll them, and pull at my pecker; the distant memory of an erection adds to the warmth. God, it feels so good! How simple life has become; happiness is a C-rat Winnie in one hand, warming balls in the other.

The Zippo lies next to the cigarettes. Its tiny eagle, globe, and anchor glint bravely in the dim light, though wear has eroded the brass to reveal the baser metal beneath. Below the emblem reads

VMO-3
7 DEC 66–7 JAN 68
VIETNAM

And on the back,

FOR THOSE WHO HAVE FOUGHT FOR IT
* FREEDOM *
HAS A TASTE
THE PROTECTED NEVER KNOW

The Zippo is the third of its kind. The first rests on Dong Ha Mountain, victim to a threadbare flightsuit. The second is in the DMZ. I cannot recall the missions when I lost them, only where, but each time it seemed I'd lost a piece of myself. I'd carved my name on each, just above the emblem.

ZACK

The lighter's flare illuminates a calendar tacked to the crate. Long columns of carefully scribed numbers shift in the flickering flame. It has no months, no days, no dates. They do not matter. Only the numbers are important.

395
394
393
.
.
.
.
1

The numbers at the top are dead. I have killed them, sliced each through and through with vicious strokes of a pen—the evening's final ceremony. I pull at the Winnie and squint through burning smoke, following the columns down, relishing the little deaths. No one can make me live those days again. I find last night's slaughter and read the next aloud, "229 and a wakeup." Two hundred and twenty-nine more nows to live until life resumes beyond the bottom of the page. Two hundred and twenty-nine more little deaths. I can do 229.

The warmth in my hand calls to me. I head instinctively for the back door of the hooch, down the steps, into the field across the road. The pisser is a rusting oil drum plunged into a quicksand of urine. A cap of wire mesh protects the wayward from drowning. In my barefooted urgency I venture too near—"Oh, shit!"—and retreat soggy footed. I aim high and arch a thin stream, missing. It splats heavily, with a satisfying patter onto the earth, reflecting moonlight like pale gold, twinkling pretty.

"Ah!" It is the last good feeling of the day.

Back in the hooch, I munch a pepperoni Slim Jim with my cigarette, then a brownie. They clean the fuzz from my teeth and hold me until breakfast in an hour. We have a small refrigerator, and yesterday's Kool Aid clears my head; it's also fine mouthwash. I collect shaving gear, slip on gook sandals, and step back into the night. Carved from old tires, the sandals were among my first purchases In Country. Padding around on Goodyear treads made me feel less like an FNG. I join other half-clad shapes shuffling to the water buffalo. There is no conversation, just clattering shaving gear and curses that merge with the unceasing drone of the generators: the familiar, dismal sounds of morning in Vietnam.

The water buffalo is an oval tank on wheels with spigots clustered on each side. One buffalo stores drinking water for half the squadron. Crewchiefs and gunners in green-dyed skivvies like rags from the tomb queue in zombie lines, waiting their turn.

Saint draws a canteen cupful of cold water, spills it onto the ground, and draws another. Everyone spills the first cup to cleanse the nightly buildup of grit, but it drains the buffalo quickly. Angry crews then lift and rock the tank to get it to flow, dislodging the slime that coats the inside like billiard felt. Saint proceeds to brush his teeth, saving as much of the paste-foamed water as possible to shave with; it takes too long to draw a refill. Saint brushes the way most people scrub toilets, then aims a stream of white froth into the sand. It splashes Zeke's feet, not dead on, just far enough away to be an accident.

"Aw, Jesus Christ! Who did that?" Zeke turns to Saint. "Did you do that, you weasely shit?"

"No," Saint lies through the toothbrush.

"Yes, you did," Zeke accuses.

"No, I didn't," Saint lies harder.

"I'm going to kick your ass."

"You and who else?"

"Will you guys give it a break? Every morning it's the same shit with you two." Smitty says.

"I'll kick your ass, too," says Zeke.

"Give it a break," moans the crowd.

Zeke and Saint walk back to the hooch.

"One of these days, I am going to kick your ass."

It takes little effort to dress. Clean skivvies and socks from the pile Mamasan left the night before. Thank God for Mamasan. Aircrew work twenty-hour days, and without the Vietnamese to do laundry and sweep the hooch, we'd wear our clothes until they rotted. My flightsuit is dirty, but not dirty enough. I check its pockets: cigarettes, matches, survival knife, dog tags, Geneva Convention, Rules of Engagement and ration cards, a few moth-eaten piasters, some MPC. I select a paperback from the shelf and stuff it into a leg pocket, then don my jungle hat. It's a camouflaged Aussie bush model with the right brim snapped up, the kind aircrew favor to look cool. I bought it when I got the sandals. It still makes me look cool, I think. I'd polished a single corporal's chevron and pinned it to the brim, just below an embroidered crescent badge that said, "Vietnam." That looks cool, too.

Time to go. I check my bunk, kick the sleeping bag back into place, and fluff the pillow. There! Bed's all made. I leave the Zippo on the crate. I had a premonition not long ago. Stupid. One morning I began to believe that the next time I lost my lighter, I'd go with it. I started leaving it behind, almost ceremoniously. I fancied the mute Zippo was waiting patiently atop the crate for me, only me, and that if I didn't return, it would be there forever. The thought was an impossibility—nothing was forever. I'd just have to return to move it. I mean, if I didn't move it, who would? When I return. I'll light another weed, and kill another day.

I join the silent crews trudging to the flight line a quarter mile away, relishing these last moments of floating numbly in my own mind. We skirt the burial mounds between the hooches and the workshops. Living in a cemetery had felt strange until I discovered that the whole damned country was a graveyard and the proximity of old or recent bodies didn't make much difference. Another graveyard lies close to the squadron s tool room: the field of dead tail rotors. A stunted forest of Huey blades sprouts in the sand; five-foot totems of black and red impaled at odd angles. Dozens line the trench we'd dug to divert the monsoon's run-off away from the shops. Gunship tail rotors are especially vulnerable to brass damage. The wind stream carries expended cartridges into the blade path. We installed new blades and made totems of the old. I don't know why; it was just something to do.

New blades are few and far between now—so are engines and aircraft, pilots and crewchiefs. Everything is used and reused to the point of exhaustion. When brass strikes, we exhume a less-damaged cousin from the graveyard and swap it out. It is a bitter irony; the most highly trained, best-educated members of the most modern fighting force in history cadging totems from a sandy graveyard to fly aircraft another day.

I head for the ready room to check crew assignments: Captain Bill and the Guinea are my pilot and copilot, and Dutch is gunner.

Bill is an Academy Man (Naval), but only the Fat Baron holds it against him. The Baron has no use for New Breed officers or Academy Men, whom he views as dilettantes. The Fat Baron is the squadron's executive officer, a "'tang" rooted in the Olde Corps. Mustangs are born enlisted and grow up to become commissioned officers. 'Tangs are almost always hard asses, and the Baron is no exception. The tough thing about 'tangs is that they've been around so long, they're almost always right. The Baron saw action in World War Two, which makes him older than rocks and hills and right nearly all the time. He is large, mean, and carries the characteristic gut, florid face, and snarl of a Marine gunny (gunnery sergeant). Officers and enlisted alike hop under cover to avoid having to salute him.

Bill is a good ol' boy with a mouth set to chew a blade of grass. I like Bill; he treats me like "Zack" instead of "Corporal." Once at Khe Sanh, a morning fog turned suddenly to monsoon and shut down flight operations. It was a rare opportunity, so Bill and the other pilots blew into the officers' club and filled up on high-test while the crews guzzled three-point-two in the hooch. Bill was three-sheets when the sun came out. Three-point-two didn't do anything except make you piss, so I was OK, but I have no idea how he stayed upright long enough to make it through the pilots' briefing when flight ops resumed. Bill leaned heavily on me as I helped him slog through Khe Sanh's red ooze on the way to the gunships.

"Corporal Zack. We have a Mission," he said ponderously. "D'ya know what the Mission is, Zack?"

The mud was sticky, and sucking sounds rose from our boots as we walked. Bill was a big guy, not tall but very, very thick, and it was hard keeping him vertical.

"No, Cap'n. What's the mission?"

"Y'know 918?"

The caves near Hill 918 reportedly held a VC base camp.

"Well," Bill continued, "Intelligence says there's a hospital inside 918." Bill began to snicker. "Zack, you got any of that CS and Willy Pete I told you to get rid of?"

CS grenades held tear gas. Some pilots preferred not to carry CS or white phosphorous in the aircraft. A stray round hitting either of them would put us into a world of hurt. Crewchiefs, however, were obsessed with weaponry and believed that too much was never enough. The grenades were concealed in my tool box, the crewchief's sanctum sanctorum, a place no pilot would dare investigate.

"Of course not, Cap'n. But I can get some quick enough," I lied.

"Well, Zack, I'll tell you what we're gonna do. I'll take us in real slow, and real low, and when you spot the opening you chuck out all the grenades you can. Then we'll make one loop around the hill and machine gun the cripples when they come running out! Har, har. Sniggle, snort, snort!"

And we tried, too. I poured Bill into the pilot's seat, and we flew out to 918, with me and my grenades all tucked up in the back. But the hill was fogged, and Bill almost creamed us trying to get in close through the mist. Eventually, he gave up and left; and even though we never went back, it made a hell of a story in the hooch.

The Guinea is deceptively quiet—very polite, with the threatening humor of a mafioso. He's New Breed 'tang, and the Baron hates him because the Guinea got his bars about a millennium faster than he did. The Guinea and Bill spend a great deal of time at Khe Sanh, where the Baron banishes officers he doesn't like. The squadron rotates crews to the northern bases in pairs, usually for one- or two-week stints. Junior to Bill, the

Guinea sometimes copilots the lead ship or captains the chase. Khe Sanh is great duty: no main-base chickenshit and lots of action. The grunts love us and we love them, except for the brass in base command. I don't know why they dislike fucking–air wing, but they do, and show it in nasty little ways. Like the time they cut off our C-rations.

Aircrew lived off Cs at Khe Sanh. Morning patrol spanned breakfast, evening patrol straddled suppertime, and during lunch we were usually in the boonies, flying cover for the grunts. Even if we made it through the chow line, a steaming platter of gruel in our hands, we'd frequently have to abandon it when a frantic grunt burst into the mess hall screaming, "Medevac. Up!" We scarfed Cs to stay alive. Now, I'll admit we might have been a little picky about what we ate, ransacking the cartons, strewing an olive-drab litter in our wake. And maybe we consumed more than two flight crews should and were a little sloppy about how we disposed of garbage. Our hooch was directly across the road from the commanding officer's (CO's) command post. OK, so one or two cans of chopped eggs did make their way into his bunker by accident. And, yeah, that shit didn't smell too good when it came fresh out of the can, let alone after it rotted in the muck for a few days. But that was no reason to cut off our food!

The Guinea dealt with it handily; he grounded our aircraft. Maintenance problems. No food, no fly. We got the food back quickly, but we sure as hell had to fly.

"I better hear you up there at 0600, sharp," the CO warned.

One day we were ordered up in the monsoon. Monsoon was a dangerous time to go aloft. Visibility was nil, and we frequently had to rely on AC/GCA (Air Control/Ground-Controlled Approach) to return to base in the fog.

It was 0550. We were turning our engine in the revetments, visibility minimal, our rotors flinging the night's dew into the soft cloud about us. The ceiling would keep us low, within easy range of any ground fire. I suspected this was exactly what the brass wanted. It was one way to cut down on the consumption of Cs.

At 0555 the Guinea announced, "We are going to do it" with conviction from the right (pilot's) seat. I exchanged glances with my gunner. Whatever he'd cooked up, it was going to be good.

The Guinea lifted, taxied to the runway, and cleared for takeoff. Soon the combat base was a couple hundred feet beneath our skids. We orbited south of the perimeter.

At 0558 the Guinea contacted the tower. "Scarface flight to Khe Sanh tower. Request permission to buzz the CP."

The voice of the controller was indifferent, as if low passes on the command post were SOP (Standard Operating Procedure). "Rog on your pass, Scarface. Advise when clear."

At 0559 we were a klick out, at 300 feet altitude. The Guinea lined up on the CP, falling into a gun dive and accelerating: 120 knots . . . 130. Vibration picking up. The copilot's eyes were on the cockpit clock. "Hold back. Thirty seconds. Hold it back a little! Twenty seconds." The Guinea eased the throttle as the second hand crawled toward 0600, then nosed back into the dive: "Mark! Mark! Mark!"

The maximum weight of a UH-1E model 540-A is 9,500 pounds. It has a rotor diameter of 44 feet, and the amount of noise and vibration generated when blades that big try to stop that much weight from plummeting out of the sky is enough to raise the dead.

"Think he hears us?" asked the Guinea.

We did it the next day, too, but then the CO sent his executive officer to our hooch, and he was really angry. . . .

Dutch, for his nineteen years, seems born to the zone. He's one of the best gunners and unquestionably the worst avionics man in the squadron, if not the Crotch.

"Fuck, man, fuckin' aircraft radio's a motherfuckin' piece of shit!" is typical Dutch. Dutch hates adjectives and most verbs, and "fuckin'" works better, especially delivered with a South Boston accent.

Dutch is in avionics only through the twisted humor of the Staff NCO in Memphis who assigned Military Occupational Specialities. Dutch's rough-hewn exterior, tenacity, viciousness, plus the fact that he was barely toilet trained, marked him as a mechanic, which meant he should have become a crewchief. The Staff must have got his laugh for the day by putting Dutch in avionics. Avionics men regard mechanics as subhuman raggies—grease-balled, multithumbed orangutans with dull minds incapable of comprehending the subtleties of advanced technology. Raggies see avionics types as tweets, twidgets, fags, or other forms of ball-less fairies who have to use tweezers to jerk off. We knew Dutch had balls because he'd scratch them between turns at picking his teeth.

Bill, the Guinea, and Dutch: a good crew. Should be a good day.

The mission schedule calls for Inserts & Extracts operating out of Phu Bai in the morning and Dong Ha after noon. I. & E.s usually guarantee action, which is good, for the time goes quickly. I check in with Gunny in the line shack. It's crowded at this early hour, and the strong smell of coffee and bodies fills the air. I grab a stained cup from its nail near the pot and draw half a mug from the huge urn. It is scalding and bitter, and the burnt oils of a thousand brews have left their signature upon it.

Conversation is limited.

"Got a cigarette?"

"Gunny, I got crabs!"

"Do you ever buy your own?"

"Camels! No filter?"

"You and your crabs are taking General Tompkins to the Mountain today."

"Got a light?"

"Gunny, you want the general to get crabs?"

"Maybe you want that I should smoke the fucking thing for you, too?"

"Hang your dick in the wind stream, let it suck the crabs off."

"Who's sucking off?"

"Dateline Phu Bai: Flash! The Gunny sucks off Frankie's crabs. Awarded one-mission Air Medal with crab cluster. Yessir, sportsfans, remember you heard it here on KVMO3, Scarface radio."

"C'mon, Gunny, cut me a huss."[1]

"Why don't you just gimme the pack? You need to cut back."

The crewchiefs draw equipment from storage bins lining one wall of the shack. The bins are plywood cubes, two feet on edge, with the owner's name stenciled in crimson military script. There are twenty-seven, stacked three high, one for each crewchief when we have that many, and when there are more the FNGs stash their trash wherever they can. Gunners don't get bins, only crewchiefs. It is a tribal perquisite indicating stature in the squadron. There are some things Marines do not like to share.

Some names have been painted over; white primer covers the original crimson. Still, the old letters peek through like bloody ghosts beneath sheets of gauze. Though new names cover the old, the shadows of those former crewchiefs remain. Names tied to faces who no longer fly, whom fewer of us remember. Four months In Country. Five? Three hundred and ninety-five minus two hundred twenty-nine is, is. . . . Up and down the row, so many ghosts live in the bins. None lurks behind my name. The bin is mine, mine alone from those first days since VMO-3 came In Country. My name stands bold in defiance: Zack.

---

1. "Huss" stems from the old designation for the model UH-34 helicopter—HUS-1. The reliable HUS was so highly regarded by Marines that it became synonymous with "goodness," and variably meant "Do me a favor" or "Give me a break."

The gear is bulky: flak vest, survival pack, machete, and flight helmet. I shoulder it and force my way through the crowd. It would be easier to leave everything in the aircraft overnight, but the equipment wouldn't last long. Though sentinels walk lonely watches on the mat, there are none to guard against the guardians.

VT-11's fuel-sample jar rests with many others in a rack on the edge of the flight line. I elevate the glass and twirl it; yesterday's fuel whirls into an amber tornado, bubbles, and clears quickly in the hooch light. A perfect sample: no water or sediment. It's SOP to draw a fresh sample each morning. A clear specimen demonstrates that the night's condensation of water has been drained from the fuel cell and that the sand, which finds its way into everything, will not get into the pumps. Taking a fuel sample is a lousy job. A spring-loaded valve is built into each cell. There's little room under the aircraft, and you need to drain both cells if you're going to do the job right. The valves are always clotted with dirt and fuck up the specimen even if the fuel is OK. If there's dirt or condensation, you wiggle out from under the aircraft, dump the jar, and start over, or empty it where you lie and risk saturating your flightsuit with jet propellant.

I study the meniscus and note where Gunny has marked yesterday's level in crayon. He checks the bottles each morning to see if the level has moved since the previous day's mark, and thinks this keeps us honest. I glance into the shack; Gunny is still chewing on Frankie. I quickly slosh an inch of fuel from VT-10's jar into my own. Gunny still hasn't figured out there are more ways to change the level in the jars than doing our job.

I walk onto the flight line, past the rows of silent aircraft. Each looms alone in an island of darkness. We disperse the ships to prevent exploding fuel and rockets from engulfing the squadron. Victor Tango 11 stands at the end of the line. The distant hooch light is a soft glow caressing her satin skin. She isn't much, but the big green machine is mine, and I have the fiercely possessive pride of ownership that fills all crewchiefs. It seems a lifetime since I earned my NATOPS (Naval Air Training and Operating Procedures Standardization) qualification, and felt the unique thrill of my responsibility. She is the *Warsaw Falcon*, third of her line, called after a potent brand of Polish pickle. I slave for her each day, cleaning, polishing, stealing parts to pull her through one more mission. Unlike fixed-wing, where each aircraft is flown by the same pilot, helicopters are wed to a single crewchief, and pilots and gunners rotate assignments daily among all the aircraft in the squadron. Divorces happen but tend to be catastrophic for one, or both, of the parties.

I approach quietly as she sleeps, outlined by another dawn. She came to me a virgin! My first brand-new aircraft, no shopworn harridan handed down, used and abused, by another. I'd thought her dowdy in her shipping cocoon, a frump restrained by laces and stays like a corseted matron past her prime.[2] I stripped her on the mat at Marble Mountain! How she thrilled me when she glistened naked under the tropical sun. She was mine!

She is tired now, and the flight hours have not been kind. Her emerald sheen is faded by engine exhaust, bleached by the sun. Her windows and bubbles are pitted and pocked by Khe Sanh clay and Dong Ha dirt, scrubbed to opacity by Phu Bai's sand, and I lack the energy to attend her as once I did, though I care for her no less. Scabs of old wounds blemish her skin. Each hasty patch of foreign metal is mute testament to some forgotten zone, some nameless, numbered hill. So many missions she has carried me safe through the zones, absorbing the rounds and splinters, rocking to the blast of high explosive that would shred my flesh. Always, always she has brought me home. She is tired, I am tired, but we survive, the *Falcon* and me. I will always love her.

"Being a crewchief was important to you, wasn't it?" interrupts Tom.

"About the most important thing in the world, I guess. It's probably still the one accomplishment I'm most proud of. It wasn't easy to pass NATOPS. I was a good crewchief," I add simply, relishing the quiet pride in the title.

"What made it so special?"

"There were three Marine Huey squadrons in Vietnam. Twenty-five or so aircraft to each squadron, one crewchief to an aircraft. Some of the senior NCOs were qualified, but they weren't assigned their own ships. At any one time, out of all the Marines In Country, there weren't many more than a hundred men qualified to do the job I did. You know how Marines are—we think we're special anyway. A hundred Huey crewchiefs—it made us more special."

"You miss it."

A quiet wash of nostalgia fills me, yet it is not a distant feeling. It comes from the past and the present all at once, comforting. I feel the space of her about me, smell her fluids, feel the morning dew of her skin against my cheek. So close. . . .

---

2. New aircraft were shipped wrapped in form-fitting gray rubber stitched together like a Frankensteinian condom.

"Yes and no. I miss the immediacy of it, but I still carry the sense of 'crewchief' inside. It's never far from me. It's who I was. It's who I am."

I peer into the engine deck to begin preflight.

"Shit! Goddamn motherfucking pig whore!"

My flashlight reveals fresh oil. I trace the leak quickly; it oozes from a filter casing secured to the engine by three screws, wired to one another for safety. I should cut the wires, replace the O-ring, torque and rewire the screws. But the leak is small, and I have little time for chow, and there is a very large screwdriver in the toolbox. If I'm careful I can crank the casing down without stripping any threads. A moment, and it is done. I make a note to keep an eye on it; along with the hydraulic dampers on the rotor head that need refilling every three hours, the #2 hangar bearing that slings grease under the tail rotor cowling, the bell crank in the aft radio compartment, the. . . .

Dutch arrives with the weapons: the two M-60s he and I fire, the barrels for the external -60s fired by the pilot and copilot, and spares for both. He locks the door guns into their mounts, removes the covers from the externals, attaches the ammo feed trays, then blasts everything with WD-40. The nonfouling lubricant is worth its weight in gold. Crewchiefs and gunners hoard private caches and share unwillingly. Preserving friendships or preserving guns is a difficult choice.

We finish, and I hurry to the line shack to sign off the yellow-sheet book. The black pressboard binder with "VT-11" stenciled in red contains the flimsy forms that give the mission book its name. Crewchiefs and pilots sign to certify the aircraft's airworthiness and record its malfunctions before and after each mission. It is also the official record of which bodies go out in an aircraft, in case the bodies themselves are incapable of reporting after the mission is done. If an aircraft is lost, its yellow-sheet book is quickly removed from the bustle of the shack. Its absence shouts as eloquently of tragedy as any memorial service.

I scrawl my signature, declaring us fit to fly. Dutch joins me as we head to chow.

OH-FIVE-TWENTY HOURS, THE MESS HALL, PHU BAI, THUA THIEN PROVINCE, R.V.N.

A murmuring line threads through the battered screen door. The dawn stirs a cooling breeze, but the cooks are bathed in sweat, manhandling the open burners since 0300. The massive griddles shed heat like a furnace. Clouds of vapor hover over the steam table, permeated with the sweet stench of food and grease and sweat. Some of my squadronmates labor over the grills, committed to a month of servitude in this hell within a hell. They move through the mist in soaked T-shirts and sullen haste and glance at us silently through tired eyes, more in accusation than recognition. I nod and say nothing, knowing better than to snap insults. They don't respond to jokes very well these days. Each morning, the aspect of mess duty is potent persuasion to stay on flight status. "Flying is voluntary," so Gunny tells us. What he doesn't say is that you end up here if you quit.

The sign at the head of the chow line declares, "Today is Monday! Take Your Malaria Pill!"

Yesterday, the sign ordered, "Today is Sunday! Go To Church!"

I have no recollection that yesterday was Sunday, nor does it necessarily follow that today is Pill Day. Sundays and church are inextricably linked by the sign, as are Mondays and malaria, but it is no given that one must follow the other. The signs tell me what I need to know when I need to know it. There are none for Tuesday, Wednesday, or the rest. They are not important. If they were, there would be signs. Each passing moment is now, and few nows need another name. Besides figuring out the order of the nows is someone else's job, and it must be very hard to do. Gunny does it for us in the line shack; it is the kind of thing gunnys are good at. Each morning he posts the date for us to copy onto the mission sheets. The numbers could be runes or hieroglyphs, meaningless symbols to trace onto paper and forget. Now is 229 and a wakeup. It is all I need to know.

The line moves diffidently along the steaming table. Messmen dump comestibles in your tray if you hold it forward or otherwise let you pass ignored. Accidents and invectives reward those who can't make up their minds. There are no surprises. Breakfast is powdered eggs, a strip of fatty bacon, steam-sodden biscuits with margarine soaking into oily smears on waxed paper. Hashed brown potatoes and Kadota figs complete the selection. No one knows what Kadota figs are, but they look like goats' balls steamed in syrupy urine and are about as popular. The alternative to figs is stewed prunes, emphatically unnecessary in a land where one prays for constipation. The messman knows me and ladles the eggs from the top of the heap instead of the bottom. They are pale yellow and dry, and could look like eggs if you concentrated. The watery mess on the bottom turns green from contact with the aluminum. Yellow or green, they taste pretty much the same, but Stateside habit inclines me to yellow.

I carry the tray into the mess and secure a place at one of the wooden tables. Checkerboard sheets of red and white vinyl capture the air of a cheap picnic. There is salt, pepper, and that most essential of military staples, catsup. If an army travels on its stomach, it is because anything that passes through an army is lubricated with catsup.

Another table supports three monstrous vats. Tepid water is a perennial offering, favored because you can see exactly what you are drinking, even if it means studying slowly drowning things with wings. Another vat holds coffee, and the third either powdered milk, tea, or green or red Kool Aid. The Kool Aid has no flavor; it is merely green or red. I seek a challenge and choose coffee. A layer of dissolved grease floats on the surface. Problem: get the coffee without the grease. I swirl the ladle and create a tiny whirlpool that centrifuges the grease up the vat's walls, then pull a clean draft from the vortex before the scum closes. Victory! I feel its warm glow as the coffee fills me.

The meal requires inspection prior to consumption. I tear the biscuit apart, looking for the small black bugs that regularly invade the flour. This one has suspicious spots, which I remove between thumb and forefinger. The eggs have begun to separate. Yellow liquid runs beneath the mound of hashed browns. The cooking oil from the potatoes floats in tiny golden globules atop the runny egg. Little rainbows play on the droplets. It might be art, if not meant for sustenance.

Hazel Baby joins us. "Hey, GI, what you flying today?"

"I & Es, GI," I throw back. "GI," the title applied to the scum-sucking dogface Army, is an epithet among Marines. "Cap'n Bill and the Guinea and the shithead here," I jab a thumb at Dutch.

Dutch grunts, "Fuckin-A, man. Gonna zap some zips. Grease some gooners. Slam some slopes. Heh, heh, heh. Dump some dinks."

Hazel Baby and I roll our eyes. Dutch has been in the zone too long.

"What you got?" I ask.

"Shit. The Baron and Combat Bob. Medevac chase."

Medevac chase is so-so duty. You work on the flight line, waiting for the siren to summon you into action, then rush into a hot zone to cover the transport as it picks up wounded men. Sometimes you shoot, sometimes you just circle low over the trees, sometimes you get your shit blown away—then it's back to the line until the next mission. Moments of action, hours of boredom is what they called it, truly enough, in the war books I'd read.

Combat Bob earned his nickname by being the biggest pussy in the squadron. Nearly every crewchief has his own tale of Bob's ineptitude under fire. On one VIP hop, he and I drew sniper fire while taking on passengers in a forward base. Bob lifted at the first shot, nearly dumping the passengers from the cabin. I was still on the ground and had to jump for the rising ship. A passenger risked his own ass by reaching out to drag mine in as Bob shot out of the zone. The word is that the Baron intends to straighten Bob out or kill him in the process.

"Who's your gunner?" I ask Hazel.

"Muffy."

"Christ," I commiserate.

"Yeah. . . ," Hazel Baby mumbles disconsolately, then chews a rectangle of fatty bacon into pulp before choking it down.

Muffy is as intelligent and arrogant an avionics man as you would never want to meet.

"Muffy stays away from me with both feet," I say.

"What the hell do you have against each other?"

"Remember Walter, back in Pendleton?"

"The fuel-truck driver."

"Yeah. Real nice guy, but slow. Muffy was ragging on him all the time. You know how Muffy is—nice smooth words with a smile on his face and a knife in your back."

"Yeah. That's Muffy," Hazel agrees, chewing.

"I can't even remember what he said to Walter, but Walter was just too slow to see that Muffy was always making a fool out of him. It pissed me off. I told him to cut it out, that Walter couldn't help it. It really pissed me off."

The memory still riles me. Muffy is intelligent, and from aristocracy, to hear him tell it. Intelligent people have a responsibility to those not so fortunate. Even if it's just not to kick them in the ass for something they can't help.

"Muffy asked me what was I going to do about it? I told him I would personally grease his gonads when we went In Country. I'd get him up in my gunship and make sure he didn't come down. It would be an easy thing to arrange over the zone. He thinks I meant it."

"Did you?"

I pick at my eggs, growing cold against the aluminum. "Back then, maybe no. But now? He's still a shit as far as I'm concerned. I like letting him worry about it anyway."

"Did he stop ragging Walter?"

"He stopped ragging, period!"

Chow is over. We exit through the rear door and join the line to the slop bucket. Food scraps go into the first can. It is seldom emptied during the course of a meal and is already overflowing. The air is thick with the stench of decaying slush. I bang the tray on edge, shedding the stickier remains.

One way of telling how long someone has been In Country is how severely the edges of his tray are curled. Three garbage cans stand beyond the slop bucket, each containing a fuel-fired immersion heater. The heaters can be lethal. Once, a messman who refilled the tank before it had had a chance to cool down lost most of his facial hair in the explosion. Suds and slop flew everywhere. He had no eyebrows for a long time. We enjoyed that; he was an avionics man.

I dip my tray in the first can, filled with sudsy water already broken down by the onslaught of grease. A clotted commode brush hangs in the next. The bristles swirl coarse patterns into the residue on the tray. The crowd behind chants, "Move on," more out of habit than impatience. There is no place to go. A rinse in the scum of the third can completes the ritual. I sling the tray across my back to dry in the breeze on the walk back to the flight line.

Hazel joins us. "Hey, only 683 more meals to go."

OH-SIX-HUNDRED HOURS, THE MAT, PHU BAI,
THUA THIEN PROVINCE, R.V.N.

We laid the mat ourselves in the early days In Country. Phu Bai's sand can bog down heavy equipment as surely as any desert. Seabees graded the surface, but we put down the steel matting that became our flight line. Marston matting is the linoleum tile of Southeast Asia, a vast jigsaw puzzle in which every piece is the same. Each mat is a rectangle of corrugated steel, perforated with holes for drainage and traction. It takes two men to lay a section; each piece interlocks into its neighbor and requires at least two blows from a sledge to drive it into place. It takes tens of pieces to park a single aircraft, hundreds to make a landing pad, tens of thousands to build the entire parking area, hundreds of thousands to make an airstrip, millions for an air base, zillions for a war. All laid piece by piece by piece. The mat is an oven in the summer, and we proved it could cook eggs by stealing one from the officers' mess and watching it slowly sizzle. The mat stores heat well into the night, and on morning preflights it is easy to stumble across a pit viper enjoying the warmth of yesterday's sun. You learn to sweep your path with the beam of your flashlight on the way to the aircraft.

The pilots arrive and walk through a quick preflight of their own. Bill asks, "Will it fly?" I answer with a shrug and the sign of the cross and hold up two fingers to ward off the evil eye. I take my customary seat on the port side, just behind the copilot. In VMO-3, crewchiefs sit on the port side. From here I can monitor the pilot's instruments and make sure the tail rotor is clear of obstructions when we taxi. Also, jams on the port M-60s are more difficult to clear than on the starboard guns. The port guns are mounted sideways to provide a smooth feed path for the belted 7.62 ammo stored under the cabin seats, and it's difficult to reach the cocking lever in flight. The crewchief is supposedly more experienced than the gunner, so we're assigned the more complicated task of maintaining the port guns.

I pull on forty pounds of flak vest and flight helmet, snap the tether of my gunner's belt into a tie down, then strap in for takeoff. The heavy web belt permits me to move around in the cabin but ensures that I'll fall only six or eight feet if the aircraft and I suddenly part company. The igniters snap rapidly in the annulus of the combustion chamber, and the Lycoming engine bursts into life. The crescendoing whine of rotors quickly joins the engine's hollow roar, blades whistling at higher and higher pitch as the turns build. The aircraft is alive behind a concentric wall of rushing wind, but the gale thrusts outward so that the eye of the man-made storm remains calm.

Dutch clears right, and I left. Bill taxis to the arming area, where the Rock and Lewish wait to check the weapons circuits There are no occupied buildings in line of sight of the arming area, just dunes to absorb misfires when the firing circuits are tested. The Rock stands off to one side, holding several crudely painted signs.

"Hands on panel," his sign instructs. The pilots place their hands atop the instrument panel, far from any triggers. Lewish, standing by the starboard bomb rack, connects his voltmeter to the guns' circuit.

"Test guns," the Rock signals, and Bill squeezes his trigger. The voltmeter shows continuity and the correct signal strength to two of the guns fired by the pilot.

"Hands on panel," orders the Rock, his faded utilities torn by the wind and sand. His eyes, creased behind flight goggles, show no humor, but they rarely do. The Rock does not enjoy standing in front of a live aircraft testing firing circuits.

Lewish switches the voltmeter to the rocket's circuit, also fired by the pilot. "Test rockets," flags the Rock, and the firing sequence is completed.

Lewish moves around the aircraft and repeats the sequence for the port armament, then the nose-mounted guns controlled by the copilot.

The procedure is carried out swiftly. Other aircraft await their turn on the taxiway.

The Rock instructs a final time, "Hands on panel." Lewish removes the test equipment and connects the firing circuits. "Armed," the Rock signals, and the pilots acknowledge with a thumbs-up. Weapons check is a serious ritual, but hardly foolproof. Last week Combat Bob got out of step in the sequence and launched a live rocket down the flight line. It exploded in a dune. Scared the shit out of everyone; we thought it was incoming. The Rock and Lewish were apeshit, especially Lewish, whose nuts got roasted in the back blast of 2.75-inch HEAT as it left the launcher. Word has it that the Baron is giving Bob one last chance before he gets shipped to the grunts. The Baron has said that Bob would make a good forward artillery observer. They have a very short life expectancy.

We taxi to the main runway to request takeoff, but Roseanne Bravo interrupts. Roseanne is the controller for aircraft operations at Phu Bai. The Marine air-traffic controller in Phu Bai tower worries about takeoffs, landings, and airspace occupancy; Roseanne worries about everything else.

"Scarface Four-One, hold your position on the ramp. Change in aircraft assignments. Tango Twenty-five will be joining you as Four-Two."

The call signs for our flight of two Huey gunships are Scarface 4-1 and 4-2. We're 4-1, the lead gunship. Tango 9, Larry Zimpfer's bird, was supposed to be 4-2. Tango 25 belongs to Hazel Baby, slated for the medevac pad.

The Fat Baron checks in, "Scarface Four-One, we are turning and will join you shortly. Tango Nine is experiencing compressor stalls."

"Roger, Four-Two. Do you want to take the lead?" Bill defers to the executive officer.

"Negat, Four-One. You are the lead."

"Shit," Bill says to the Guinea, "that bastard will be critiquing every move I make. This is going to be like training."

When 4-2 joins us on the ramp, Bill raises Hue–Phu Bai tower.

"Hue tower, Scarface Four-One, two for takeoff."

"Roger, Four-One. Taxi runway two seven, winds from the west, seven knots."

Bill moves onto the runway with the Baron not far behind. "Scarface Four-One, this is -Two," calls the Baron on the tower channel.

"Four-One copies," answers Bill, trying to sound professional.

"Four-One, you have not set your altimeter. That is incorrect procedure. The tower did not provide it, and you did not correct his error. Request altimeter, Four-One."

"See! He's starting already," Bill cries to the Guinea.

"Altimeter two niner six seven," interrupts the controller, a nervous edge to his voice. "Clear for immediate takeoff," he urges.

Bill noses over and shoves us down the runway, his jaw clenched. This day may not be so good after all.

The mission is simple enough. Escort Superchiefs 5-1 and 5-2, two UH-34Ds from HMM-163, on a troop insertion. If everything goes well, this will be number eleven toward my eighth Air Medal. If it doesn't go well, it'll be numbers eleven and twelve.

The mission system is complicated. The crew earns one mission for each operation or troop unit we support within the same day. If we fly in ten different operations, we earn ten missions. If we make ten sorties in support of the same unit or the same operation, we earn one mission. If we get shot at, it counts double. Somehow, getting shot at by ten different gooks counts more than getting shot at ten times by the same gook or his brother in the same zone. On my best day, I earned fourteen missions by flying sorties in support of seven different units and getting shot at each time! On my worst day, I flew twelve fucking hours for a lousy credit of two missions because the same damned gooks shot at us all day. Missions are like green stamps; twenty credits on your mission sheet earn you an Air Medal, a sixteen-pointed piece of bronze with an eagle hurtling balls-to-the-wall into the deck. Each subsequent medal earns a bronze star to stick on a yellow and blue ribbon. Five bronzes equal one silver. The ribbon can physically hold four silver stars, which (plus the ribbon itself) represents twenty-one Air Medals, the most you can stuff onto one chest.

A citation accompanies each award. It begins,

> For Meritorious Achievement In Aerial Flight As A Designated
> Crew Chief In Marine Observation Squadron 3
> During combat support operations in support of the Republic of Vietnam
> the insurgent Communist guerrilla forces (Viet Cong) from. . . .

We get nothing for the missions but the citation and the chance to fly more of them. No time off for good behavior, no early return to the World. But, as tired and jaded as I am, collecting missions is still important to me. It is what I do.

OH-SIX-FORTY HOURS, THE CO BI–THANH TAN,
THUA THIEN PROVINCE, R.V.N.

We approach the cratered hills of the Co Bi–Thanh Tan and locate the special one for today. The -34s orbit high over the flatlands that flow from the hills to the distant coast, waiting for us to scout the territory for hostile Indians. A morning mist fills the narrow valleys, and the hilltops float like green islands in a silent sea of fog, so dense you could vanish inside it forever. The grunts are on a snoop and poop, so there will be no zone preparation or unnecessary firing to draw attention to their position. The insertion must be accomplished quickly or the team is compromised. Bill climbs to 1,500 feet while the Guinea pinpoints the landing zone (LZ). Covering us, 4-2 is ready to roll in should we draw fire.

Bill on final: "Scarface Four-One, starting my run." The Baron responds with a double click on the radio. There is a brief, nauseating moment of weightlessness as the aircraft pauses at altitude, suddenly quiet after the steady drum of the rotors. We drop precipitously, like the first moment beyond the great climb of a roller coaster. The leaden breakfast of grease and potatoes produces a brief queasiness, a taste of bile. I choke it down, but rising bacon isn't half bad, and I welcome the chance to burp.

Our airspeed is increasing—130 knots, 140—as we race into the dive, engine and rotor tachs showing max rpm. The aircraft falls beneath me; I hold tightly to the gun mount as negative Gs force me off balance, upward and backward. The Huey rarely allows us to fight from comfortable sitting or standing positions. I man the internal gun from a crouch, balanced on the balls of my feet, the muscles in my legs screaming, pivoting the weapon forward or aft while I scan the rushing earth. Our job is to deliver flanking and rearward fire as the aircraft sweeps through the zone. The deck plates vibrate beneath my boots, a constant thrum through leather and bone. A tie-down ring hurts my foot through the thin sole, and I shift position to relieve the pain. Airspeed 150, 160, red line: the tachs quiver just below max. Bill holds the aircraft in its dive, and the windshield fills with the mottled greens and browns of earth.

Everything in the cabin vibrates forward: spare gun barrels, ammo cans, C-rations, canteens, smoke grenades: the flight deck is a cluttered mess. We've never figured out how to keep everything in place, but nothing ever seems to fall out. It all piles up under the pilots' armored seats, and pisses them off if it jiggles far enough forward to fall into the nose bubbles.

The wind howls through the rocket pods. I lean into the stream to scan the zone; the gale forces itself beneath my visor, swirls over my eyes, cooling, watering. The chin strap on my helmet works loose, lashing furiously at my neck and cheek. I try to ignore it but cannot, and take my hand from the -60 to tuck the stinging whip away. Such a short moment, but it puts me off balance and will add seconds if I need to return fire from the zone. I return my hand to the gun, feeling guilty.

We close on the zone. Breath comes hard inside the hardened carapace of the flak vest. The chest plate digs bluntly into my abdomen and bladder. I don't know what's worse, the pain or the urge to piss. It also bores against the big Talon zipper on my flight suit, forcing it into my sternum. I shrink from it, but pushing backward draws the bottom of the vest into my bladder even more, and the back plate grinds through the thin cloth over my spine. It is a dead weight, leaden against each vertebra as if specially designed to crush only the most sensitive nerves. I fix upon the Talon, and hate it with an intensity unmerited by a penny's worth of metal, but I endure. The zone is now large across the gun sight. I can do nothing else but endure.

"What are you smiling about?" asks Tom.

"I was just thinking, the one good thing about being crouched behind that damned gun after breakfast, with that damned flak vest digging into your guts, was that your asshole got stretched from here to hell and gone, and you could cut the most satisfying farts since Christ was a corporal."

Tom laughs, then turns serious.

"What did you think about in the dive. What did you feel?"

"Think?" I snort. "You don't think in a gun dive—you just are, and Marines aren't allowed to feel. You ought to know that. We have that cut out in boot camp."

"Follow yourself down. What do you remember?"

It is easy to see the aircraft. It has grown in the room around me as I've talked about the mission. I sense the space of the gunship, the scream of the wind, the ache in my stomach, the tension in my chest. It is my element, and I know it better than the job I worked this day. There I am, only the grimness of my lips showing beneath the smoked visor. The helmet is so large, and I am so young! One hundred and ten pounds I was, only twice the weight of the vest and paraphernalia of war that girded me.

I am an older ghost, whispering into my own young ear across the years. Tell me, young Zack, what is it that you feel?

He does not hear me in the wind, or cannot respond. He is a good crewchief, so intent upon the zone. No matter. I know the answer. I am he, and we have carried it through the years.

Feeling has no place on the flight deck of a gunship. Surviving in combat does not permit it. Zack knew it, and I know it. We are crewchief. The mind melds with the machine, welded together by speed and purpose. You do not "feel" in the dive. The eye, the brain process sterile data on terrain and visibility and places hiding sudden death, all filtered by training and the experience of a hundred other dives. The result of the process, what action you will take next, is computed each instant as new data comes in. The formula of the dive has no place for emotion, as it is known and felt in the World. Oh, a surface thread of apprehension or excitement tingles on the perimeter of the mind, and might become emotion with time. But there is no time over the zone, which is good, for the formula breaks down over time. Given too much time, apprehension evolves into fear. We learned about fear in training, and we learned to fear it in itself. The formula uses fear to counter fear. "Will I fail?" is the base fear. "Will I let my buddies down?" It must be so. There is no survival if the formula fails.

Yet there is something. Tom's questioning voice urges me to look more deeply. I feel it in the memory of the dive. I see it in the set of young Zack's lips. During all those dives in all those faceless zones, it was there, and it too drives the formula to survive.

It is power! Raw in its simplicity, constructed irresistibly of fighting armor and wind and air-cooled steel. Oh! to hurl like a god at the rushing earth. I had forgotten the taste of it. It is the power of the executioner, merciless, absent of the heat of hatred, who knows not his victim, nor cares to, nor will remember him after the deed.

"I'd forgotten how powerful you could feel, going into combat. I haven't felt that in a long, long time," I say slowly. The memory is comforting, the recollection of stronger days, before PTSD and therapy and dissecting my life to see what I might salvage, what I must purge.

"Powerful, not frightened?" asks Tom.

"I think the sense of power controlled the fear. Usually, I didn't have time to be scared."

"Power, not bravery?" he explores.

"Shit!" The question is absurd. "You feel powerful, you feel scared, you feel pissed, or you don't fucking feel. There's no such thing as feeling brave. Bravery is attributed. Brave is how old guys remember themselves when they were young. Honor, pride—those are feelings. Put brave in one hand and brown skivvies in the other and see which weighs more."

"You miss it," says Tom. It is no question.

"I was a crewchief," I explain again, proudly, the word its own definition.

"How long did the feeling of power last?" asks Tom.

"About two seconds into the dive. You got busy real fast; too fast to feel much of anything for very long. Everything was so compressed, though, like a lifetime was compressed into an instant."

"Do you miss it?"

"The power? I haven't thought of that in years. Maybe. Mostly, I miss the sense of doing something important. And of belonging. We were one hell of a group. It was important to belong."

I see the aircrews of VMO-3. The image is warm. It was when I belonged. But the memory fades, replaced by the grayness of the Center. I whisper no more into young Zack's ear. I am only me, and no longer crewchief.

"Do you know what the hell I do all day?" I spit, suddenly angry.

"You work with computers," Tom answers evenly.

"I work with fucking computers. When you've dropped out of the sky at nineteen with eight machine guns and fourteen rockets, with the lives of a squad or platoon counting on you to do every fucking thing right, working with fucking computers is a little anticlimactic. Life is anticlimactic. Nothing that I do now compares with what I did then. I was a crewchief! People depended on me. Now, what I do doesn't mean a thing. It's kind of hard to go on when you feel like you've topped out at nineteen. Yeah, I miss it."

I do not like the turn of this conversation. The fullness of the distant mission reminds me of how empty I feel.

"What about after a mission?" asks Tom.

"What about it?" I snap.

"What did you feel then? Did you still have the sense of power or importance? Were you excited, relieved, afraid? What?"

"There wasn't anything after a mission but more bullshit and other missions. I never thought about a mission after it was over, other than what went wrong with the aircraft or if I could have done a better job. I told you. It wasn't a good idea to feel, and besides, we weren't allowed."

Bill pulls in heavy pitch to break the dive. The Gs mount rapidly, grinding me farther into the flight deck. The M-60's pistol grip, the handhold above the door, and my aching legs form a vibrating triangle that keeps me positioned behind the gun as we flatten out over the zone. We bank left, then right, then left again, dodging and twisting at high speed, trying to draw their fire and avoid it if it comes. The Guinea sweeps his TAT across the rushing landscape.[3] I snake a hand into the pyrotechnics carrier and grab red smoke to toss if the zone turns hot.

We circle. Once. Twice. Ten seconds, twenty. The hill is covered in elephant grass, surrounded by other hills ringed by trees and scrub. So many places to hide.

A final pass. Bill barks, "Clear zone," and skims away above the fog. Dutch and I swivel aft, scanning the fleeing land for movement.

Quiet zone. No trouble. No nothing, which is just what the checkout dive proved. The few times that Charlie takes potshots are justification for checking out a landing zone before the transports come in, but most times

3. Emerson Electric's Tactical Armament Turret mounted two M-60s and 1,000 rounds of 7.62-mm on the Huey's chin. The guns were synchronized to follow the motion of a swivel-mounted electric sight suspended above the copilot.

Charlie stays cool. He has all the time in the world and knows we'll be back with the lumbering transports. It's Charlie's country: his game, his rules.

We rejoin the -34s. Superchief 5-1 leads the way; 5-2 lags a half mile behind. It is a narrow hilltop with room for only one aircraft. We cover the lead transport, and 4-2 brings up the rear. The ballet is practiced and precise and I watch it play out over the long tube of my M-60. Flaring to drop airspeed, 5-1 closes on the zone. The rotor wash mats the grass into undulating waves of green. Bill pulls into a tight circle centered on the -34 as her rear wheel touches earth. If Charlie is here he will open up now, with the grunts helpless in the womb of the aircraft. The grunts spill out as 5-1's forward gear settles.

Bill banks tighter in our loop to place my gun and the TAT on the zone. The Guinea is glued to his sight, playing its cross hairs over the brush outside the circle of the -34's rotor wash. I can see the door gunner poised over his -60, covering the grunts as they wade into the shimmering grass. They fan out and secure the local zone. The Superchief rocks forward, building a cushion for takeoff, broadcasts, "Lifting, lifting," then claws its way into the sky. When 5-2 closes, Bill breaks into a longer loop, allowing 5-2 entry into the zone. Scarface 4-2 remains high as we cover the second transport. The second Superchief disgorges its load like a beast giving birth to bundles of camouflaged green. The grunts dissolve into the jungle as the helicopters flee the LZ.

The radio operator broadcasts a quick situation report.

"Scarface Four-One, this is Hightower. Do you copy, over?"

"Roger, Hightower. Say sit-rep," orders Bill.

"Sit-rep Jane, I say again, sit-rep Jane," responds the team.

The zone is secure. Bill wishes them luck, and we break contact.

Checkout, insertion, sit-rep: two minutes, perhaps three. The mission is over.

OH-SEVEN-THIRTY HOURS, THE MAT, PHU BAI,
THUA THIEN PROVINCE, R.V.N.

We land in time for squadron muster. Top Daniels stands in front of S-1 (Squadron Administration), and glares at Dutch and me as we saunter in from the flight line. Top is the squadron's first sergeant and like all First Shirts is the unit's epicenter of chickenshit. He rags us daily about keeping clean, fraternizing with the mamasans, wearing improper military attire, having unmilitary attitudes, giving slovenly salutes, and wearing your cover in the mess hall. When Top was Top in the World, he could threaten loss of liberty or skinhead haircuts to keep us in line. This is not the World. This is fucking Vietnam with fourteen-to-twenty–hour days, where you shave your head because of the heat and you get no liberty because there is no liberty. There is only one thing that Top does control. R & R. He gets to say who goes, who doesn't, where, and when. Top holds R & R in one hand and your balls in the other. So we come to muster to listen to him talk about V.D. and the mamasans, and to learn about acting like Marines, even if we are In Country.

Top is bitching about his saw again. When we arrived In Country, we were incredibly ignorant of how few tools would be available to do the simple woodworking necessary to turn an empty hooch into a home. Top had been around and came prepared with his own carpenter's rig. Someone stole his saw the first week, then lent it to someone, who lent it to me. It was so long ago, the trail is lost in antiquity. I oiled it when I thought of it and let it rust when I didn't. Top often demanded its return, even threatening a shakedown like he used to do in the World if something was stolen. He'd order the entire outfit to attention in front of their bunks while searching lockers for contraband. But a shakedown would disrupt the flight schedule, so he couldn't deliver on his threat. I hid the saw behind the desk in my hooch and vowed to keep it forever. It was a small pleasure, but screwing the Top made me feel good, so I did it. You took what joy you could get, In Country.

Muster over, Dutch heads to avionics to mutilate black boxes, and I go to work on the aircraft in my section. This morning, VT-4 needs a new bell crank in the elevator linkage. There are no new bell cranks, so I head to the deadline to cannibalize the part. VMO-3's deadline stands dismally at the farthest corner of the mat. Forlorn but never forgotten, aircraft out of commission for parts are parked unprotected beneath sun and monsoon awaiting the components they need to fly. Once consigned to the deadline, a bird is pitilessly plucked for spares.

VT-12 is in the middle of the line. She was airlifted there the day the 12.7s shattered the Jew's left eardrum and peppered Bill's bull neck with shrapnel. I know the story well. Bill and the Jew had just lifted from an outpost well out in the boonies, to take a group of grunt officers on a scouting mission. She was hardly off the ground when the 12.7s smashed her down. A round crashed diagonally through the Plexiglas beside the Jew, pierced his flight helmet above the left ear, and exited through the roof. The shock wave destroyed his hearing.

I'd not have been so lucky had I been in his place. The ship was below the canopy when she was hit, and my habit was to stick my head out the door and watch the tail rotor until I was certain we were clear. I know the 12.7s would have kissed me squarely in the forehead had I been there. When I got back from R & R, I quizzed the Jew again and again about what had happened. I had to know.

You see, VT-12 was mine, the first *Warsaw Falcon*, bureau number 152438. The Jew had taken her only that morning, the first day of my week of R & R.

She survived the crash, with damage to the stabilizer bar, the cabin, Bill, and the Jew. She was a good ship and could take a lot, but the damaged stabilizer kept her grounded. Knowing that getting a new one would take a long time, the squadron began stripping her for parts. Now she needs an engine, main mast and blades, tail rotor, all her radios, and the door the shattered window was in. She is an empty shell, but I remember how beautiful she once was. I take my old place, port side, and slide the cabin door closed. I will take the bell crank later; there is time. No one can see me, so far out on the mat, and they would tease me if they knew how I miss her.

I will stay with her, for just a while.

Ten-Thirty Hours, The Hooch, Phu Bai, Thua Thien Province, Republic of Vietnam

By midmorning the tropical sun has become uncomfortable, well on the rise to intolerable. I steal time from the aircraft to refill my canteens at the hooch's refrigerator. Phelps is there, and Give-A-Fuck and Smitty, and Babysan. Babysan's a good kid, pleasant and pixieish, and very gullible.

She stands angrily in the entrance, hands on slender hips, her delicate face framed by her coolie hat. Tiited back, held beneath her chin by black satin, it surrounds her midnight tresses like a straw halo. She stamps a tiny foot.

"No! No! Numbah ten!"

"What's with Babysan?" I ask Phelps.

"We just told her the squadron was moving to the other side of the strip, and that she'd have to drag our laundry over here every day. She says, 'No way, GI.'"

The squadron was soon to move, huts and all, north of the airstrip, and no provision had been made for laundry facilities. We'd barely have enough water for showers, and our drinking water was already limited to what the buffalo could hold. The mamasans would have to lug our laundry to the far side of the base until a well was drilled near our new digs. Babysan had good reason to be pissed.

Give-A-Fuck perches cross-legged and barefoot on the divider between the cubicles, a one-by-twelve shelf five feet off the deck; he begins to snarl and clutch at Babysan. Give-A-Fuck thinks he's a werewolf, sometimes. His face contorts into a bestial grimace, baring fangs. The funniest part of the act is his feet—the more he growls, the tighter they curl, like prehensile claws. He swipes the air at Babysan, who pulls back in alarm. Give-A-Fuck leaps from his perch, still snarling.

"If you don't do the work, we'll send you to Dong Ha," he threatens.

"Dong Ha . . . heavy shit at Dong Ha. . . ," we agree solemnly.

Babysan shakes her head again, but dubiously. Dong Ha is getting hit heavily—not a nice place to be at the moment.

Behind her, Ron pulls his rifle from storage. "You're going to need to learn how to use this. Lots of VC at Dong Ha. Let me show you."

Ron pulls an M-14 round from his pocket, shows it to Babysan. "This is the casing, and this is the projectile. Open the bolt like this—pull it back firmly; it's kind of hard. Slip the round into place here, then close the bolt."

Ron has manipulated the weapon into Babysan's hands, and stands behind her. She stares at the closed bolt, puzzled. Ron has always been straight with her, and I can see doubt grow as he lifts the weapon to her shoulder and talks her into sighting down the barrel.

At the far end of the hooch, Smitty walks in, his mess kit slung over his back. "Hi, guys! Hi, Babysan!" he calls happily.

"This is the safety, right here." Ron holds Babysan's slender hand in his. "Push it this way to 'fire' and this way to 'safe.' Now watch!"

Phelps swivels, forcing Babysan to face Smitty. She stares down the long barrel of the rifle as he approaches, smiling. Ron pushes the safety to "fire" and pulls the trigger. *Crack!* Smitty flies back like a rag doll, smashing into a cot, then flopping like a dead thing on the hooch's deck.

"See!" Ron notes, plainly.

Babysan works her mouth three, four, five times, but nothing comes out; her eyes grow wider and wider as she stares at Smitty's limp form. Ron loosens his grasp, and the rifle falls at her feet. At last a screeching "Eeeeeeeeeeeeeeeeeeeeeeee" pierces the hut, and she flees.

Three seconds later, Babysan peeks suspiciously around the corner of the door. Smitty lies as he fell, arms and legs awkward in death. Phelps says, "Ooh, Babysan—what you did."

"Eeeeeeeeeeee, eeeeeeee," she screams, feet flying. We can still hear her, even as she rounds the corner at the end of the line of hooches. "Eeeeeeee. . . ."

Give-A-Fuck bends to retrieve the weapon and falls on the floor. "Har, har, har. Christ, I'm gonna piss myself! You see her mouth—'Abba, abba, abba!' Har, har, har!"

Ron helps Smitty up. "Ooh . . . I think I broke my fuckin' back."

Give-A-Fuck ejects the round: a blank. They'd emptied the powder, and Babysan couldn't know that only the primer had fired.

"She's gonna be pissed as hell at me," Ron says.

"Har, har, har," rasps Give-A-Fuck.

"I'm gonna tell her it was all you guys' idea. I'll say you made me do it. You too!" He jabs a finger into my chest.

"Me! No fuckin' way."

"She'll believe me. I'm such a nice guy."

"Bull!"

Phelps jabs a finger at each of us in turn.

"Well you're the one who suckered her into biting a lemon; you slurp your *Playboy*s to gross her out; you cop feelies on her teeny little titties; and you're the one who wired her chopsticks together! And what do I do? I teach her English!"

"You bastard! We all set this gig up," Smitty says.

"Babysan will never believe it. I'm such a nice guy!"

TWELVE-HUNDRED HOURS, DONG HA, QUANG TRI PROVINCE, R.V.N.

We land, park on the rolling slopes near the airstrip, and head to the mess hall. There is little time to eat. At Phu Bai, flight crews move to the head of the chow line, but this is a grunt base. Attempts to cut in are met with angry thumbs jabbed toward the rear.

There is a solution for every problem. Most of the grunts wear rectangles of red cloth sewn into their trouser legs and centered on their utility covers. We tell the mamasans these are V.D. badges to mark the diseased. In truth, it is the proud and simple badge of the Sparrowhawk. A Sparrowhawk helicopter assault team is always on call for rapid insertion into the hottest of hot zones. The grunts who man these teams take life seriously and drop everything to scramble to the loading zones when the cry "Sparrowhawk, Up!" echoes through the compound.

Dutch sneaks behind a nearby hooch and bellows the magic words. "Sparrowhawk, Up!"

Nearly every man in line splits, races to his quarters to retrieve weapons and don flak gear, then runs to the loading zone.

Dutch wanders in from the corner of the hooch. "Dumb fucking grunts. Works every time."

Being from out of town, we borrow mess gear from the truculent mess sergeant pacing the chow line like a slave master. He doesn't like fucking–air wing who show up to borrow his gear and consume the supplies he metes out with miserly animosity. The more we eat, the less he can sell on the black market. Lunch at Dong Ha is less attractive than breakfast at Phu Bai: boiled and skinny hot dogs floating like turds in a vat of steaming water. Pea-green globules of fat glisten on them as the messman serves up two per man, slithering from the meat and pooling into a congealing scum in the mess tray. The messman dumps a spoonful of baked beans next to the dogs.

"Too fuckin' much. You're givin' 'em too fuckin much," the mess sergeant roars, and grabs the spoon to remove the excess beans faster than I can escape with the tray.

The sergeant demonstrates proper portion control. "This is how you fuckin' do it" punctuates each crash of his serving spoon into the next few trays. He jabs the spoon back at the hapless messman. "Now do it right!"

A sheet of cornbread rests at the end of the line under a threatening sign that warns, "Take 1!" I study them carefully, searching for one larger than the rest, with perfect sides that will not crumble when I pick it up. I extract a single cake from the center, where the sheet is thickest. It is two inches square and an inch thick, and I walk off feeling proud of myself.

The beverage selection is familiar. The Kool Aid is red, the coffee has a slick, and the water looks like it came from a pond. All of it gives me the shits. Consuming anything at Dong Ha gives me the shits. I think that breathing Dong Ha air gives me the shits.

The staple dessert of Vietnam fills a nearby vat. Unlike hot dogs, eggs, or bread, we can have all we want. We can dump it on the earth or even eat the stuff. It is the only commodity of greater supply In Country than missions or the Top's chickenshit. Canned fruit cocktail. Peaches and pears and seedless grapes that float like jaundiced ova in a vat of clear goo. Some dip their canteen cups in and pull enormous, runny portions to slurp on the way back to their seats. The swamp of syrup pools on the vinyl tablecloths, attracting myriads of small bugs with short but annoying lives. I hate canned fruit cocktail. It makes me want to puke. When I get out of this place, I will never allow it in my presence again.

Dutch, Hazel Baby, Muffy, and I share a table. Muffy would rather not be near me; but the hall is full, and it's me or the grunts.

Dutch pokes into his dogs and beans. "Know what this reminds me of?"

"Cut it out," I order.

"No. You know what this reminds me of? I fed a hot dog to a Great Dane once; he swallowed it whole and then threw it up right away. It came out in one piece, and it looked just like this." Dutch holds up a wiener. "Lookit the way the beans cling to it. Just like barf."

"Cut it out."

"The mutt ate it right back up again."

"How're things going with the Baron?" I turn to Hazel.

"I feel like I'm flying with a ramrod up my rectum," he answers. "When I tried to tell him where the -34s were on the insert this morning, he told me to stay off the radio. When we flew up here, he ragged me for not calling in any spot reports. There wasn't anything to spot! You can't win."

"The Baron didn't say anything to us on the run this morning," I note, "Bill was flying textbook Ma-reen. I think it freaked him out, having the old bastard orbiting high and watching his every turn like a bug under glass."

"The Baron caught Bill and the Guinea in the ready room and chewed them out for taking too long to find the hill."

"Shit. Every damned hill looked like a big green tit sticking up out of the fog. We found it fast enough. What an asshole. Is this a war or is this a war?"

"I think Major Diaz is a good pilot," says Muffy.

"You would, you fucking kiss-ass piece of shit," I hiss. The Baron is a good pilot, but I don't like agreeing with Muffy.

Muffy turns to Dutch and Hazel, "See this guy. He threatened to kill me back in Pendleton."

"No, in Pendleton I threatened to kill you here," I explain between chews. "There's a difference. We're not in Pendleton, now. When're you gonna go flying with me, Muffy?" I tease.

"I'm getting some fruit cocktail," he says stiffly and walks to the vat.

"You know, Muffy is a pretty good gunner, Zack. He's a shit, but a good gunner. You ought to give him a chance," says Hazel.

"I'll give that bastard a long walk on an unhooked gunner's belt. That's what I'll give him."

An angry mob wearing red patches is gathering outside. Guttural threats filter through the screens, "If I find the motherfucker I'm gonna tear him a new asshole." The grunts have clearly left their sense of humor in the World. We finish quickly and move on. The slop line is a near replica of Phu Bai's; we clean and return our gear to a messman, since the sergeant makes fucking–air wing clean it twice out of pure meanness. For the record, Hazel Baby notes that there are only 682 meals to go.

The pilots are waiting at the aircraft. They get the same fare we do, but don't have to resort to devious tactics to gain admission to chow, nor do they have to wash their own gear. Rank hath its privileges. We turn up and taxi the short distance to Dong Ha's fuel pits to top off. Line pressure is erratic; the flow varies from barely a trickle to a burst with enough force to tear the nozzle from your hands. My first experience left me bathed in JP (jet propellant). The pit jockey trots over with his clipboard for me to sign for the fuel. The Crotch makes you sign for everything. except chickenshit and fruit cocktail. Why sign for fuel? Is there a quota? If you reach it, do you stop flying? Will the war end when every outfit reaches their limit? I sign the chit "John B. Handcock, Gunnery Sergeant, VMO-2." It gets harder and harder to be creative.

The villes and paddies surrounding Dong Ha yield to the desolate beauty of the highlands. It is easy to imagine that we are flying back to the dawn of time. The terrain is primordial; a prehistoric beast lazily munching on the treetops would not seem out of place. The highlands are fierce and defiant, though scarred by the endless assaults of defoliant and artillery. Valleys twist and turn amidst the mountains, a maze of green upon green that outline and isolate the battle zones where we have fought.

We near the Rockpile, which commands a region as wild and beautiful as it is deadly. Valleys like rivers of lush green flow from the twisted hills, joining, growing ever wider like streams merging to build the river. The Rockpile stands at the confluence, dominating all approaches, soaring without preamble from the valley floor. These jagged stones pile higher and higher like clutching giants struggling in one another's grasp. The mountain seems at war with the land itself, as if the earth could not bear so terrible a monstrosity in her womb. Marines man an observation post on the peak, calling in H & I (harassment and interdiction fire) from the forward artillery position at Camp Carroll, even upon the flanks of the mountain itself when enemy forces seek to force them from their aerie. A wooden landing pad is the only flat surface on the mountain. The UH-34Ds that resupply the grunts tenuously anchor their forward landing gear upon it while the tail hovers in space as the aircraft is unloaded. Grunts coming off the Rockpile have told me about having to battle rock apes for the choicer turf. I have never seen a rock ape and do not know if they exist, but other rumors hold that they are simply grunts left too long on the mountain.

Bill circles the Rockpile. A grunt with an M-14 perches on a ledge seven hundred feet above the valley floor. Unshaved and forlorn, his flak jacket hangs open on a naked chest. He picks at a C-ration, then throws the can off the precipice. He could be a rock ape, I think. It tumbles down, creating a small avalanche, and disappears into the rocks and trees below. We wave to him happily, and he flips us the finger. We wave with even greater fervor and throw our own fingers out the starboard door. He points the M-14 threateningly, but he will not shoot. Fucking–air wing, which delivered him into this land, alone will retrieve him.

The Guinea points to one of the green valleys, and Bill climbs to follow it as it narrows and branches into a maze of climbing rills. A pair of troop transports from Superchief circle in the distance, looking every bit as prehistoric as the surroundings. The old -34s are the Model-T of helicopters in Vietnam: reliable, slow, able to absorb withering punishment, ugly as hell. A -34 looks like a bloated grasshopper with twirly antennae and an oversized ass.

The fixed-wing aircraft assigned for the zone prep check in: two Phantoms, call sign Condor. Fixed-wing zone preps are SOP on troop insertions this deep in Indian country. The hills are honeycombed with caves and spider holes. Bunkers, invisible beneath the canopy and grass, deliver interlocking fire covering many of the ridges. The LZ is swathed in elephant grass, and we roll in quickly to mark the target. It's an easy shot. Bill stands well off, dives into rocket range, and launches two Willy Petes, then pulls away, coming no nearer the target than necessary. The missiles streak in, trajectories marked by dirty white contrails, and burst in a dense and lethal cloud.

Condor leader and his playmate are orbiting well out of sight. The caves conceal 40-mm flak batteries, radar-controlled quad 50s, and SAMs that give the fixed-wing nightmares. NVA gunners simply roll their weapons out, unleash their firepower, and return to cover before their launch point can be detected. The guerrilla war near the Rockpile is surprisingly high tech despite the primitive surroundings.

The Condor flight dives to earth far from the zone and flies along the contours to avoid radar lock-on. The Phantoms pop up a half mile from their target and close quickly, arcing daisy cutters into the hilltop. They skim across the zone, ahead of their own ordnance. It detonates in the jets' wake, scant seconds after the aircraft have passed. The daisy cutter is a pregnant canister of explosive with a spear for a snout. The spear detonates the canister several feet above the ground; its shrapnel spreads sideways to clear the zone of obstructions and mines without leaving a crater. A distant *Crump* echoes through the valleys, soon swallowed by the endless canopy. The Phantoms circle and return for another drop, although it hardly seems necessary. The hilltop is obscured in a haze of phosphorous, pulverized rock, burning foliage, and dust. The new cutters disappear into the rising cloud, burst orange and yellow, and toss more debris aloft. The Phantoms return to strafe. Jet jockeys love to strafe. They dive on the smoking zone and pepper the surrounding hills with 20-mm. We watch the guns flicker in distant silence, bright pinpoints of light twinking on the Phantoms' waggling wings and exploding moments later, just as silently, in brilliant stroboscopic bursts amidst the rocks and trees.

The Phantoms' deadly ballet transfixes us. From the safety of altitude, I feel once more the surge of power. I think that nothing can withstand the assault, but in the valleys lie the rotting hulks of aircraft.

The Condor flight breaks off, the first phase of their mission accomplished. They orbit high to cover the troop insertion and to hear Bill's BDA. The battle damage assessment is the price gunship crews pay for the air show. Jet jockeys have a fixation with BDAs; it's their version of body count and fuels the war stories they take back to their ready rooms. Sometimes, when the gunship pilot rates the zone prep poorly, the jockeys grow belligerent, to the point where air-to-air hostility becomes a possibility .

Bill dives, far from the zone, and flies contours just as the fixed-wing did. We race to the ball of dust marking the target. Scarface 4-2 covers us high, and the Superchiefs stand off, waiting to be called in. We burst over the hill, banking and turning wildly, circling the brush fires started by the cutters and WP. Ten seconds, twenty seconds, Bill zigzags, holding a course no more than a few moments. The wind has swept the cutters' smoke from the zone, but Willy Peter is dense and lingers. Its searing vapor moves slowly through the grass and into the canopy. Anyone above ground will be flushed from hiding. The zone seems quiet. Dutch is braced in his doorway, facing aft, one finger locked in the trigger guard of the M-60, an extension of the gun, the other poised on the ICS button to instantly communicate anything he sees to the pilot. Thirty seconds. I grasp the machine gun, steadying myself against Bill's turns, holding red smoke in my left hand with the pin out. My ICS switch is pinned to my flak vest; I cannot communicate while I hold the grenade and the weapon, but Bill will know I've seen something if I open fire.

We continue to circle. Forty-five seconds, a long time to scope out the zone, but the prep, while necessary, has nullified the element of surprise.

Bill eases into a shallow starboard turn for the final sweep. The prep is thorough, 90 percent coverage on the cutters' bursts. The fixed-wing are pleased with the BDA. They've blasted a clearing large enough for a single -34, maybe two if the pilots want to push it. The surrounding hills provide ample cover for the enemy, but the fixed-wing peppered them thoroughly on the last pass. There are no discernible bunkers or spider holes, and no trails that we can see. I decide to swap green for the red smoke, knowing that Bill will want to mark the LZ for the -34s. The dust has settled, and the -34s will need to know the wind direction to make their approach. I slip the pin back into the red grenade and lean away from the gun to grab a green one.

Shots on the starboard side!

"Taking fire at four o'clock," Dutch speaks as he shoots.

My pulse quickens, breathing stops, blood pounding in my ears. I see colors without form, swimming, blurring into one another, mostly green, white. A half second of disorientation. The cabin shimmers.

One second, and training cuts in. The cabin becomes crisp and clear. I stabilize my body against the gun, become part of the aircraft. Sporadic shots. "Fire at six o'clock." I hear myself, a steady voice, professional, unfeeling. I am trained not to feel, to be outside my body, to judge. I am the crewchief.

Bill rolls to starboard to bring Dutch's gun to bear, diving at the same time to flee the zone. The Guinea traverses his TAT and makes noise, but his rounds cannot reach the target. The turn steepens; the leaden weight of the flak vest pulls me back. I'm off balance, only one hand on the gun. I drop the smoke into the clutter of the flight deck and grab for the gun mount to keep from going down. My weapon points into the sky, useless; Dutch is firing aft, and the cartridges ejecting from his gun pepper my legs. The flight deck is filling with hot brass; if I move I'll land on my ass. The Baron calls, "Four-Two in hot," and the blast of HEAT echoes behind us, then the rattle of aerial -60s. I can't see shit! Bill finds a valley, flattens the dive, and puts the hill behind us. The wind stream is solid at my back. It swirls up into my flight helmet, cool against the oily sweat on my headband, fanning chilly-fresh inside my flightsuit, rippling the hair in my ass— damp, tickly, good. I brace between gun and gunship, grip my weapon, push the safety off, and fire, fire, fire, sending long lines of tracers down the pylon. No target, any target, two hundred rounds: fire, fire, fire!

We clear the zone and climb to a safe altitude. "Everyone OK back there?" asks the pilot. Bill turns in his seat and surveys the mess in the cabin, smiling. Dutch grins. I shrug and give a big thumbs-up. The shock of enemy contact is gone, a memory of a memory. We laugh, like kids chased by grown-ups for misbehaving.

The Superchiefs abort and head for home. We do not want this hill enough to land the grunts against enemy fire, but the zone is now a target to do with as we please. Bill contacts the Condor flight and turns them loose. In moments, the Phantoms' remaining ordnance sails in. A billowing plume rises higher and higher, making it difficult to see exactly what hill we're lambasting. It doesn't matter; any of them will do. We stand off and practice long-range aerial gunnery, so far from the zone that the tracers burn out before they land. I play with the gun and try to keep my rounds within the same acre. It is useless, but we like to shoot, and the guns are already dirty and have to be cleaned, so what the hell. The Guinea fires his TAT dry. Bill flies by on starboard wing and lets Dutch empty his gun, then on port to give me my turn. He comes to a near hover, aims, and fires the fixed guns until they all run empty or jam. The fixed-wing depart, thanking us for the gun time. They are happy. We are happy. In five minutes we've expended enough ordnance to reduce a fair-sized neighborhood to rubble, but we do not investigate the results of our bombardment. There is no point. The snipers went to ground in the minutes it took to clear the airspace so the Phantoms could attack. It is doubtful that we got anyone, but we saturated the zone thoroughly enough that someone will claim a probable KBA (killed by air) in his ready room this evening.

The mission is over, the twelfth and thirteenth against my eighth Air Medal.

Flying past the Rockpile, the grunts wave and cheer—they enjoyed the show. We get a big thumbs-up. I wave back, but I'm already beginning to regret the firing mission. Now I'll have to clean my guns.

THIRTEEN-TWENTY HOURS, QUANG TRI CITY,
QUANG TRI PROVINCE, R.V.N.

On the way back to Dong Ha, Bill detours to Quang Tri. The Air Force O' club near the Jolly Green airstrip is about the only place in I Corps that consistently has hard booze. We touch down in an open field near the compound. The pilots head for the club, leaving us to guard the aircraft. It's not long before we draw a crowd of kids begging cigarettes and candy; selling Cokes, their sisters, their mothers—anything. Hazel Baby and I intercept them while Dutch covers us. There is no way we're letting any of the little bastards near the aircraft.

As a cherry crewchief flying VIPs to an orphanage, I felt sorry for the emaciated little fuckers ooh-ing and ah-ing around my flying machine. I was Captain America, "winning hearts and minds" with chewing gum, until they swarmed over the ship like roaches. A toothless crone grinned and nodded insanely while her charges ripped off smoke grenades, first-aid kit, C-rations, and Plexiglas cleaner. I drew a weapon and drove them off, feeling like the original ugly American, but as I checked the aircraft I knew I was damned lucky they hadn't planted explosives. My pilot tore me a new asshole, but I'd learned a valuable lesson—trust nothing In Country.

We stop the mob fifty feet from the gunships, gesturing with weapons and nodding "No." They're dressed in rags, the discards of uniforms from several nations, and their limbs protrude like filthy sticks from the baggy, oversized garments.

"Hey, GI," one ragged specimen of twelve or fifteen calls out. "You want my sistah? She virgin!" A slender, dark-eyed kid who might have been pretty in another world peeks coyly over his shoulder. She is half a foot shorter and could be eight or ten. In a few more years and with a pound or two on each tit she might make a good piece of ass. The miniature pimp thumbs over his shoulder and brags, "She give numbah one blow job. Suck-a-hatchie numbah one!"

Hazel Baby and I shake our heads, refusing the offer. The pimp elbows "sister." She licks her lips on cue, running a small pink tongue over the "O" of her mouth. She is indeed young, her teeth not yet blackened by the betel nut that addicts so many of the older villagers. "Fucking kid. . . ," I shake my head. The scene no longer shocks me. No longer do I pity them. The months In Country have left me with only a remote sense of disgust and the growing conviction that these people for whom we fight are not entirely human.

"You want my mothah?" the kid calls, seeing no response. "She virgin too. Numbah one boom-boom!"

"Hey, kid," Hazel Baby shouts, "your mother suck dick?"

"That ought to win their goddamned hearts and minds," I mutter.

"Sure, suck good dick, one t'ousand P," the pimp offers.

Hazel wonders aloud if the pimp will instantly turn his "sistah" into his "mothah" for a thousand piasters.

"I'll give you ten P," offers Hazel, meaning it.

"You cheap Charlie!" the kid shouts back. "You numbah ten, you numbah ten thou'!" He steps toward us, suddenly menacing.

Hazel and I slowly back away.

"Fuck you! ya little shit," Dutch shouts from the aircraft. I sense real anger in his voice. "Get your ass outa here or I'll blow you away." Dutch levels the -60 and flips its gun sight into firing position; the kid stops. I turn from the kid to Dutch, and back again; it is difficult to say whose eyes hold the most hatred, whose clenched jaw is more grimly set. The kid stands his ground, fists clenched. Hazel and I move slowly apart, opening a firing lane for Dutch. A few timid kids shift nervously, but all hold their ground.

Suddenly, the pimp begins a chant. "Cheap Charlie, Cheap Charlie!" The others join in; "Cheap Charlie, Cheap Charlie!" they singsong.

Dutch joins in from the aircraft, laughing, "Cheap Charlie, my ass!" then relaxes his weapon.

The chant fades, and a new one takes its place. "Ciga-lette! Ciga-lette! You give one Ciga-lette!" they yell in cadence. Hazel Baby and I pull packs of dried C-rat Camels and Chesterfields from our flightsuits, toss them onto the sand, and back away to the aircraft. The ground swarms as the ragged troupe fights for the tiny four-packs. "Hey, you numbah one GI," they shout.

The pilots have been gone half an hour, and the midday heat is a shimmering blanket on the LZ, rising in waves from the sand. In the days of the summer monsoon, everything is hot, from the air that clots your nostrils to the sweat that tickles through the short hairs under your balls and pools deep in the crack of your ass. You get used to anything but the heat seems harder to bear than anything else. It is maddening, preying on the mind. The simple act of breathing becomes obsessive. Carry a load, like ammo, or steel matting, or a man, and your lungs scream with the craving for cool oxygen.

Hazel and I decide to reconnoiter the mess hall to see if there's anything cool or at least wet to drink. Leaving Muffy and Dutch on guard, we sneak into the compound. The heat has driven everyone indoors—the Air Force and our pilots to the O' club and the ARVN to their barracks. We locate the mess hall, force the back door, and invade the kitchen. There's little worth stealing. Spam. Fruit cocktail. Some dink food that we wouldn't touch. An olive drab gallon of pears, which we liberate.

Dutch wears a John Wayne opener on the chain with his dog tags—most of us can't bear to wear them—and peels back the lid. We pass the can around, drinking the syrup and scarfing the fruit straight from the container. The warm syrup is best, smooth and soothing. Gooey-fingered and sticky-lipped, we wash ourselves sparingly with water from our canteens.

Dutch carefully closes the can, walks toward the Vietnamese kids, and sets it in the sand. "Chow time," he calls. They swarm onto the field as Dutch retreats hurriedly. The pimp leader picks it up, peers into the empty interior, and throws it at us, lapsing into native curses that can only mean "Fuck you!"

I turn to Hazel. "How many hearts and minds d'ya figure we won today?"

FOURTEEN-FORTY HOURS, THE DONG HA LSA,
QUANG TRI PROVINCE, R.V.N.

It is fucking hot. My head aches under waves of heavy pressure, like some-one is plunging both ears with a plumber's helper to the rhythm of the "Anvil Chorus."

Parked next to Dong Ha's Logistic Support Area, we are buffeted by the rotor wash of passing -34s and -46s that drives the dust through the aircraft with the vigor of a blast furnace. Dutch and I shelter in the pilots' seats each time one passes, baking in pools of sweat beneath the greenhouse of Plexiglas. It'd be lots better if we parked in the revetments, away from oper-ating aircraft, but we're on short call to take some horse's ass Bird colonel on a series of observation hops, and we can't expect the Bird to walk all the way to the revetments. Must be a whole fucking two hundred yards. Oh, no! Fucking Birds can't walk. Goddamn Birds. The cabin and weapons fill with dirt, more to clean at day's end while the horse's ass fucks off in the O' club. God, I hate Birds.

An aircraft clears the LSA, and we return to the cabin. I watch a mob of sweat-soaked grunts labor under the gaze of a humorless gunny. They move fifty-five–gallon drums from one side of the LSA to the other, then fill the empty spaces with other drums identical to the first, as if they're rotating stock in a grocery store. Gunnys suck. What the hell is it with them, anyway? Why can't they stand to cut anyone a little slack?

Jesus, my head hurts. I glance at the cockpit thermometer. Forty-seven degrees Centigrade. What the hell is Centigrade, and what is it with this military metric bullshit anyway? When we fly, we have to do everything in meters. When I report the location of our chase aircraft during flight, I tell the pilot, "He's fifty meters at eight o'clock," but inside I know I mean yards. I only say "meters" because everyone says "meters." Yards, meters, what fucking difference does it make? A meter is longer than a yard. Let's see, that means he's closer than I report. No, farther. . . . Closer. . . . I strug-gle with the math in my head. Who gives a shit?

It is so fucking hot. Forty-seven fucking degrees. Centigrade to Fahren-heit. When the hell do I take out the thirty-two? Nine-fifths, five-ninths. I check the conversion table in my crewchief's pocket checklist. Jesus Christ,

117 degrees American in the goddamn aircraft—117 goddamn miserable lousy fucking degrees.

It is so fucking hot! I even keep my hair short. Some guys shave theirs. In the World, you had to beat me to cut my hair this short. I run my fingers through a quarter inch of bristle—even my hair feels hot. Hair can't feel. Hair isn't supposed to feel. If my hair is this hot it must be because my goddamn brain is baking.

Dutch is loudly singing the black-humor ditty about a singing telegram delivered by dress-blued visitors to the parents of a fallen son.

> Your son is Dead, they say.
> He bought the Farm, today.
> He lost his turns and crashed in the Dong Ha LSA.
> He really Hit the ground.
> He's scattered All around.
> What more can I say?

"Shut up." I order.

"What?"

"Shut up, you south Boston sack of shit!"

It hurts to talk. The pain in my ears flashes forward in a giant blocking tackle, settling into my eyes, swelling. I raise my hands to blot out the sun, trying to contain the pressure building inside.

"What the hell did you say?" Dutch threatens.

I cannot see him. The pain is building so fast. I warn him, choking the words from behind my hands, "I think I'm going to puke."

"Christ, get out of the aircraft."

Dutch grabs my arm and helps me from the cabin. I cannot see. He leads me to the edge of the LSA. The full heat of the sun weighs down into the green of my flightsuit. "Oh shit. . . ." Dutch bends me over, and I empty the day into the dust. Powdered eggs. Hot dogs. Bits of pear. Gone. All that precious food.

Dutch produces a canteen and rags, dampens one and presses it to my forehead like my mother used to do when I was a kid. I open my eyes tentatively, and wipe the salt tears away with the cuff of my flightsuit, gasping for breath. It seems much cooler now; the glare not so painful as before. I breathe deeply until the aching pressure begins to wane. Dutch raises the canteen to the lips; I rinse the foul taste gratefully and spit bits of food into the dust. Dutch helps me to the gunship, sweeps the troop seat clear, and applies a fresh rag as I lie down.

"This heat, man. . . ," Dutch shakes his head.

I smile weakly. "Heat, shit. I was doing OK until you started singing. You sing like you have a mouth full of Boston beans."

"Screw yourself," says Dutch, but he stays by me, dampening the cloth from my canteen, then his. He gives me some aspirin from the first-aid kit. In a little while the pain passes.

Fifteen-Thirty Hours, The Demilitarized Zone, Quang Tri Province, R.V.N.

Flying below the DMZ feels different from cruising the rest of the country. North Vietnam broods menacingly beyond the scar of the firebreak that separates the two nations. The threat is real enough everywhere in Vietnam, but on the DMZ the thread of life stretches especially thin.

I pause, remembering the desolation of the DMZ. What an evil place it was, as if it were alive but could abide no life within itself; it seemed to embody the death of all things.

"What do you see?" asks Tom.

"Oh, nothing, I guess. Just remembering. The DMZ." I answer.

The sound of the letters is itself chilling. What more does one need to say?

"What was special about the DMZ?"

"It was so . . . desolate. I flew there summer and winter, but all I can remember is gray and tan. Funny, I don't ever recall it being green. I know it must have been green. Everything was green. They spent enough time killing back the jungle to create the firebreak."

The firebreak was a six hundred–meter by eight mile swath of bulldozed earth Secretary of Defense Robert McNamara ordered dug below the seventeenth parallel—a free-fire zone to stop anything that came out of the north. It worked as well as anything we tried to do in Vietnam.

"I used to be afraid of it," I laugh. "When I first started flying there, I'd worry that we'd stray too close to the border by accident, and the bogeyman would reach up and get me. It sure felt scary. But I got used to it."

"Used to what?" asks Tom.

"Used to everything. You can get used to anything, after a while. I was a crewchief," I explain.

"Think. About this day you're putting back together. What're you feeling about the DMZ? About the day?"

"I don't feel anything about the DMZ," I say simply, "and you wanted a day, so I'm giving you a day. Nothing special about this day. Well, I guess I didn't puke every day, but I don't feel anything. What is it you think I should feel? I can't put in what isn't there."

"OK," Tom says slowly, "Let's just go on."

There is little conversation in the aircraft; the Bird colonel sitting between Dutch and me cramps everyone's style. We've given him the extra radio headset and disconnected Dutch because there are only four comm receptacles. Dutch sits hunched against his gun, visor down. The bastard—I think he's asleep. The Bird shifts back and forth in the cabin, looking out the port door, then the starboard, then craning his neck over the instrument console for a forward view, trying to keep his terrain map from shredding in the wind. Dutch and I wear the only two gunner's belts aboard, and I am not entirely unhappy that the Bird ignores my gestures to remain strapped in. Perhaps he will fall out.

I see Gio Linh in the distance. The fire base is taking its perennial dose of incoming. We're too far away to hear the crash of artillery, but we can plainly see plumes of dirt splash skyward, as shock waves drive a dusty haze through the perimeter. It reminds me of dropping rocks into a well. Ploop, ploop, the rounds fall in. Fish in a barrel. Ploop, some distant, temporary structure decomposes in a flash of light and dirt. Nothing we can do about it—the rounds come from North Vietnam. Nothing anyone can do about it, except tough it out. I wonder about the rock apes on the Rockpile, but the grunts of Gio Linh lead the lives of gophers. We fly by, close enough to hear the *whuuuump* of large caliber landing. *Whuuuump*, ploop. *Whuuuump*, ploop. More geysers of dirt.

The Bird watches the fire base without expression. He is large and lean, and the eagles on his collars are impressive. Too bad they have to waste such a good emblem on such a shit rank, I think. Captains and brigadiers seem to be the best ranks, at least that I can tell. Our captains seem like they have a hell of a good time—all those lieutenants they have to shit on—and most of the brigadiers I fly seem to be real decent heads. In between, most officers turn into a bunch of goddamn admin weenies. I glance back at Gio Linh. Ploop, ploop. God, what a boring fucking war.

We leave the fire base and turn west to Con Thien, the second of the two positions guarding the southern boundary of the DMZ near the coast. Con Thien is Vietnamese for "Place of Angels," which is a nice bit of irony, depending on your point of view. Mired in dust or mud, living underground, few Marines appear less angelic or are closer to becoming angels than the grunts of Con Thien and Gio Linh, the two bases that vie for honors in the Incoming-of-the-Day Club.

Con Thien is taking incoming as well, and Bill advises the Bird that there is danger of an air burst from artillery arriving from the north. The Bird disregards Bill's warning and orders us to fly a pattern between the fire bases.

Back and forth. I don't know what we're looking for and do not ask. It makes no difference. I study the riddled land beneath us, looking for signs of enemy activity or anything suspicious, and I'll report anything I find.

The Bird sees something. He leans past Dutch, who has the good sense to wake up, points at a piece of earth, and shoves the map under Bill's nose. "That's it," he shouts. Bill nods. I look out. Craters, just lousy craters in a land of craters. Big deal. We circle once, the Bird folds his map, Bill nods again, then we head home. Mission 154 is over. In the distance, incoming still occupies Gio Linh. *Whuuuump, ploop. Whuuuump, ploop.* Fucking place never changes.

Seventeen-Hundred Hours, The Mat, Phu Bai, Thua Thien Province, R.V.N.

We shut down for the day. Bill and the Guinea wave a brief good-bye and go wherever it is officers go after flight ops are finished. It isn't work because we won't see them on the flight line until morning. The mess hall opens at 1700 hours, but the line is usually long, so we plan to make chow in an hour or so.

Dutch checks out the ordnance shack.

"The Rock isn't there, only Lewish," he says.

The Rock has one commandment: You clean what you fire. Squadron policy is to store the crewchief's and gunner's M-60s and the barrels for the aircraft's fixed guns in ordnance when flight ops finish. The ordnance man is supposed to inspect the weapons, making certain they've been cleaned. It takes an hour for two men to clean eight machine guns, an hour better spent doing most anything else if you can get away with it.

"Great," I answer. When the Rock is out of sight, Lewish will accept anything. If the Rock finds the dirty weapons, Lewish will claim we dumped them in the racks while he was out arming an aircraft. Lewish takes the -60s without comment and places them in VT-11's rack. Dutch and I leave in haste, busting internal organs in glee at having dodged the Rock. We secure the aircraft, spray the fixed guns with WD-40, and tie on a tarp to shield them from the night's dew. I head for the line shack, and Dutch to avionics.

The Rock is in the shack.

"You forget something, Corporal?"

"Who? Me?" I look around; there are lots of corporals in the squadron.

"You," he jabs.

"Oh no, Sarge. I just dropped the guns in the shack so I could go to chow. I was coming back to clean them."

The Rock is having none of it. His withering look asks, ' Do I look like I was born yesterday?"

"I checked your fucking guns, and your fucking guns are cruddy. Get your ass into that fucking shed and clean them fucking guns. You shot them, You clean them!"

The Rock stomps away. My ass puckers when it's chewed, so I concentrate on letting it go slack. I notice that I have not been breathing either, and work on that as well.

Dutch is not in avionics. No one knows where he is. I leave word to get his ass to ordnance and trudge back to the cleaning shed. Lewish has the weapons waiting—a large chunk of his ass is missing too. Lewish has earned himself a heavy dose of EMI—extra military instruction—for his favor. EMI means doing whatever shit detail the Rock can construct for as many hours as Rock desires: scrubbing the P-40 ammo carrier, sweeping rat shit. Useful stuff like that.

The cleaning shed is a tin roof on stilts over wooden workbenches. In the center, four 55-gallon drums sliced to serve as basins hold a noxious brew of jet propellant and hydraulic fluid. Seldom emptied, the liquid is black and gritty with the residue of thousands of missions. I strip my guns and dump everything in to soak. Hazel Baby and Muffy occupy the second set of basins and are half finished with their chore.

"The Rock got you, huh?" asks Hazel.

I shrug. "Hey, it's always worth a try. What's he going to do? Cut my hair off and send me to Vietnam? He can't give me EMI while I'm on flight ops, and I have to work in the hangar tonight anyway.'

I roll up my sleeves and scrub at the caked carbon with a brass brush. You have to use soft metal against the steel. Steel brushes work better but remove the bluing, and the gun rusts faster. The carbon is as hard as the metal itself and yields slowly. I pick at it with a chipped fingernail.

"How're you feeling, Zack? I saw you toss your cookies in the LSA."

"I'm OK. Must have forgotten to take my salt tablets. I got a little weird," I apologize.

The carbon is stubborn. I give up doing it the right way and scrape it with a knife.

"You had a bunch of mail again. And a care package. Timmy put it in the hooch," says Hazel.

"You get anything from Hazel?" I ask Hazel Baby.

"Yeah," he answers tightly. "I don't know why I stick with that broad."

"Uh, oh," I think, "Hazel is at it again." Hazel is nearly thirty, has a son about ten or eleven, and rides off and on with the Hell's Angels. Her picture shows a hard prettiness, a lean woman dealt too much life. Hazel Baby has been waging a long-distance battle for her affections since Pendleton. Hazel's allegiance seems torn between the nineteen-year-old sloshing gun parts in the tank next to me, and her old man in the Angels. I think she's left Hazel Baby and come back twice by now, all through the mail. It's easy to track the status of this strange relationship. Hazel Baby names his aircraft *Hazel Baby*. Each time they break up, he paints her name out, and each time they reconcile. . . .

"What did she say?"

"I don't want to talk about it."

"Come on."

"She's riding again. I keep telling her that guy is no good for her. No good for that kid of hers, either."

"Not much you can do from here, though."

"When I get home, I'm going to kick some ass," says Hazel, but his tone is helpless, not defiant, and much can happen here and in the World before he goes home.

"How about if I get you someone nice to write to?" I offer. My girlfriend has recruited a number of pen pals among her college roommates. Several letters arrive each month from Doyle Hall, College of Notre Dame of Maryland.

"I don't want someone nice. I want Hazel."

I see Muffy smirking as he cleans his weapon. "Muffy, you got a girl?"

"Of course I have a girl," Muffy answers, as if the thought that he not have a girl, and a good one, is preposterous.

"Got a picture?"

"Not with me."

"Want to buy one?" I ask slyly.

"You are a bastard."

I know my shot is out of line, but I can't resist with Muffy. While conversation regarding women preferentially runs to the crude, a guy's girl is sacrosanct. Show a picture of your girl to someone else, and the only response desired or given is a low, slow whistle of approval, a muttered "Nice . . .

really nice"—even if she is uglier than a pit bull and twice as mean.

"C'mon, Muffy. Tell us about her."

"Fuck you."

"C'mon," I persist, knowing that he will. Each of us wants few things more than to talk of home and those we love, to give form to our memories and feelings, if only for a moment, in words chosen with exquisite care; to sculpt, in air, a vision that we alone can see.

"Well," he begins slowly, "she's blond, wears it up in a kind of flip in the front, but it's long down the back. She has blue eyes, and naturally dark lashes. Doesn't wear a lot of makeup. I told her not to, and she doesn't. Nice family in Georgia. Her daddy's a lawyer. She's really smart. National Honor Society. Studying nursing . . . uh. . . . We. . . ."

It is happening, as always, when we speak of those whom we love. Words melt into thoughts; the thoughts run warm and silent inside, and you really don't remember when you stopped talking. Others stand and stare, but they understand that the vision is more real than the weapons shed and the guns.

My own thoughts fill me as Muffy speaks. She is small and dark, with laughing eyes in an elfin face. She has a pixie smile atop a tiny chin, and a tiny nose that wiggles like a rabbit's when she is happy.

The memory is warming, but visions such as these cannot endure for long in the weapons shed.

"Getting married?" Hazel interrupts.

"Of course!" Muffy answers.

"Going to have kids?"

"Of course."

"I didn't think avionics men could have kids. I thought all they had were transistors, what with those little teeny peckers and all."

"Y'all go fuck yourselves."

SEVENTEEN-TWENTY HOURS, VMO-3 ORDNANCE, PHU BAI, THUA THIEN PROVINCE, R.V.N.

Gunny bursts into the shed, puffing hard. "Emergency extract. Zack and Dutch. Where the hell is Dutch?"

"MIA," answers Hazel.

"OK," he points to Muffy, "you're it!"

"Why us? Who's on the hot pad?" I protest. Each night a different crew is assigned to be on call for emergency operations. This is not my night.

"The hot pad comes on at eighteen hundred. They're still at chow. You're it."

"What about him?" I point to Hazel.

"He's got perimeter guard. Move it!"

"I'll go find Dutch," protests Muffy, desperately.

"No time. Move it!"

We drop our weapons, wipe the fuel from our hands, and rush to the ordnance shack. "Hey, Muffy, looks like your ass is finally mine."

He glances nervously at me but says nothing.

The Rock rolls up in the ammo truck with fresh guns. We race to the aircraft, bouncing wildly on the corrugated matting. Muffy and I strip the covers from the fixed guns, Rock loads the extra ammo, and Lewish arrives from the line shack with our helmets and flak vests. Bill and the Guinea lope across the flight line, struggling back into their flak gear. Looks like they didn't have the sense to clear out either. Muffy removes the main rotor tie down and tosses it to me. The pilots drop into their seats and pop igniters as they complete the takeoff checklist. The crew of Tango 10 races to their ship. Lewish circles the aircraft with the arming meters. He and the Rock check the firing circuits as Bill gets the mission coordinates from Roseanne Bravo. This is an emergency extract; there is no time to taxi to the arming area.

From the conversation with Bravo, I glean that Hightower, the patrol we inserted into the Co Bi–Thanh Tan this morning, has run into some heavy shit.

Superchief, farther down the flight line, is talking to Roseanne as well and reports the -34s ready to lift.

Lewish runs to Tango 10, and the Rock finishes arming our guns and follows him.

Muffy and I strap into our gunners belts.

Bill gains tower clearance, broadcasts, "Scarface Four-One, lifting," and we rise above the aircraft parked on the mat. The downdraft of our rotors buffets the gunships and slicks beneath us as we pass. We dash to the runway, turn right, and nose into the takeoff run.

I see Dutch standing in a group of Marines, his mess kit dangling from his back. Dutch grins and thrusts up a thumb. "Gung ho!" I should give him the finger, the bastard, but am too caught up in the race to become airborne to deliver insults. "Gung ho!" I flash back as the runway falls beneath us. I am crewchief! It is ridiculous, and romantic, and from someone else's war, but it makes me feel good as I finish strapping in, and I mean it.

SEVENTEEN-TWENTY-SEVEN HOURS: RUNWAY TWO SEVEN,
PHU BAI, THUA THIEN PROVINCE, R.V.N.

We establish translational lift and pass the Superchief transports rising from their own flight line. Bill informs the Superchiefs that the gunships will head directly for the zone; the slower -34s will rendezvous as soon as possible. Their job is to extract the grunts under fire. Ours is to ensure that there will be grunts left to extract when they arrive.

Bill checks in with Landshark a hundred feet over Highway One.[4] The flight will be low level all the way—no time to waste in establishing a safe cruising altitude. Landshark confirms Sav-A-Planes firing across our path from the firebase at Scat 1-4 into the Co Bi–Thanh Tan.[5] Bill has the artillery missions cleared to grant us a straight flight into the zone, then switches to UHF Guard and broadcasts in the blind.[6] "Any aircraft, any aircraft, this is Scarface Four-One broadcasting on Guard. Emergency extract Yankee-Delta 592234. Any aircraft respond button Vermilion, over." He repeats the transmission, then switches to the UHF frequency coded "Vermilion" in hope that some other aircraft with ordnance and fuel to burn will respond.

We get lucky. The radio squawks a distant signal, "Scarface Four-One, this is Deadlock Two-One and Two-Two. We copy your transmission. ETA Yankee-Delta 592234, two five minutes."

Bill rogers the call, glad to have the help of two VMO-2 gunships.

The Deadlock flight is the only one to respond and will arrive over the zone at about the same time as the Superchiefs. We'll have to hold until they come.

SEVENTEEN-FORTY-FIVE HOURS: THE CO BI–THANH TAN,
THUA THIEN PROVINCE, R.V.N.

Bill raises the grunt operator on Fox-Mike.

"Hightower Four, Hightower Four, this is Scarface Four-One. Do you copy?"

---

4. *Landshark* was the call sign for the regional tactical air controller. Landsharks Alpha, Bravo, Charlie, and Delta kept tabs on all aircraft and missions operating within their areas of responsibility, as well as artillery-firing missions. Aircraft checked in with the local Landshark controller when entering or exiting the controller's airspace to avoid flying into a barrage.

5. *Sav-A-Planes* were the reports Landshark provided on artillery missions, which aircraft obviously wanted to avoid.

6. Guard was the UHF emergency band that all aircraft monitored. If you needed help from anyone in a hurry, you broadcast your message on Guard, and the world heard it.

"Hightower copies five-by-five," whispers the radio operator. Grunts always whisper. It makes little difference whether they're concealed under heavy canopy with VC feet away or exposed in the midst of a firefight. The operator whispers to Bill although we can hear the crash of RPG (rocket-propelled grenades) through his transmitter.

"Hightower, we are two Huey guns, ETA two minutes. Extract following, ETA nine minutes. What is your situation?"

"Scarface, sit-rep Mary, I say again, sit-rep Mary. We are pinned by mortars and RPG from tree lines 200 meters south and 250 meters west our position. We have wounded in the zone. I think I have you in sight, Scarface, we are at your . . . ten o'clock. Say again extract ETA?"

"Hightower, extract ETA now eight minutes. Scarface flight is circling north of your position. We will attack west and south."

We can see the smoke rising from the zone. The grunts are trapped in an open field of low scrub, near the vertex of a ragged "L" formed by intersecting tree lines—a perfect crossfire. The dark green bodies of Marines lie prone on the deck, some in shallow trenches scraped into the rocky soil, a few taking refuge in craters. Others lie in the open, their lifeless forms needing no further protection.

Our plan is straightforward. The zone offers no opportunity for strategy or finesse. The gunships will assault the tree lines, inserting themselves between the grunts and the North Vietnamese gunners. Coming out of the north, Scarface 4-1 will dive obliquely and strafe the western tree line with measured bursts to conserve ammo until the -34s and Deadlock ships arrive. Moving counterclockwise through the angle of the trees, we'll bring our fire to bear on the southern positions, raking them as we escape to the east. Muffy is on the outside of the turn and will fire point blank into the enemy as we pass. I'll train the port gun aft to cover our ass while trying not to hit the grunts in the center of the zone. Scarface 4-2 will phase her attack 180 degrees out from us. It's a very long daisy-chain, too long for the aircraft to effectively cover each other, but it's the best we can do to keep constant fire in the zone. And the attack is too predictable—the enemy gunners will anticipate our approach after the first run. Still, it will keep at least one gunship between the grunts and the tree lines and should keep the VC down. It takes very large balls to stand up to a Huey gunship firing flat out on the deck.

North of the target, Bill drops to ground level. At 300 meters from the trees the firing shifts in our direction. Muzzle flashes wink distantly, and tracers flash past the aircraft as we race headlong into the fire. Rounds ricochet from boulders ahead and behind us, but we close on the target unhit; the narrow profile of the onrushing Huey is our best protection. Bill unleashes two HEATs into the western tree line. They burst on the edge of the woods, erupting in a dirty cloud of earth and foliage. It's impossible to see whether the shot is effective—the cloud from our rockets obscures any muzzle flashes. We pick up flanking fire as we close on the western woods and circle to rake the inside of the "L" with 7.62. Muffy squats behind his -60, firing measured bursts. Exploding RPG erupts in our wake, tracking as we arc along the periphery of the zone. The attack has its intended effect: nearly every NVA gunner struggles to keep their weapons on us, and for a few moments they leave the grunts alone.

Bill blasts two more HEAT into the southern tree line. We're carrying seven shot pods and cannot spend our stores too quickly. The Guinea swivels his TAT, methodically squeezing the trigger to send tracers into the dust and haze oozing from the forest.

Thirty seconds into the attack, every gun on the aircraft is in action except my own. I wait until the continuing loop of our flight path brings the western trees into my line of fire. I kneel in the doorway, facing aft, wedged between the gun and forward door frame, braced against the aircraft's gyrations. I lean outward against the weapon, depending upon the traverse stop welded on the gun mount to hold me in position. The penny's weld of steel is the only thing that prevents me from falling the length of my tether as we hurtle through the zone. The tree line creeps into my sights, and I throw short bursts against muzzle flashes winking through the clearing smoke.

In the aircraft we are silent and intense, swallowed by the din of machine gun, rotors, and wind stream. Muffy fires coolly, continuously. His gun rapidly fills the deck with bouncing, jiggling cylinders of searing brass. Every now and then a burning cartridge flies through the cabin to land in a soft fold of my flightsuit. They're hot enough to raise welts if you fail to brush them away. Muffy tracks his gun aft to concentrate fire on some target in the aircraft's wake. As he pivots, ejected brass flies forward in the cabin and bounces off the rear of the copilot's seat.

One burning cartridge sails high and vanishes into the Guinea's collar. The Guinea's yelp nearly drowns the chatter of the guns. 'Crewchief, get it out!"

I am firing into the receding tree line, and his scream breaks my concentration. I turn to see the Guinea struggling against his shoulder harness, trying to dislocate his arms to get at the cartridge. His foot is jammed on the cockpit ICS switch—the noise of the wind and a string of "Get it out!'s" fill my flight helmet.

I draw away from my gun but see new lights winking and pull my trigger once more.

"Crewchief, get it the fuck out!" the Guinea orders.

All the firing is aft now; I motion to Muffy to cover us and leave the weapon. The Guinea writhes in his seat, trying to escape the cartridge trapped in his underwear. Each motion wiggles it farther down his spine.

Muffy covers us, but plays at aiming his gun to carom a constant stream of brass off the Guinea's flight helmet. A smile of pure evil peaks out under the smoked visor. Some of the brass comes my way, but I can dodge it. I think that anyone with the balls to pull this kind of shit in a hot zone can't be such a bad guy after all. He plays the gun back and forth, peppering the Guinea, then returns to firing purposefully into the zone.

I remove my flight glove and stick my hand down his back, searching for the cartridge.

"Farther, farther," he cries.

I lean forward, driving my arm down the small of his back.

"Farther!"

I stop, and find the comm switch with my free hand. "Just wait a damned minute. Just how far do I have to go?"

"Get it!"

The cartridge, hot enough to burn my finger tips, is stuck in the crack of his ass. I pry it out with two fingers and drop it into his lap. A souvenir. The Guinea shrinks back into his seat and smiles wanly in gratitude. Muffy laughs soundlessly into the wind stream. He'll have a ball with this when we get back.

Behind us, Scarface 4-2 begins its run. Bill throws a final burst of machine-gun fire into the trees and breaks away, staying low, dashing north by east to begin another pass.

Muffy and I check the pilot's weapons. One of the port-side -60s is jammed; I lean into the wind stream to clear it. I release the barrel lock and grasp the flash suppressor to pull the barrel free. The flight glove gives little protection against the heat, and I toss it quickly into the litter of the cabin. I pull a bent link from the chamber, open the feed cover, and snake the cocking tool in place to push back the bolt. In less than thirty seconds the gun is rearmed. Muffy and I tear open new boxes of 7.62 and sweep the empty boxes and expended cartridges into the airstream with our arms. The brass falls into the reddish hue of the evening, sparkling merrily as it tumbles end over end. Rearmed, Muffy and I brace into position and watch the grunts as Bill prepares to reenter the fire.

"What's the matter?" asks Tom.

"Oh, I was just thinking about them."

It is suddenly difficult to speak. My voice is strained, with too little air to drive it, like it's trapped inside and can't escape.

"Who?"

"The grunts."

The tightness grows. My eyes ache. I swallow hard. but a lump pains high in my throat. Hot tears flash. I force them away. I am practiced at this. It happens often when I speak of the grunts. I cannot allow them to come.

"You know," I say slowly, "a lot of people say that Marines are brainwashed. They're right, you know. We are."

Tom watches silently, letting me speak.

"In boot camp, they taught us one thing, over and over again. Marines take care of their own. It's like a religion.

"I never saw any of those guys in the zone before, and I doubt that I ever saw any of them after that mission, but it was the most important thing in the world for me to help them. We would have done anything for them. Hell, we did do anything for them. The risks we took . . . God! I was nineteen! You could say it was only our job, and we had to do it, but the feeling ran deeper than something you do because it's a job. Maybe it was because I knew—I really knew—that they'd do the same for me. I could count on them. I had to deliver when they were counting on me."

"That's kind of what being a Marine is like, isn't it?" asks Tom.

"Yeah," I answer, missing the old days, remembering what it felt like to trust and to be trusted. To feel that I was doing something important.

"Were you afraid?"

I look back into the zone, with eyes that feel so old, and I do not know. It seems that I project feelings that I should have felt, feelings that normal people should have felt, rather than what I did feel. I can't remember feeling. I can only project that which logic demands be felt. What was it that I should have felt?

"I was afraid for them. No one knows what it's like unless you've been there—the fear. To be trapped in the zone. . . ."

"Did you think about that in the aircraft, the fear?" Tom asks.

I try to remember, but cannot. At last I answer, "If I did, it didn't last long. I was too busy."

"Try to remember," Tom says pointedly. "What did you feel?"

I try, and try and try, but I can't remember feeling anything. I see myself in the gunship, functioning like a machine. If there is feeling, it seems only the anguish that I feel today, projected back into the old memory. Any man, any normal human being would feel anguish for the suffering in that blasted field, and fear for his own survival as well. Did I? What did I feel?

The lump in my throat rises higher. A great bursting sensation overwhelms me, and I see the zone too clearly in the dim light of the Center.

No! No no no!

I squeeze my eyes shut, slam closed the doors that seem to be opening.

I answer Tom with words like cold steel. "I told you. We weren't allowed to feel."

We roar through the attack, our second pass a duplicate of the first.

Near the end of the run rapid shocks course through the deck plates and up through the soles of my flight boots. A hit! Probably in the tail cone, but I can't know for certain. I glance back in time to see a scattered sweep of tracer in our wake. The fire tracks our flight path; probably a lone automatic who doesn't know how to lead. Except for a lucky burst, he stays consistently a few yards off target. The angle of fire places him somewhere between the tree line and the trapped grunts. I scan the ground but see nothing.

Bill contacts the Superchiefs. Their ETA is under five, but we don't know how much longer we can cover the grunts. The other aircraft finishes its run, and Bill makes a dummy pass to conserve ammo for the extract. I remember the story of one of the Fat Baron's missions. When his guns ran dry, he made dummy runs to keep the VC down. When they wised up and charged, he swooped through the advancing mob and rammed them with his skids. Hell of a way to fuck up a good aircraft; they said it was a bitch for the crewchief to get the blood out of the radio compartment.

Another dummy run. The gooks realize what's going on and concentrate their fire on the grunts.

The automatic seems to be gaining experience; his rounds track closer on each pass. Finally, I spot him, hiding in a clutter of boulders strung between the Marines and the main enemy force, just under our flight path. Outcroppings of rock conceal him during our approach, but he is able to shoot freely as we pass overhead. For a split second I see him in the shadows of his natural bunker, a scrawny little dink in black pajamas and straw hat hunkering behind an AK.

I shout at Bill, "Taking fire six-o'clock. Pull in hard to port!"

Bill increases our banking turn and drops my gun nicely on target. I let go a burst, but the dink drops out of sight behind the boulders.

The turn is holding us too long over the zone. The ground fire and RPG are heating up, and we have to break out soon. Bill holds the bank to give me more time. I spray aft, playing the rounds across the boulders, trying for the slit. Ricochets spin off wildly. It would be a difficult shot under any circumstances, but the Gs and gyrations of our turn keep throwing me off target. I lock my shoulder into the weapon and hold the trigger down, pouring continuous fire. You can shoot an M-60 for a long time in the wind stream before it overheats. Shards of rock burst from the boulder, and a low haze of dust clouds the slit. Finally, the tracers pierce the opening over his head, disappear into the shadow of the outcropping, and flash off the stone behind him. Rock shards and shattered lead blow into his back, tear him from the ground, and throw him face forward up and over the boulders. His hips lodge in the slit, head, chest, and arms draped brokenly across the pockmarked stone. I continue to fire into the body, which jerks like a puppet as the rounds penetrate, then ricochet back from underneath. Bill breaks our bank and dashes at ground level to escape. I fire aft long after the lonely little dink has vanished from sight.

The cavalry arrives in the form of two fresh Deadlock aircraft as we begin our next pass. Bill fits them into the daisy-chain immediately. Four gunships now circle the zone. It begins to look like the grunts might have a chance.

At last the Superchiefs arrive. Bill coordinates the extract. The four gunships tighten their cover, pulling in close to the Marines while the -34s make straight-ins from the northeast. The lead transport contacts the grunts, his voice sounding flat and sprinkled with static over the Fox-Mike.

"Hightower, Hightower, this is Superchief medevac. Do you copy?"

"Roger, Superchief, we have you in sight."

"Hightower, get your wounded up! Superchief Five-One will land to your north, Five-Two to the east. Smoke your zone, Hightower. I say again, smoke your zone!"

"Roger, Superchief, popping yellow smoke."

"Hightower, get your wounded up! We are in hot."

The Superchiefs drop in tandem, their door gunners raking suspected enemy positions as they flare to land. Every airborne gun is in action. But for the lone soldier in the rock bunker, we have seen no enemy. We kill trees, rocks, bushes, the earth itself. The Scarfaces and Deadlocks pickle their last rockets, which tear through the smoke and dust shrouding the tree lines and shred more of the land.

The grunts are up and running as the -34s land. Crewchiefs and gunners leap from the aircraft and race to help the struggling Marines.

From my gunship I see the wounded and the lifeless forms of the dead half-carried, half-dragged through the dirt, then stuffed through the dark maws of the medevac ships. The living will not leave the dead. The legend in the flesh: Marines take care of their own.

A minute, maybe less, and the extract is complete. The Superchiefs lift amidst the continuing barrage and dash away at low level. The gunships break off, eager to escape as well. We circle east and rejoin the medevacs laboring toward Phu Bai. Superchief 5-1 contacts Roseanne to alert Bravo Med to stand by with stretchers. Closer to the combat base, Phu Bai tower grants the Superchiefs' request for an emergency straight-in to the medevac pad. We hear the tower order aircraft on approach to abort and orbit the airfield until the medevac clears. The Superchiefs land next to the old civilian terminal at Phu Bai. I know the stretcher bearers will swarm over the aircraft and hustle the wounded into the labyrinth of the Med. I have seen it many times.

The Hueys proceed to our own parking area and shut down; the fuel truck rolls out to greet us, and the Rock and Lewish load fresh rockets from the P-40 into our empty pods. Mission 155 is complete.

Oh, we got shot at. Make that 156.

"Did the wounded make it?" asks Tom.

"Don't know," I shrug.

I am tired now. So tired. The session does not feel good anymore, and I want it to end.

"Didn't you try to find out?"

"No."

"Why not?"

"We never knew who lived and who died on these things. They were voices on the radio and faces in our aircraft, and we cared about them while they were in trouble and tried to save them, but when it was all over there was nothing left for us to do. There were so many. I don't remember ever having the time to find out. I don't remember trying, either. Maybe we didn't want to know. I don't know."

"What did you do after a mission like that?"

"Flew other missions. Or worked on the aircraft. Or jerked off."

"Didn't you talk about the missions with anyone?"

"Maybe for a few minutes in the line shack, but everyone had their missions. Most were nothing special. This one had a lot of gun time, so maybe I talked about it for four minutes instead of two. But then I'd forget it. I forgot this mission. It was nothing special."

"You didn't forget this mission."

"Yes, I did."

"No, you didn't."

"Yes, I fucking well did. I only remembered it because you asked me. I'd forgotten it on my own. It never would have come back unless you'd asked me."

"No, I asked you to tell me about a day. You picked the missions to make the day. You picked this mission. You never forgot."

Silence, and anger. I feel manipulated. I do not like to feel manipulated, but I cannot deny the truth in his words. How is it that I remember something today when I know I'd forgotten it within an hour of the event itself?

"What did you do after this mission?" Tom continues.

"I went back and cleaned my guns."

"Is that all?"

"Well, I went to get something to eat first."

Tom studies me, waiting for something else. I am confused. I am doing what he asked, putting together a day. What more is there? A day is a day is a day. I am doing my best. I cannot put in what was never there, or what I can no longer find. At least he does not ask me how I feel. I am tired of being asked how I feel.

"By the way, whatever happened to the gunner, Muffy?" Tom asks suddenly.

"Oh, yeah, I forgot to tell you. On the way back to Phu Bai, I reached down and unhooked his gunner's belt. Then I tapped him on the shoulder and pointed to it. He had no way of knowing whether I did it during the fight or not. Boy, did he go pale. When we landed, he asked why I didn't push him out. I told him Hazel Baby was right, he was a damned good gunner. If he could be as good a human being as he was a gunner, he'd be right fine. Then I told him why he pissed me off so much—ragging on old Walter."

"And?"

"He became a right fine human being after that. We were friends until we left Country, and I always liked to fly with him. Y'know, he was one of the best gunners there ever was."

EIGHTEEN-THIRTY HOURS: THE MAT, PHU BAI,
THUA THIEN PROVINCE, R.V.N.

The Scarface crews head to the mess hall, the officers to their side and we to ours. Bill had asked Roseanne to hold chow, and the unwilling sergeant has, predictably, done the least he could. Spam sandwiches on coarse white bread, green Kool Aid, and all the fruit cocktail we wanted. The "Monday Pill" sign is gone.

We take our food in silence, bone tired: a still life painted of crud and cordite in sweat-soaked flightsuits, matted hair and empty eyes staring at nothing. Masticating the tepid meat dumbly, swilling green fluid. So fucking tired.

We return to the flight line for a closer look at the aircraft. A single main rotor is holed—itself enough to guarantee the ship won't fly the morning missions. We will replace the blade during the night, but it will have to wait until full light to be tracked and test-flown.[7] There are a few tail-cone punctures for the metalsmiths to heal, souvenirs from the gook in the boulders. More serious is the hole in the nose hatch, which lines up with a hole in the UHF radio. That will keep us grounded if avionics doesn't have or can't steal a replacement. The tail rotor is brass-damaged. Fucking tail rotor. If none better are in the graveyard, I'll have to rebalance it.

First, there are guns to clean. Muffy and I return to the ordnance shack. The disassembled remains of the day's firing remain as I left them. No good fairy or cleaning elves have wandered through. Muffy offers to help me with the double load of weapons, and I accept gratefully.

Afterward, I grab a truck and driver, tow the *Falcon* into the hangar, and join the night crew to replace the main blade, patch the tail with epoxy, and rebalance it. All the blades in the ditch are worse than mine. Dutch wrestles a new radio into place. The metalsmiths rivet patches of aluminum and spray everything green. The hours pass, neither quickly nor slowly—just numbly.

Inspecting the aircraft more closely, we find damage to the tail-rotor cable.

"Hey, guys, look at this," I call eagerly, in a voice I used in show and tell.

We study it curiously. One strand of the wire rope is severed, one is damaged, and one remains whole. It appears the least of the damage, but it is the one hit that could have brought us down. The tail rotor counteracts the torque of the main blades and keeps them rotating around the aircraft, instead of the other way around.

Everyone scrapes his finger across the frayed strands; everyone grunts

---

7. To reduce vibration, helicopter rotors are adjusted, or "tracked," so that the blades rotate within the same plane. The crewchief carefully swings a long pole, its end wrapped with rubber and masking tape, into the blade path of an operating aircraft until the ends of the blade begin to cut the tape. The aircraft is shut down, and the blades are adjusted until each one cuts in the same track.

professionally. I show off the near-fatal wound the same way patients fresh from surgery entertain audiences with their scars.

The cable has to be restrung, but the hour is late, and Gunny tells me to secure. I'll fly Scotty's bird tomorrow while he goes on R & R.

TWENTY-THREE HUNDRED HOURS: THE HOOCH, PHU BAI, THUA THIEN PROVINCE, R.V.N.

Alone, I walk toward the hooch. The night is cool, and the sound of the generators keeps me company on my way back. A few of the guys are still up, like myself just getting off the line.

"Got some good gun time today," I mention, dumping my boots loudly on the deck, but no one wants to hear about missions.

"You going to open your care package tonight?" Saint peeks around the "wall" between our cubicles, a green blanket suspended in the crude frame that has separated his feet from my head for many months.

Saint is a good-natured hulk from St. Louis: five feet ten, two and a half feet wide, and a foot and a half thick. His crinkly eyes and broad mouth are set in a perpetual grin beneath a shock of unkempt hair. He rises from his cot and ambles over with the roly-poly gait of a mariner, as if his feet were set in buckets of cement. I have never heard him speak an evil word about a single human being, except for squids and dogfaces, who don't count anyway.

"Oooo, oooo." Saint gets excited when food is present. "I'll trade you some of my stuff."

"You don't have anything that human beings eat," I counter, "just asparagus, or ham and motherfuckers. You are a pig. You eat pig food. Fodder. I should get you hay. Go away."

Saint's packages from home contain mostly inedibles, among them asparagus. Lots of asparagus. Saint loves asparagus. He slurps it cold from the can and drinks the juice. Asparagus does funny things when it reacts inside a human body under a tropical climate. Saint stinks; his pecker tracks stink; he has sweat that gives sweat a bad name. He bathes as frequently or infrequently as the rest of us, but he still stinks. I know it is the asparagus.

Ham and motherfuckers are C-rat units of chopped ham and lima beans held together by a paste of green goo. The origin of 'motherfuckers" is unknown, but it fits. Most normal people discard these units or drop them on the enemy, but Saint eats them warm, cold, chopped, or mixed with other inedibles like asparagus. Saint is strange.

"Have you written to Grace?" he asks.

I'd let him speak long distance to my girl one of the last nights in the World. Worried, she had asked him to look out for me. A few days later, just nanoseconds before we left the World, Saint was "Dear Johnned." He took it bad—really bad—and became possessed with the conviction that he would do everything he could to prevent it from happening to me. The bastard wrote letters to my girl, reporting on me. She wrote him back, and neither one of them would tell me what they said. Saint was much bigger than I, and I couldn't beat it out of him. The nights we'd wrestle over a letter, he'd just pick me up off the floor and hold me until I passed out from the asparagus.

And every fucking day he'd badger me, "Did you write?"

"Last night. Mom and Dad tonight."

"You're supposed to write to her every night. You better keep hold of her. You got something good going. Don't fuck it up."

"I write to her almost every night. I'm beat."

"I told her I'd look out for you. She'll think I'm not doing it."

"When I write to her I'll tell her you're as big a pain in the ass as ever. Now just fuck off."

"There you go again. Always negative. You should look at the positive aspects of life. Look at what you got going for you."

Saint is always lecturing me for being morose, telling me I'm lucky to be a crewchief, telling me I have a good education, a good girl, telling me that when I get out of the Crotch I'm really going to be something. *Piss off.* I like being morose; it suits me. I would be OK if it weren't for assholes shoving their sunny good humor in my face. Saint is a metalsmith and gunner, and flies missions a month on and a month off. He has no reason to be happy about anything, but he's been an incurable optimist since I've known him. The easiest way to shut him up is to stuff his face. Perhaps he is not so stupid after all. I turn to the care package.

The box is a treasure. There is fudge and a salami covered in green mold. There are paperbacks and writing paper and clean handkerchiefs. I blow my nose, and note that the white cloth comes away black—the smoke of the day's firing is still on my face. There are tiny cans of hot dogs like cute versions of the turds I had for lunch. Why do they think I want canned meat? There is toothpaste and a brush and soap. Everything is packed in popcorn, which we munch while I unpack. I remove each gift carefully, one at a time, wanting to make it last, relishing the touch of all these things from the World. The very air in the box comes from the World! I visualize the box being packed in the kitchen of my home or in some place I used to know better than the hooch. Around me is the stench of piss and cordite, and of myself and my mates. The cans and fudge and salami bear the finger oils and breath of someone who could never, ever imagine me cascading shards of rock into a gook's body. Sometimes I no longer know who I am, but the package lets me remember who I used to be.

I feed some fudge to Saint to make him go away and begin my letter. I write every night. It is how I escape.

> Dear Mom and Dad,
>
> I'm doing OK. Things are pretty quiet around here. I got your package today. Thanks for all the stuff. The salami was covered with mold but we just wash it off with dirty water and it's as good as new. How about some of your home made wine to go with it, next time? I need some T shirts, and if you wrap the bottle in them it'll come through fine. Please DO NOT SEND CANNED MEAT. Brownies are good. I have a friend who likes asparagus. If you send canned asparagus, I don't have to give him anything good. Went flying today. Observation hops. Nothing much going on. Only 228 days left. Worked late on someone's aircraft tonight. Mechanical problems. Well, I have to hit the rack. Have to go flying again tomorrow. Probably more observation hops. They're pretty boring. I'll write again in a few days.

I reread the letter. Lies and blather and stuff that makes no sense. Only the part about the asparagus is true. It is not me, not what I want to write—need to write. It is hard to write to parents. Easier to write to friends. I cannot tell them what I did today. I tried writing the truth once, but they told me it upset my mother—said not to write about "those things." Killing. Death. They couldn't even print the words. "Those things." Shit. My old man—he was in World War II—but even he doesn't seem to want to hear anything. Why doesn't he understand? I need to write, want to write, but the letters are for them, not for me. I give them what they want.

I sign it "Love . . ." and use the name they called me in the World. They would not know "Zack." They do not want to. The letter goes in an envelope with a red dragon and map of Vietnam embellished on the front. I scrawl, "Free" in place of the stamp and labor to address it.

Stepping out of my flightsuit, I pull on my Goodyears and pad to the mailbox in my skivvies. The sand feels good squishing between my toes. I post the letter and walk back slowly, enjoying the night sky. There are many stars; I watch them a while. So many shooting stars. What is the legend? Each one a soul going to heaven—or is it hell? The hut is dark when I return, and I pick my way down the center aisle past the quiet forms of my friends. They stink. I stink. Everything stinks. I will take a shower in the A.M. I'm glad for my cubicle in the corner by the door and for the occasional breeze that comes in off the perimeter. It hits me before it picks up the stench of another, and I'd hate to dwell downwind of myself.

Mamasan has piled fresh laundry on my cot. I arrange it carefully in stacks: underwear, trousers, blouse, and socks. I need order in my fucked-up existence and cannot abide a messy cubicle. I was never this way in the World, but now the compulsion suits me.

Three minutes to blow enough air into the mattress to suspend me until I fall asleep. I can spare no more, but more would be wasted.

I climb in and stretch toward the calendar.

Day 229 is dead.

OH-FOUR-SEVENTEEN HOURS: THE HOOCH, PHU BAI, THUA THIEN PROVINCE, R.V.N.

The night watch touches me lightly, chanting over and over, "It's oh-four hundred . . . it's oh-four hundred. . . ."

"Fuck you, asshole," I mumble, but I am a light sleeper, and aircrew with 228 days left In Country wake easily.

"How do you feel now?" asks Tom.

"Drained. Tired. Dead."

"Typical day, huh?" he shakes his head slowly, with a kind of quiet wonder. I wonder at the tale myself, that a day so beyond the ken of normal folk was—is—typical to me.

I stretch and yawn. "Close enough for government work." It is years since I've used the cliché, but it fits in the Center, and the irony pleases me.

The day litters the room. So much lifeless rubble. I kick at it absently, lacking the energy to put it back from whence it came. It is strange how much more easily I bore the burden in my youth. Now the memory alone fatigues me. I want only to leave it, to dissolve in the grayness of the Center.

Tom will not let the rubble rest. He searches the pile of missions and fatigue, looking for what I do not know.

"I know you're tired. But what do you feel?"

The debris flutters, like trash luffing in a languid wind. "Feel, schmeel, peel. I told you. I don't feel."

"You felt something during that last mission."

"No, I felt something telling about the mission, but it went away. I felt nothing during the mission. Except how funny it was when the Guinea's ass got blistered. At least Muffy didn't tell anyone how I got stinky-fingered. I don't . . . didn't . . . feel anything!"

Tom cocks an eye, and I cannot conceal that I note the slip as well. I am suddenly angry at slipping. It is a chink in my armor. I know there is danger in these wakening memories when I mix past with present. I level a stare that says, "Go away!" then add, "Look, is this going to be one of those get-in-touch-with-your-feelings deals? If it is, I don't wanna play."

"Why?"

"I don't wanna play that, either. You wanted a day, you got a day. Fuck off."

"Why are you so angry?"

"I dunno," I answer sullenly.

But I do know. Tom watches me; I do not meet his eye. The old anger distantly churns. I think of how I was, those many years ago, of those shifting images—the flight line, the gunship. Proud. Assured. Brave. Yes, brave! Without humility, I will attribute that to young Zack. He was brave.

Suddenly, I remember why I am in this place. The Center, the damned Center. The day he made me rebuild. For a while I remembered who I had been. *Why did you make me remember? Why do you make me feel? It only reminds me of what I've lost!*

I glance about me; the litter in the room is suddenly annoying. It is part of why I am angry. I hate the clutter. I hate these pieces of myself lying like the rubble of a mission on a gunship's deck. It is untidy, unprofessional, out of control, dangerous. No crewchief would permit this. I want to put it back, where it belongs, but I am too tired. It slips through my fingers, like air through air, and my incompetence shames me. Oh, the difference between then and now. *Brave young Zack!*

"Do . . . do you know what it's like, not to be a . . . not to be. . . ?" The word is impossible to form; it defines what I have lost. "I wasn't always like this, you know. I was whole! I was a p . . . I was a. . . .

"I have this . . . this image of myself, like once I was whole, and then something important—something really, really important was torn out, and I will never, ever be whole again. The things I saw, and did. Oh, God, what it did to me! This isn't just that loss-of-innocence bullshit people talk about after they've been to war. This goes deeper. I used to be a p . . . p. . . . I used to be a person!"

I rail at Tom, "You know what a person is? A person is human. A person has a soul. A person feels for other people. I was a person before the war. You keep asking me about feeling. I knew how to feel, once. But the missions—people depended upon me, others of my own kind. They needed me. A crewchief mustn't feel, a crewchief couldn't feel! You

couldn't do your job and feel. I did my job. Maybe . . . maybe I wasn't a p
. . . anymore. Maybe I wasn't whole, but I had value, to others and to
myself! I was crewchief! But now I am . . . I am . . . what? No crewchief.
Only the memory of crewchief. No feeling. A fucking empty shell! I'm . . .
not . . . anything."

What am I?

I smash hand into fist, desperately needing someone to blame for what
I am.

"And the worst thing, the worst: I volunteered for Vietnam. There's the
shame, and the stupidity. How proud I was. Corporal Zack, Crewchief,
USMC. Professional! Warrior! I wanted to be a good Marine. They provided the mold, but I was willing clay. But God, oh, God. I was only nineteen. I couldn't know how much I would lose. No one told me. Someone
should have told me, made me listen. . . ."

I conjure another image: the sun of the dry monsoon painted in dazzling
whites and yellows too painful to behold, beating on the aircraft through
shimmering waves of heat. Now I see myself moving in blurred strokes of
tension, fatigue, boredom, my eyes etched in madness. Nineteen. How
could I have known?

Tom leans closer. "What would have happened if you had felt during
those missions?"

Why is he so dense?

"The grunts would have died. I'd have died. I couldn't have done my
job. We were professionals. I told you, we weren't allowed to feel."

"What will happen if you begin to feel?" he asks quietly. "Will," not
"would," not "might." I fear the difference.

There is a wall between me and the answer. The image of a great white
barrier stands before me, stretching infinitely to the right and left. I could
walk forever and never circle it. It is high, but somehow I know that it has
a peak that merges into an infinite sky at a point I cannot see. I could try
to climb it, but that seems the harder way. I touch it; it is hard, slippery,
and my hand skitters off. Is the wall my shield or my prison?

"What will happen?" he asks.

I touch it again; my hand penetrates! "Oh, no. No!" The tide of fear is
sudden, overwhelming, and it shocks me. I speak deliberately, in a child's
voice, needing to follow a simple line of logic. Logic is the finger in the
dike. I cannot take it out. "That can't happen. You see, if I started to feel
again, it would all come back. All those things I should have felt then, but
didn't. And if it came back, I don't think I could handle it. I have to keep
this locked up, you see. I have to. I . . ."

"Are you listening to what you're telling me?" Tom interrupts.

"I have to . . ."

"Listen to what you're saying." His voice is strong, stronger than the panic rising within.

"What?"

"You do feel, and did feel. You were trained not to let emotions rule you. You went from one mission to another, either something trivial and boring, or one like you just flew in here. Either way, you had to function. You were a good crewchief. You knew you couldn't function if you let your feelings take control. You cared too much about your squadron mates and the grunts. That doesn't mean you didn't feel. You just locked everything away. Until now."

Feeling aches so, high up in the throat and behind the eyes. They mist, and I blink away the watery haze and swallow the rising pain. Get a grip. Steady . . . steady. "From the halls . . . From the halls. . . ." I am crewchief, inured to the cry of the wounded, the despair of the zone. Zap some zips. Grease the gooners. Heh, heh, heh. "From the halls of Montezuma. . . ." I did not feel then. I will not show that I feel now.

"Look, I can't handle this. All the pain and fear in the world was in those zones. Maybe I had to keep it outside my aircraft. Maybe when I get too close, it wants to come back. It shouldn't be that way. That was then, this is now. But now was always now. Now never lasted beyond now. Now never became then. There was no then. Are you following all of this? Even if there was a then, I should remember then, not feel then."

I move as if to stand, but the litter in the room bars my way.

"I don't want to go through this anymore. Uh, uh. No. Fuck this, fuck . . ."

Tom leans forward suddenly, rubs his hands, slaps one onto each knee, and hitches himself to the edge of the chair, then removes them just as quickly and rocks back, arms folded high across his chest. The motion startles me, and I stop, watching. He inhales, then exhales fiercely. I feel his breath stir in the closeness of the room. It is an unflattering motion, unbecoming for a shrink, and it instinctively sets me on guard. I watch him warily and breathe in deep, measured drafts, hyperventilating to regain control as I was trained. I am in control, now, on guard. Feeling on guard guards against feeling.

"Do you know how memory works?" he asks.

"No," I answer after a pause, suspicious. The question distracts me, but I recover quickly and heighten my guard. All questions have purpose in the Center.

"There's a theory about remembering traumatic events."

He watches silently.

"OK," I answer nervously.

He cocks a serious eye, gauging, "You sure you want to hear this?"

The warning is clear. This "theory" may lead down paths untraveled. He asks not if I want to hear, but if I wish to turn back.

"I want to go on," I answer the hidden question. Who can tell what will come, but at least he shall speak, not me, and in silence there is greater control.

Tom assumes a curiously confidential air, as if sharing a secret of his trade. "We think there are three types of memory: very short term, short term, and long term. Very short term memory is like the memory of your last heartbeat or breath, the kind of memory you use to hold onto a telephone number from the moment you look it up to the moment you finish dialing. It lasts only an instant or a few seconds at most, then it's gone. The mind discards these kinds of 'very short term' events—it knows it doesn't need to remember every breath you take, and you only remember phone numbers if you take some special action to memorize them—which means you push the information along from very short term memory into the next stage."

Tom pauses, measuring my level of interest. I enjoy these glimpses of the working mind, and motion him to continue.

"Events that people deal with in the cognitive sense get processed into short-term memory. It spans from, say, thirty seconds to a few hours after the event takes place. Short-term memory preserves the event itself, as well as the feelings that you experienced during the event. For example, if someone sneaks up behind you and goes 'BOO,' you feel frightened, and the sense of being startled lasts a short amount of time after the actual event is over. Short-term memory is very vivid and carries a lot of feeling with it.

"Normally, your mind does one of two things with the stuff that makes it into short-term memory. Information that you don't need for more than a few minutes or hours gets flushed. For example, you only need to remember where you parked your car from the time you leave it until the time you drive it away. You don't need to remember where you parked your car five years ago. Information that you need to remember for a long time, such as where you live, gets processed into long-term memory. To do this the mind assimilates and organizes the information in some manner that's structured and accessible, like entering data into a computer. The structure varies from person to person, but it's whatever is consistent with your view of the world. As information is assimilated into long-term memory, it loses the vividness and depth of emotion characteristic of short-term memory. When you recall happy or sad events long after they occurred, you remember them as being happy or sad, but you don't re-experience the intense feeling that accompanied the original event. The vividness was stripped away during the synthesizing process the brain employs to organize information it needs to keep for a long time. What you have left is a sort of nostalgic recollection of the feeling, not the feeling itself. If the mind didn't do that, we'd all be basket cases re-experiencing intense emotion every time we remembered something that happened years ago."

My guard is down. The flow of memory he paints is fascinating. I feel taken into a confidence, as if for once we sit on the same side of the room.

I interrupt, wanting to understand. "I don't get this 'structure,' and what do you mean 'my view of the world'?" His words provide a path to order and reason that I crave. "Structure is whatever makes sense to you. It's the rules you use to build your world—what you've learned to expect and accept as normal, or 'reality.' People need structure in order to organize their private universes, otherwise all you'd have is chaos. Whatever doesn't comply with someone's definition of 'normality' doesn't fit into their universe. It never gets properly stored in long-term memory."

"Flying in helicopters and ricocheting rock shards into an enemy doesn't normally fit anyone's model of the universe," I interject.

"You got it," Tom answers.

"So how do you make things fit?"

"You usually find some way to explain an event in terms of existing experience, something that's already familiar, that fits the structure of your world. In therapy, we attempt to uncover relationships that aren't apparent, in order to find some avenue for the assimilation process to continue.

"For example. Say all you have to live on is Saint's asparagus, and you hate it. When you get hungry enough, you might find some way to make it palatable by discovering that it tastes like some other veggie that you do like, and then concentrate on that element of the flavor. Asparagus gets assimilated into the structure of your universe. Sometimes it can require a great deal of time, but you can either spend your whole life and a lot of energy fighting with something you can't get away from, or learn to deal with it."

"It's a nice theory," I note evenly.

"I like it," Tom says, satisfied, then adds pointedly, "When you were a crewchief you had to kill, to save your buddies and yourself. That's another ball game. Getting used to killing is a hell of a lot tougher to accept than learning to like asparagus, no matter how much you've been trained."

"I hate asparagus," I respond in a tense voice. My chest quivers; the words come choking. "I always have and always will. There never was a thing you could do to it to hide the taste."

I shake off the dread, consuming moments of silence to regain control. I killed without anger or pity. But the horror of the images I carry. . . . I have never told Tom how they haunt me, that pale company of the dead. Faceless then, why have they faces now? Their deaths belong to the war. Why should I know them? It was the war that stole their lives, and the feeling from my soul. Must I scavenge among the fragments of my past, face the ghost-faces, and feel the dreaded feelings to be rid of them?

Tom studies me, curious, but I will not tell him what I see. He would make me face it.

"So what happens to the shit that doesn't fit?" I ask.

"We're not exactly sure."

"Oh, great. Fucking great." Sarcasm is a powerful shield in the Center. It holds the faces and the feelings at bay.

"But we have a theory."

"Great. Another theory. You are just full of fucking theories. My tax dollars at work. Oh boy."

Tom ignores the bite of my words and continues. "We think, in PTSD, that traumatic events aren't assimilated into long-term memory because they contradict the basic rules of your universe. You can't organize the event in any way that makes sense to you, and it's so contrary to your rules that you can't find a way to make it fit or explain it in terms of something that does fit. It just can't go in.

"All your life you've viewed yourself as essentially a good person. One of your rules is that 'Ron is a good person.' Why else would you have stuck up for Walter? Good people don't kill. Soldiers kill; Marines kill. In boot camp, you're taught to avoid seeing enemies as people. That's why we call them dinks, slopes, gooks, etc. It devalues them as people. It makes them 'the enemy,' and the enemy is something you can be trained to kill, especially if your life or the lives of your friends are threatened. The problem is that 'desensitizing' works only for so long. Eventually, you can't escape the fact that we're all people. Soldiers and Marines are people, and who you're killing are people too. Being in uniform didn't change you as a person.

"You had no time to deal with any of this during combat. If you'd tried, what would have happened?"

"I'd have hesitated, or become confused. I would have died. The grunts would have died," I answer, feeling the doom. To hesitate was unthinkable.

"Even when you returned from a mission, there was no time. When one mission was finished, you had to go out immediately on another. In any case, the theory is that the traumatic event was never assimilated into long-term memory. There's the crux of it. Since it was never assimilated, the event wasn't entirely stripped of its short-term characteristics, the sense of vividness and emotion, for example. The feeling."

"So I'm stuck with a shit pot of old short-term memories that're constantly juicing my hormones, and I live in the Twilight Zone. Is that what you're getting at?"

"Not exactly."

"Well, what the fuck, exactly?" I ask, exasperated.

"We're not sure where the memory of traumatic events lives. It seems to occupy a place all its own. It's not in short-term memory, because it's not held for a short period of time. Still, the memory carries much of the vividness, the detail, and some of the feelings characteristic of short-term memory. It's not in long-term memory because the traumatic events won't fit into any normal structure. The only way to deal with the memory is to access it, but for memories to be accessible, they have to fit into a structure. It's a chicken-and-egg paradox: you can't deal with the memory because you can't get to it, and since you can't readily get to it you don't deal with it."

"You are really fucking confusing me."

"Look, isn't there a name for a temporary place in a computer where you store information when you're messing around with it, before you file it off into some permanent place? Something like a scratch pad?"

"Yes. We call it a buffer. There are large buffers and small buffers, and middle-sized buffers. Everything is relative. A buffer is just some amount of the computer's memory that you use on a temporary basis. The computer keeps track of the memory that's being used—the unused portion is called free memory. You grab the first piece of free memory you can find, load it with data, manipulate it, and then write it out to permanent storage when you're done."

"What happens to the buffer when you're done with it? Do you erase it?"

"Not usually, you just release the memory back to the system. When someone else gets the same space later, they erase whatever used to be in the buffer by putting their own stuff in on top. You can never trust what you find in buffers. They're like sheets of toilet paper that you use, then leave on the roll when you're done instead of flushing. The next asshole better tear off the old shit before he uses the paper, or he'll have a real crappy time," I laugh, enjoying the analogy.

"Suppose, for some reason, you don't save the information and just leave it in the buffer. Can you find it again?"

"Not on purpose. When you need a buffer, you get the first one that's free, which is usually not the last one you had. Like I said, you can't predict what shit is in it. The information in the old buffer is lost."

"How long does information stay in a lost buffer?" Tom asks.

"I told you, until the data is overwritten. Or. . . ."

"What?" he asks.

I respond slowly, feeling the impact of the words. "Until the computer dies."

The metaphor is complete. Lost buffers. The fragments of the past where lie the feelings I seek and deny. We sit for long moments. Again I feel the shards of the day about me, only now I see in them wet, glistening things, like flesh torn from raw and ugly wounds. Are these the feelings that I seek? I poke at one, as if with a naked finger. It seems to cling to me, neither hot nor cold but of a temperature with my own body. It is ugly and should repel, but it does not. It has no features, no form—it's just a wet, glistening thing, I cannot say from where. It feels no pain as I touch it. It has no nerves to feel pain. The pain lies in the wound from which it was torn. Everything Tom has said tells me that I must find the wound and open it, and put the thing back so that it may be healed, if not made whole. I fear the search for the wound, and the pain in healing, and know that no surgeon shall ever be able to hide the scar.

"It really is a nice theory," I observe absently, poking again at the thing. Funny, it has stuck itself to my finger. Once touched it will not come off. It is me, but not a part of me.

Tom smiles, satisfied. A shrink shall be a shrink.

"OK. What do we do now?"

Tom's answer is no surprise.

"It's up to you. Remember, I can never make you do anything that you don't want to do. You were feeling pretty comfortable when you came in today. You can decide to stay where you are, or you can decide to go on. You know where the feelings are."

I chew a finger, play with the stubble of the day on my chin, pull my nose, watch him watching me as I decide.

What decision is there? I cannot ignore what has been revealed, either by purpose or by chance. I have been crewchief, and my fiber is to move in one direction only. Tom knows this. He knows me, and knows what I shall do as well, but the choice is mine. I feel maneuvered, but not manipulated, and I marvel with some humor at the game we have played this day. "A day," he demanded. I gave it to him, one of my own making, and now I stand positioned to see and understand the choices that lie ahead if I wish to remember a day, and not relive it.

I remember, among his first words, "the ability to choose."

I stand to leave, and he with me. The rubble no longer bars my way. We shake hands, and for a moment I recall the surge of power felt long ago over the zone.

"When do we begin?"

At each moment we are the sum of our existence. I have been crewchief, and am.

"We already have."

Chapter 4

Hope is a waking dream.

—Aristotle, as cited by Diogenes Laertius
in *Lives of Eminent Philosophers*

Faces of Fear

"I had the dream again last night."

The Center is noisy today. I hear the chimes from the arcade next door. And hot. Fucking government can't buy decent air conditioners. It is difficult to concentrate.

"It came three times. Can you believe it? I couldn't sleep worth a damn."

Tom looks cool in spite of the heat. I hate people who don't sweat.

"What's been going on? Can you connect it to anything?"

"Shit, no. Sometimes it comes when I'm under stress, or pissed off, but sometimes it comes out of nowhere. I can't connect it to anything. I don't know. Hell. . . ."

I stare into space. Half-formed images shift like ghosts in the tired void that is my mind, lacking purpose and direction.

The dream is always the same, beginning with darkness, something that is more than just the absence of light. The darkness is palpable, like a smothering shroud, and beyond the shroud is a great evil. It is close at hand, close enough to sense, but not close enough to touch, and it can see me. It can always see me. It grows hot in the darkness, stifling, difficult to

132

breathe. I fear the fear; the dread of it builds and builds. Suddenly, searing light rends the gloom. It blinds and exposes me, but the evil hides within its harsh glare. Then darkness consumes the light, except for the twinking stars, and the evil closes in, but it has no face. Flashes-light-darkness, flashes-light-darkness, over and over. If I could see what was there, I might know how to fight it. But I cannot, and my mind cannot stand long against the terror. It flees in panic, seeking safety in consciousness. Panic drives me to wake up, but the dream's grip is terrible. It keeps me in a halfway world of flashes-light-darkness and the need to escape. I choke, retch, scream, "Wake me up wake me up wake me up!"

Tom leans away and arches a single eyebrow. He takes a long breath and lets it go slowly, searching for the right words. The quintessential shrink in action. I know when he is planning something, about to seize the opportunity to break new ground.

"What are you afraid of?"

The shifting images stop. Most fade from my mind's eye, forgotten. But those that remain take focus from the question, and a path begins to form. The question is artful and gains my attention.

"What do you mean, 'are'? Don't you mean 'were'?"

"Is there a difference?"

The thought amuses me. Vietnam, so long ago but never far from my thoughts. There is no difference after all.

"What are you afraid of?" Tom asks again, insisting. He is not going to let this pass.

There were many things that frightened me In Country, but I know that's not what he's asking.

"I saw a lot of bad shit, flying missions, but nothing frightened me more than one duty I had to pull over and over again."

I stop. It is eerie. I can feel the dream, and the fear that washes ahead of it like a foretelling tide flows into the room.

"Which was?" Tom urges.

"The wire. I was terrified of the wire."

"Do you see a connection?"

"No."

But there is a connection. I fidget in the steel chair. It's stupid. Embarrassing. I don't want to talk about it. I mean, I've told him a lot of dumb things, but this one takes the cake. I stare into the titles on the bookshelves behind him, polysyllabic contrivances that pass for words. It helps me not to think about the dream. Too much thinking brings the dream on. It takes so little to feel the breath of its terror, even in daylight.

Sometimes, in the halfway world between panic and consciousness, I've struggled to remain in the dream. When the evil is retreating, I can almost see it. If only I could see more clearly, I would know. I did see it once, in that halfway world between the soundless fear and the piercing plea of my wakening scream. I grappled at it with fumbling mind-fingers, but it fled, and the memory of it ran like a slippery rope through my hands. I fought to stay in the dream so that I might know its face, but it melted away with my screams ringing clearly in my ears. Only the empty memory of having seen it remained. I felt sad. If only I could capture it, just once, then it would never come again.

I hate the dream. I hate it, I hate it, I hate it. If I could only see.

"What is it that you're thinking, just now? Out with it."

"I don't know, it's hard to concentrate."

"You were thinking about something. What image passed through your mind?"

I pause, shifting in my chair, not meeting his eyes. It is stupid. So stupid.

"Do you know, I really hate full moons."

The dream began soon after I left Vietnam. I never dreamed In Country. Perhaps I didn't have the need. I worked hard at trying to understand the dream when it began, when I saw that it was the same dream again and still again, and yet had no face that I could recognize. It seemed that there was a simple thing that I had to do. If only I could withstand the terror long enough, if I could seize and gaze upon the thing of darkness and recognize its face, if I could understand my fear of it, it should go away.

There were many things to be afraid of In Country, but only a finite number of essential fears. Cowardice. Disfigurement. Death. The special loneliness of the zone—the archetype of Marine fears—was a fourth to some. To me, at least. There may have been more, but these were mine. However few the fears, the paths to them were infinite, and each twisting journey contributed its residue of terror to the chamber in my mind where the night horrors are born. Against my will the chamber still holds the memories of those journeys. I travel them too often, and when I do, they turn back upon themselves, twisted as a Mobius in darkness, so that I cannot tell where they begin, where they end, and where they lead. I know only that they twist among the old memories, sealed in nesting boxes, that lie scattered like rubble throughout the crypt of my fears. Boxes within boxes, each containing a sarcophagal fragment of some old terror.

Two years, now, in the Vet Center, spent unsealing the nesting boxes. One by one, over and over. I rummaged within them and poked about, as with a stick, and tried to discern some order from the remembrances the boxes concealed. I learned and relearned many things, and used them profitably in rebuilding a life. But for years the dream did not change, in spite of all that I had learned. It visited me at its whim like a persecuting demon. I believed it lived on some untraveled byway of my memory. I believed there would come a time when the dream would be turned out of its hiding place and I would recognize its face. Then it would go away.

At last I traveled an old path, one traveled many times before, and in the confusion of nesting boxes I recognized a fragment of the dream. Then I realized that the dream was not simple at all, that it lived in many places. The portrait had not a single face but many, and I recognized none while searching for only one.

An enlightenment.

The dream's bones hid in the nesting boxes! Oh, not in all that cluttered the paths of my memory, only some. Each shell yielded its fragment, the fragments became faces, and the faces lent form and feature to the composite.

I found many fragments. In the Center I turned them like tiles in a mosaic. I moved them across the mural of my life until the puzzle-pieces fell into place. At last I began to see the faces.

And then the dream began to change.

SESSION: ON THE WIRE

Corporals and below could expect about one night per week on perimeter guard, assigned to either the Reactionary Force or one of the perimeter bunkers. I hated the bunkers, hated them more than any other duty I pulled in Vietnam. The memory of those nights I spent on the outer perimeter of Phu Bai, on the wire, is painted of shadows and flares and piss-corroded bunkers. And fear.

Sergeants and above escaped perimeter duty and, enjoying the perquisites of rank, missed few opportunities to rub it in.

"Hey, Zack, looks like you got the wire tonight. Ho, ho, ho," Gunny chortled. The fat bastard.

"Ho, ho, fucking ho," I muttered, and tagged a "Fuck you, dickhead" on the end, a safe ten decibels below old Gunny's hearing range.

Each day, Gunny posted the guard roster, and each time my name appeared I felt as if I'd just had the duty a day or two before. Few chores made the space between days pass so quickly. After flight ops closed, the luckless and unwilling mob consigned to the perimeter changed from flight-suits to grunt gear, with real-Marine rifles, steel pots, and ammo belts, and trudged toward Phu Bai's outer wire. We remained "air wing," however, and our transformation to "real-Marine" was superficial as we marched, dragging our feet through the dirty sand, bitching all the way.

"Fucking goddamned duty."

"You bet your sweet ass."

"There's a goddamned USO show in, too. Filipino pussy."

"I was gonna jerk off tonight."

"Save it for tomorrow."

"Pecker might not be there tomorrow. Shit."

"Then do it in the bunker."

"Not in my goddamned bunker, you goddamned peckerwood."

"Shit."

"Air wing"—"fucking–air wing," in the world of the wire—was painfully out of its element on perimeter guard. Life in the air was city-fast, factory-loud. "Air wing" was cool. We crossed the country in giant strides as with seven-league boots, flying over a macroscopic world of rivers and hills and valleys that we turned into zones when we felt the need. The engines, the rotors, the ganged weaponry empowered us, as if machinery and limitless airspace could magnify life itself.

But the wire . . . the wire was the province of the grunt, the "real-Marine": muted motion, muted sound, each breath deliberate and measured. It was a microscopic world of dunes and grave mounds, and the nascent zone lay biding its time like a vigilant beast. Humanity itself seemed diminished as we approached the unfriendly world of the perimeter. Conversation and give-a-fuck attitudes declined logarithmically; it was hard to be cool on the wire.

The Sergeant of the Guard met us, arms akimbo and mouth drawn in distaste in the manner of all noncoms. We named his kind "Rock of the Marines." Fucking–air wing referred to all hard-ass noncoms as "Rock." Rock was a dour specimen of real-Marine who hated having us as much as we hated being there. The air wing's side of Phu Bai had once been guarded solely by real-Marines, but the grunt detachment had been reduced some months before, leaving only a small complement of real-Marines to show us how it was done. "Guard your own shit!" the grunts called to us as they moved out. Rock well knew that you couldn't even beat fucking–air wing into real-Marines. On the wire we were unreliable and dangerous—to himself, to ourselves, and to anyone else within rifle range except for (possibly) the enemy. He was right, but was as trapped as we were by Marine traditions that flaunted high-tech specialization. Tradition ordains that all Marines be riflemen. Like any legend, there is an element of truth.

All Marines do fire a rifle—about five hundred rounds in boot camp at paper targets that don't shoot back. The notion that firing a rifle makes one a rifleman is a piece of circular logic that could have been fatal on the wire.

Rock's only ally was the Corporal of the Guard, "Baby Rock." Cast from the same mold, they were indistinguishable, other than that one was a sergeant and the other was not. I could discern no justification for there being two guard NCOs, unless someone considered leaving one to fend alone amongst fucking–air wing to be cruel and unusual punishment. Both were grunt-variety real-Marines, which meant that both were assholes.

There purported to be an Officer of the Guard, an occasional, disembodied voice on the field radio who conferred only with the sergeant or corporal. We never saw him, so to us he did not exist.

Rock took control of our herd.

"All right you fucking people, come to attention. I said, 'Attention!' Line up for rifle inspection."

Sergeants of the Guard live for rifle inspection. It is one of two rare moments when the grimace they wear churns itself into a smile. The other is when they scream, "Get down and give me twenty!" for pissing them off.

Rifle inspection was always painful. In flight I fired an M-60 machine gun. I carried Willy Peter, smoke, gas, and concussion grenades, a combat knife, a Filipino butterfly knife, and a machete to chop the tops from C-ration cans or threaten prisoners. All were pristine. You could eat out of the chamber of my -60 if you liked the taste of fresh oil. But my rifle—one semiautomatic, gas-operated, magazine-fed, air-cooled shoulder weapon fired from the closed-bolt position—hung rusting and gathering dust from a nail above my cot, when it wasn't rusting and gathering sand under the crew's seat in my aircraft. The only time I touched it was to play real-Marine on the perimeter.

Rock snatches it from my grasp. The manly report of flesh on solid wood echoes even here, on the wire. Razzle-dazzle on parade. The weapon is a blur, twirling in the setting sun. Clockwise, counterclockwise. Swish, *smack!*, swish . . . I feel the wind of its passage brush the short hairs in my nose, tickling. Do not sneeze. Never sneeze! Rock's eyes lock with mine, showing me he can do this shit blindfolded. Swish, *smack!* Halt! Weapon rigid, slanted forty-five degrees to the deck, thumb in the chamber, he sights down the bore and reflects the dim redness of the perimeter sun from his thumbnail upward to illuminate the rifled spirals of the barrel's lands and grooves. It is the moment of truth.

The rifle is filthy. Surprise, surprise.

"This rifle is fucking filthy!"

Silence. I know the drill. It is best to be silent.

"What's going to happen if we get hit and you have to use this weapon, Marine?"

Silence.

"You are going to Die, and you are going to fucking let your buddies down and they are going to fucking Die because Your Fucking Weapon Is Filthy!"

Silence. I used to think he'd strike me, but he never did. After a while, I knew he never would. That's boot-camp stuff. Real field Marines don't do that. But there's always twenty.

"How are you going to kill the enemy and save your buddies with this?" he asks, shoving the chamber into my face. I peer into it. It's the only time I ever do—look into it, that is—at rifle inspection on the wire. It is truly filthy.

When I was still an FNG, I made the mistake of answering—worse still, explaining. "I would kill them with my M-60, or the grenades in the bunker, or even blow them away with a pop-up flare, and besides, I've heard that you can bury an M-14 in sand, shake it out, and fire it all day long. Anyway, I'm a machine-gunner, Sarge." It had all sounded quite reasonable to me at the time. I remember that I smiled.

He made me give him twenty, right there in the fucking sand, and that grimace churned itself into a grin the whole goddamned time. Shit.

Silence.

Baby Rock produces a bore brush, cleaning rod, and a bottle of Hoppes as if by magic. I believed it was magic, the first time I saw it, but soon I discovered it was simply part of the one-act play carried out each evening on the wire.

"You will clean this weapon, Now, and I don't Ever want to see your weapon in This condition Again," directs the sergeant. He hurls the M-14 at me at the proper angle for port arms. It shoots across the narrow space between us, and I make certain to let out a *whoooof* of air, as if the force of the throw is too much for fucking candy-assed air wing to handle. The *whoooof* works. Rock smirks and moves on to abuse the next man in rank.

I shuffle to the side and bore-clean my weapon, then scrub the chamber and knock some rust off the sights. Baby Rock produces light oil, and I wipe on a thin film, not too much so that it won't collect sand. I take my time because it's the only chance I ever get to clean the damned thing, and as long as I keep looking busy, Rock and son-of-Rock will leave me alone. Soon, other fucking–air wingers line up behind me, freshly loaded with their own dose of crap, waiting for the cleaning tools. One or two are giving twenty. It is the start of a normal evening, on the wire.

After rifle inspection, Rock divides us into two groups: the larger to the Reactionary Force, the smaller to the bunkers. The Reactionary Force is the preferred choice, although the Rock does not solicit preferences. The Force is the second line of defense for the base and spends most of the night sleeping in a tent a hundred yards behind the line of bunkers. Each man draws walking guard duty—a two-hour shift of lurking in the shadows, a loaded rifle in his sweaty hands, hoping beyond hope that no one attacks the bunkers. The Reactionary Force, responding to an attack, will be the second group of Marines to die if the wire is breached. The first will be the bunker guard.

I have tried to second-guess Rock's selection criteria to avoid the bunkers. I've stood prominently in the front rank, militarily erect, hoping he'd grab the slovenly shirkers in the rear. I've shirked slovenly in the rear, hoping he'd prefer the heroes up front for the outer guard. It's made no difference. His selection is decidedly random, and even odds never seem to favor me evenly.

"Answer up when I call your name," Rock rasps. The words come in standard Marine drawl, a guttural wind forced through vocal chords abused by too much mess hall Tabasco.

"Cantrelle!"

"Yo!"

"Say, 'Present,' dickhead."

"Present," and in faint register, "dickhead."

Snuffles of laughter in the ranks.

"Something funny, ladies?"

Silence.

"Tent."

Lucky Frenchy—asshole always gets the light duty.

"Finch!"

"Present!"

"Tent."

Damn! Give-A-Fuck never gets the bunkers anymore. Just go a little bit apeshit and see what it can buy you. The last time Give-A-Fuck drew bunker duty, Rock—it was a different Rock back then—briefed him on the evening's password, normal procedure when posting guard. "If someone approaches, make sure you're in cover, then say, 'Halt' three times. If they don't give the password, blow their shit away." The Sergeant of the Guard himself was checking posts that night and tried to sneak up on Give-A-

Fuck. Give-A-Fuck took about a millisecond to scream, "Halt! Halt! Halt!" and blew twenty fat, splatty .45-caliber rounds out his grease gun. It was monsoon too, so the Sarge went ass-deep in a flooded culvert to dodge the slugs. A grease gun might not be able to hit the fat lady at a circus unless you shove the barrel a couple of inches up her ass, but it sure is loud. Give-A-Fuck claimed he couldn't hear the sergeant screaming the password over the gun's din. He got his butt reamed, of course, but they never put him on the bunker line again, either. Damn!

The formation ran on. One by one, everyone got his assignment. Everyone but me.

Finally, "Zak-zek, Zik-zak. Whatever the fuck you pronounce it!"

"I pronounce it Zay-zack, Sergeant."

"What're you, a fucking Polack?"

Silence, but I hope he checks my post tonight!

"You got super-nuts."

Holy hell, the Jackpot! Super-nuts for the night! The Supernumerary of the Guard is the extra man on duty, called in to relieve another only if that man can't stand his post. But to get out of standing your post, you first have to convince the Rock that you have an unresolvable problem, like being dead; and real-Marines don't believe in unresolvable problems, so you're shit out of luck almost any way you look at it if you draw a standing post. Super-nuts gets to sleep in the Reactionary Force tent all night, which is as good as it gets on the wire. I've never pulled Super-nuts before. It's about time some luck blew my way.

As soon as the Guard formation is dismissed, the Reactionary Force stampedes into a large and musty tent. The goal of the rush is twenty-odd mildewed and rotting cots, the rejects of the base, that lie scattered in disarray on a floor of musty sand. Older hands lead the stampede, knowing that fewer than half the cots have both the head and foot cross braces required to keep the canvas taut at both ends. FNGs, slow starters, or cripples are left with the most dilapidated specimens, and will spend the remainder of the night tossing and turning to decide which half of their body they prefer to support. I throw myself on a good one and turn to watch the show.

"This is my fucking cot, goddamn you."

"I sat on it first, shithead."

"The hell you did. I threw my cover on it first. See!"

"Makes no difference. I was here first."

"I was first."

"I was first."

"How long you been In Country?"

"Long enough to wear out more socks than you have seabags, boot."

A duet: "Get off my fucking cot!"

Neither will rise from the cot to fight. Someone else will take it. By now all the good cots are occupied; only a few remain, missing both head and foot crossbars. Whoever loses will sleep with his feet planted firmly on the sand, hoping the damned thing doesn't fold up on him if he moves during the night.

"I'll flip you for it."

Frenchy produces a twenty-five-cent MPC, crumples and palms it, rolls the paper money across the canvas of his cot, and covers it with his hand.

"Call it."

"Tits."

"OK, you got tits, you got ass."

The note is unfolded under careful scrutiny. The head of an old Miss America smiles up. "Tits it is." The single syllable rolls off Frenchy's tongue. No one can say, "Tits and ass" like a Cajun.

"Fuck. Fuck. Fuck. Take the fucking thing. Fuck."

We quickly settle in. The mess hall has sent sandwiches in waxed paper, the guard's "midnight snack." Spam, always Spam. It's the only thing that travels well, and is safe from the tent's nocturnal invaders. No living organism but a Marine would consume warm Spam with salad mustard that runs like baby's diarrhea. Knowing the sandwich won't improve with age, I consume mine in a few large bites, washing it down with tepid water from my canteen. Even the burp is unsatisfactory. Most of the others do the same; a few sell theirs to more desperate masochists. Midnight tummies will have to take care of themselves.

I lie on my cot and pull a novel from my pocket, Tolkien's *Fellowship of the Ring*. It's still light enough to read, even in the fetid atmosphere growing in the tent, and I enjoy the tale of elves and hobbits and the black horsemen of the Dark Lord. The stylized battle of good against evil could have been set in Vietnam. Frodo, Sam, Merry, and Pippin are battling the Ringwraiths on Weathertop, when an acid voice calls from outside, "Whichever of you assholes is supernumerary, get your butt out here."

Rock stands at the entrance, and next to him Timmerman, who is pulling ready-room watch back on the flight line this evening.

Rock nods toward Timmy, "This man says Corporal Musante has to fly up to Dong Ha tonight. He's in Bunker Two. Go out and relieve him and send him in. Take your weapon with you."

Damn, damn, damn! Moose is supposed to go on detachment to Dong

Ha tomorrow. I stare at Timmy in disbelief. Timmy smiles back. For an instant I hate him. Actually, more than an instant. Fuck you, I think. Wipe that fucking smile off your face.

I trudge to the tent, pick up my gear, and trudge to the bunker. Moose has already settled in, staring out into the twilight of the wire.

"You're relieved," I growl. "Your pilots want to go to Dong Ha tonight. Timmy's waiting for you back at the tent."

Moose wastes no time grabbing his stuff. His smile stretches from ear to ear. "Hey, you have a real good time out here tonight, Zack," he says. He picks up his sandwich.

"You leave that shit here," I say. "That's mine."

Moose mouths, "Fuck you" with a grin, but leaves the sandwich and disappears through the bunker maze. I drop my equipment, peel back the wax paper, and begin eating. Damn.

Phu Bai's bunkers circle the base like an obscene necklace. A number of crumbling concrete structures remain from the French occupation but are left unused, as if building upon the remnants of their failed enterprise would doom our own. The cornered squeals of the rats that inhabit them scared the hell out of me the first and only time I entered one. Among the silent grave mounds and rusting concertina, the old bunkers mutely mock our industry, in what is only the latest war to pass this way.

The Marine bunkers are ten feet on a side and seven feet tall: squat and windowless mausoleums with triple-thick walls of decaying sandbags.

Building the bunkers sucked. The base provided the bags, the sand came from beneath our feet, and the rest was up to us. The early sandbags were canvas and had the short life span of any natural fiber exposed to a tropical climate. Later, we used synthetic bags. They are soft green, and although they won't rot, they tend to unweave themselves badly when torn. But it's the canvas bags I remember most: decayed, mold-ridden, urine-soaked canvas.

Filling them was a simple, mind-deadening operation. The bags had to be turned inside out to put the seam on the interior so that the loose thread of unhemmed canvas wouldn't unravel and dump out all the sand we so laboriously put in. Turn inside out, fill, tie, stack. Turn inside out, fill, tie, stack. One Marine held the bag, the other scooped sand with a mess tray—easier than using an entrenching tool. Both were short-tempered and foul after a day's missions and evening's labor.

"Put in more sand."

"If I put in more sand, you can't tie it."

"It can take more sand."

"No, it can't."

"Yes, it can."

"No, it fucking can't, you stupid son of a bitch!"

"Fill it your fucking self."

"Get back here."

"Fuck you!"

"Goddamn it, I'll call the first sergeant. Get your goddamned ass back here."

"Fuck you. . . ."

"Goddamit, get your goddamn ass back here or I'll report you to the Top!"

Each day the winter monsoon poured on us as we stuffed the bags. Zillions of fucking bags. Enough bags that if we dropped them on Hanoi we'd have buried the place and won the war. Wet sand into wet bags, with the skin of our hands turned white and sloughing off on the canvas. Then grunting them into place, higher, higher, bursting our balls under the strain. Jesus! And always the monsoon: soaking the sand, soaking the canvas, soaking everything.

"Why the fuck can't we get them already turned right side out to begin with?"

"Why the fuck don't they all have tie strings?"

"Why the fuck can't we have a filling machine like the fucking-Army?"

"Why the fuck can't we pay the zips to fill them? Fucking-Army does it."

"Why the fuck do flight crew have to do this bullshit?"

"Why the fuck me?"

"Why the fuck?"

"Fuck!"

The bunker is barely large enough to hold a cot and several crates of machine-gun ammo, with just enough headroom for a midget to stand erect. A single open doorway pierces the wall that faces the perimeter, and is shielded from the wire by a sandbag barricade that wraps partially around each side. The guard walks his post within the narrow "U" in front of the bunker. A firing step for the guard to stand on to peer out over the wire runs the inside length of the barricade. Sandbags layered atop steel matting roof the bunker and extend far enough forward to overhang the barricade. The walls and roof are thick enough to stop small arms and grenades, and maybe a near miss by an 82-mm mortar. Maybe.

Each bunker is a miniature weapons platform, and an important order of business is to inventory and inspect the weapons and pyrotechnics. An

M-60 machine gun rests on its bipod atop the barricade and is the main defensive weapon. The crates of belted 7.62-mm ammunition are enough to keep the gun firing all night if need be—it would be extremely dangerous to resupply a bunker during an attack.

A 66-mm Light Antitank Weapon leans against the front wall of the bunker. The LAW is a collapsible, one-shot rocket launcher similar to a bazooka. The problem is that you have to leave the bunker to fire it, or at least use it from one of the less-protected wings of an entrance. It has a limited but lethal back blast that requires a clear space in which to dissipate. The LAW is a reserve weapon, to use in case the wire is breached because it can also blow a hole through the rows of concertina.

An array of grenades is stored next to the machine gun and in crates behind the barricade. Fragmentation grenades are reserved for infiltrators beyond the wire. Concussion grenades are for closer work. The white phosphorous variety, so the Guard Sergeant says, is intended to throw at the first wave of an onrushing enemy. These will not damage the wire and are not intended to kill, at least not quickly. The unquenchable particles of Willy Peter would erupt in a cloud of white smoke and burn their way hideously through the clothing and flesh of the vanguard. The searing and lethal haze would confuse an attack as crushing waves of men trip over bodies smoking and thrashing in pain, further entangling the wire. The screams of those caught within the roiling white cloud would, it is thought, unnerve the rear ranks, causing them to break and run, to be cut down by the interlocking fire of machine guns from adjacent bunkers.

In and beyond the wire, a maze of antipersonnel mines and assorted booby traps is buried in the sand, fuzed to detonate from trip wires. Claymores can be fired electrically from within the bunker, although the success that VC infiltrators have had in sneaking undetected to the wire and turning these directional explosives around has convinced us not to use them unless absolutely necessary. The wire itself is a tangled maze of rusting concertina, strung here and there with tin cans containing rocks that clink and clank if disturbed. During daylight, we've pissed on the barbs in teenaged malevolence, hoping to contribute to the infection an infiltrator would get from a cut. It seems impossible for anyone to navigate the yards of sand over which the wire is strung without revealing themselves, but on two occasions infiltrators have penetrated the base. One even reached the reserve fuel bladders: multithousand-gallon balloons dug into sand pits near the squadron's living area. He was caught leaving the base, and the satchel charge he planted in the bladders fortunately proved to be a dud.

We also have assorted pyrotechnics to illuminate or mark targets. Magnesium pop-up flares are extravagantly tossed aloft in spite of edicts from the guard commander. He has warned us that they expose the bunker's position and shouldn't be used "unnecessarily," and that they cost a lot and are running up the price of the war. His orders are to call the command post if we suspect movement in the wire, and request permission to throw up a flare. It's an order seldom obeyed.

Green and yellow smoke grenades are available to mark the position of friendly troops or to indicate wind direction to approaching aircraft. Marking smoke is seldom required at a main base like Phu Bai, which has good visibility and open fields of fire. Once in a while, someone will toss a canister into the concertina, to watch belching clouds of sunflower or emerald dissipate slowly in the wafting breeze. It is very pretty.

The red smoke grenade is the most ominous pyrotechnic stored in the bunker. Its presence adds more than a touch of melodrama to being on the wire. More than anything else, red smoke emphasizes the fact that bunker duty is no joke. A single grenade is kept in a tiny niche next to the field telephone, away from all other signals so it won't be used in error. There is, of course, more than one canister of red smoke in the bunker, but one good red smoke is all you'd ever need. Red smoke is the last thing you'd throw at them, if they break through the wire and neither Claymore nor LAW nor machine gun nor frag can stop them. Red smoke would summon artillery from the heart of the base, and mark the zone for the gunships scrambling from the flight line. It would bring down the myriad terrors of friendly fire that will stop them when you are past caring.

It is strange, sitting in the Center, sensing the closeness of the bunker, the proximity of the wire. It was so many years ago, yet it is familiar, immediate.

"What did it feel like, in the bunker?" Tom interrupts.

The question distracts and annoys me. I take weapons check very seriously, and must complete it before the night comes. Everything must be ready for the night.

But I am no longer in the bunker.

"What . . . what did you say?"

"What did it feel like, in the bunker?"

I break my tale unwillingly, trying to hide my resentment at the interruption.

"It feels like now."

"Feels, or felt?"

"Did I say 'feels'?"

"Yes."

"No, I didn't, I said 'felt.'"

"Feels."

"What the fuck is the difference?"

"You tell me."

I hate it when he wants to play mind games. Feel, felt. What the hell difference does it make?

Tom persists. "OK. What did it feel like, then?"

"Like being trapped on a tiny island, armed to the teeth, but knowing that anything that was big enough to fuck with you was big enough to blow you away. And it felt alone."

"Were you scared?"

"Me?" I blustered. "I was a fuckin' Marine, don't you know. Me, scared? Shit!"

Tom asks softly, "Were you scared?"

I stare into the memory of weapons check, recalling how impotent it felt, despite the mass of firepower I possessed or could summon. When it was gone, I would be alone.

"What do you think?"

I memorize the killing field. The perimeter is largely flat, dotted here and there with low dunes and grave mounds that can provide cover. I wonder why we haven't leveled them. We honor so little, and nothing is sacred in this place, why not bulldoze the graves to improve the field of fire? Beyond the wire, stunted grass and low scrub sprout from the sand and stretch to a distant tree line, in which an enemy can assemble unseen.

Two guards are posted to each bunker. A third might be assigned if S-2 (Intelligence) suspects unusual enemy activity and wants to run more than 50 percent alert. The night is split into two 4-hour watches, a strain after a full day's work, but four contiguous hours of sleep are better than breaking up the night's rest by running more-frequent, shorter watches. The bunker is manned just before sundown; both guards usually stay awake until the first watch begins at 2100 hours. The second watch takes over at 0100 and ends at dawn, although crews on an early flight schedule leave at 0400 to ready their aircraft.

I share the bunker with Bickman, a metalsmith from my outfit. We toss for first pick of watches, and I win. It's about fucking time. I plan my strategy carefully. I have early flight ops, which means I'll leave the bunker at 0400—which means I want the second watch, forcing Bickman to stand the last hour or more, even though he'll have had only three hours of sleep.

"I'll take second," I say with a straight face. I won't tell him about flight ops until he wakes me at the change of watch. Why blow the whole evening for him?

We finish inspecting weapons, then test the line to the Command Post, the bunker's umbilical to the outside world.

"Bunker Two, phone check."

"Give me a short count, Two."

"1 . . . 2 . . . 3 . . . 4 . . . 5."

"Roger. Read you five by five, Bunker Two."

"Bunker Two, out."

I stand next to Bickman, feeling the day come to an end. It is a good feeling to have another day behind me In Country.

Evening can be a special time, if you let it. As night creeps toward the base, there is time to unwind from the day and prepare for the long watch. It is quiet on the wire during the early hours of evening, and not yet deadly. Though I hate the bunkers, I savor this interlude away from the flight line, away from the medevacs and the emergency extracts, away from the screams and the curses and the crushing boredom that meld together to define the war. In the waning light of the day, I can still see beyond the wire and, facing the darkening sky, let my mind escape the confines of the base. Bunker Two is on Phu Bai's eastern perimeter, far from the squadron's living areas. The road I traveled from the World lies in the east, and in the east lie hope and escape, the path of my return. Beyond the tree line the unseen sea washes the distant shores of another life.

But the cobalt night creeps from the east, then crosses and envelopes the shores that lead to home. It silhouettes, then swallows the distant trees, bearing within itself the terrors of darkness. The east owns memories of home, but the fears of the present rush toward me with the night, and I shudder to look into it. Though night falls no more rapidly on the wire than in any other part of the World, the dread of its coming seems to quicken the sun's fall below the horizon and hastens the change of heaven's tincture from azure to crimson, to violet, to black.

So I turn my back on the hated wire and the threatening night, and gaze westward. The western sky bathes the base in a warm glow, slanting tongues of fire along the tin roofs even as the reaching arms of darkness sweep forward to quench them. Home beckons always in the setting sun and has nothing to do with geography or the logistics of military transportation. For a few moments, I can ignore the brutal harshness of the hooches and the gunships on the distant flight line. I can follow the sun with my eyes and pretend that sunset at Phu Bai is only a brief moment apart from sunrise at home, instead of a world and a lifetime away. I know that someone who knows my name looks upon the same sun rising, even as mine falls. Someone at home.

Home. The World. The land of the big PX and all-night generators. A place that is not here. A wonderful fairyland of a place that evokes fond memories that have nothing to do with whatever reality home actually was.

Home is a place I long for but can't clearly remember. Veiled faces from another life, smiling. Distant recollections of warmth, security, and love washing cruelly against the cold realization that it has been, can be, will be lost forever.

Home.

Not as real as the bunker and the hooches and the flight line and the red-crusted medevacs, but more than just a place or even a memory. A target, a survival goal, a tangible emotion that tears my guts in yearning for some-place that is not here.

I long less for what it is, more for what I wish it to be. A fond memory tucked away, opened with care to be gazed upon when the need comes. A dream. The bunker, even in its dingy squalor, can still be a place for dream-ing. I gaze into the setting sun and try to remember other places, try to hope for better times.

But the immensity of the base floods in, the squalor, the pound of rotors, the curses, erasing all the hopes, the memories bound by tenuous thread to the sun. At last it falls, swallowed by grim rooflines of plywood and metal, victim to the vicious night, which floods in as if seeking to extinguish life as well as light. A part of my being follows the sun, stretching taut, snap-ping, and when it is gone I feel alone, on the wire. Then I stand in empty silence, crushed and abandoned under the weight of the night.

Night leaves little room for thoughts of home. During the day, the sprawling combat island that is Phu Bai dominates all terrain to the hori-zons. The umbrella of fixed-wing and helicopters casts a broad shadow that makes us feel powerful. Although darkness cannot shrink the range of our firepower, the lengthening shadows seem to compress it to within the diam-eter weakly circumscribed by the rusting coils of wire. With night comes physical and emotional consolidation, the marshaling of forces to with-stand the assault of darkness. The blackness beyond the wire dwarfs the ter-rible power of our technology. And the bunkers, which hold the dangers of the night at bay, feel the fear first.

I stand with Bickman, staring into the fading glow of day, and wait as the night claims us.

Tom's voice is slow. "You look sad."

"Do I?" The old memory lingers. I see us standing within the barricade, watching the sun go down. Both of us so frail—ninety-pound weaklings. The grunt gear hangs loosely from our shoulders. Hardly picture-perfect Marines.

"Yes."

"I was thinking about this one night, on the wire. A conversation. Funny how it still sticks with me. We only talked about two things in the bunker: pussy and home. Couldn't talk about pussy with Bickman. It embarrassed him. And he was so private, he didn't want to talk about home. He was tough to talk to."

"Who was Bickman?"

I laugh at the memory. "No one, no one."

Robert Bickman is an anomaly in the Marine Corps, a painfully shy Marine. More than shy, introverted—a character trait the DIs didn't normally allow to survive boot camp.

I can sympathize with Bickman. I've known the world through his eyes, the pain of primping Brylcream into my hair before the Hop, then sitting tentatively with the hard edge of a metal seat cutting into my behind, too afraid to ask a girl to dance, too afraid of being turned down and taste once more the embarrassment I've tasted so often. I know what it felt like at holiday gift exchanges when someone who drew my name from Sister's fishbowl said, "Oh, no! I have to get a gift for him!" I know how it felt to cringe under the sniggers that followed. I know how it felt when I finally did something right, and then it was put down because it was only me who did it, and if I could do it, anyone could.

It all changed in boot camp. Drill instructors single out the runts of the litter, the ones who most want to hide in the rear rank—those most likely to be like me—and make them be something, anything. "House mouses," whose lowly state in life is to take the DIs' abuse and clean their living quarters. "Hatch bodies," who give meaning to their miserable lives by holding open doors as the platoon races to and fro. I was the platoon's "admin body": clerk, runner, general factotum, and "taker of abuse." I took abuse from my own DIs because God had bestowed upon them the right. I took it from other DIs when my own fed me into their hostile dens for purpose or for fun, like the weekly Octagon soap episode.

Pound, pound, pound. My hand aches as I smash it into the post at the entrance of an enemy platoon's barracks. They hate me because I am not one of them. It is the order of our world. My DI has sacrificed me into another DI's arms, bearing my platoon's weekly tribute of Octagon soap coupons. The enemy DI's mother collects them, and so all the platoons in the series buy enough Octagon soap to scrub a battleship. This DI is very good to his mother.

"Sir, Private Zaczek reporting to the Drill Instructor as ordered," I scream the litany.

"I can't hear you, turd."

Pound, pound, pound. "Sir, Private Zaczek reporting to the Drill Instructor as ordered!"

The DI goads his platoon, "Hey. This looks like one of those weasely shits from Plah-toon 175. What should we do with him?"

"Kill kill kill kill kill. . . ."

"I can't hear you, ladies."

"Kill Kill Kill Kill Kill. . . !"

"Enter! What do you want?"

"Sir, the Drill Instructor of Platoon 175 sends these coupons with his compliments to the Drill Instructor's mother." I offer the coupons with a tentative smile. Please let me go, please just let me go.

"What are you smiling at, shitbird?"

"Nothing, sir."

"You thinking about my mother, shitbird?"

"No, sir"

"You thinking about my mother with your dirty fucking mind, shitbird?"

"No, sir. I never think about the Drill Instructor's mother."

"Don't you like my mother?"

"No, sir. Yes, sir. Sir, I like the Drill Instructor's mother, sir."

"You want to fuck my mother, you shitbird!?!" The furnace draft of his roar sucks the air from my lungs. I am dead meat.

"Sir, sir. . . ."

"You weasely shit! Platoon! What are you going to do with this shitbird who's thinking about fucking your Drill Instructor's mother?"

"KILL KILL KILL KILL KILL. . . ."

I took the abuse, and after a while learned to give it back, both to my own platoon mates and to members of an enemy platoon when we met in hand-to-hand combat. I learned to chew ass and relish the taste. In the end, twelve weeks after placing my head between the lion's jaws, I was cured. I'd never be shy again.

With Bickman, the cure did not take. He keeps to himself in the squadron, wrapping his days in metalsmith's masking tape and zinc-chromate primer, and his nights in his own thoughts. Still, I find it strange that he's been In Country for five months and in that time has not earned his combat aircrewman's insignia. The silver and gold eagle's wings bear aircrew's pride, and I wear the badge whenever I can—even on the wire—especially on the wire, since it is one thing I have that Sergeant Rock doesn't, and can't, unless he becomes fucking–air wing like me.

I look down at my chest; the fading perimeter sun flashes crimson from the precious metal. Each sculpted shaft and feather, the brilliant anchor emblem, and the scrolled stars of bronze proclaim who I am, what I have achieved. I glance at Bickman. "Hey, Bickman, when are you going to get your flight skins, pick up your wings?"

He stares into the wire, looking nervous, always nervous—not meeting my eyes, but then he never does. "I'm not."

"Don't you want to have something to show off when you go home?"

"No. I don't want to be noticed. I don't like to be noticed."

"C'mon. Everyone likes to be noticed. Everyone wants to be remembered. Why did you join the Crotch?"

His hand flutters upward to his lips, as if to shield his face behind a screen. They tremble, most un-Marine-like, and I can see the hurt and embarrassment. He says only, "No, I don't. I just want to go home. I don't want anyone to remember me." The lonely pain in his words is real, accentuated by the loneliness of the wire.

I fall mute at his halting words. How sad to live such a life, yet I can understand. I search for words to help, and as I speak Bickman stares into the darkening wire without a word.

"I've always believed that everyone wants to be noticed while we're here, and remembered after we're gone. It doesn't matter for what and by whom, just that some person, no matter how insignificant they are, remembers us, no matter how insignificant we are. Even if they say only, 'Hey, remember Joe? What an asshole.' Even if it's only that, it's something. When you've lived your life, and look back, at least you can say, 'I made a difference.' If no one remembers you, were you even here?"

He stares into the wire, saying nothing, and though I can't know if he is listening, I go on.

"In high school, they made us read Thoreau. *Walden*. Ever read *Walden*?"

I try to remember the words blind Mr. Flynn made us recite nervously in eleventh-grade English. They take fresh meaning on the wire, no longer just words, memorized under duress. Perhaps old blind Flynn saw more clearly than I ever knew: "I went to the woods because I wished to live deliberately, to front the essential facts of life, and see if I could not learn what it had to teach, and not, when I came to die, discover that I had not lived."

I sweep my arm through the gloom. "This is the biggest adventure of our lives. Probably the most important things we'll ever do, we'll do here. We have to live it, we'll never get another chance. To say, 'I don't want anyone to remember me' is like wishing not to have lived at all. God, how awful; what a waste it would be to pass this way and gain nothing, or leave nothing. How can anyone want that?"

He says nothing, but turns away, staring out over the corner of the barricade, and in my attempt to help, and motivate, I know I have only caused more pain.

We spoke no more that night, and shared no more nights on the wire thereafter. The conversations we had during my final months In Country were strictly crewchief to metalsmith. He'd fix the rents and tears in my aircraft quietly and efficiently. I'd thank him, and he'd smile that thin, nervous smile, not meeting my eyes, and then just walk away.

The Center is at last silent. The arcade players have departed. Good. The memory saddens me, and needs the respect of silence. To leave nothing of yourself behind . . .

"He failed, you know," Tom says softly.

"What do you mean?"

"He failed. All he wanted was to fade into the background, not to stand out. To be forgotten. He failed."

"How?"

"You remember him."

I hit the rack about 2130, leaving Bickman in the bunker's barricade. Shielded by the dank walls, I flash my Zippo, but the flickering flame does little to penetrate the gloom. Yellow light licks at the mold coating the rotting canvas bags oozing damp sand like leprous sores through green skin. The floor is littered with the refuse of a thousand sentries: rusting C-rat cans, waxed paper, but mostly butts. Enough torn cigarette filters to stuff a mattress. I flare a C-rat Chesterfield, all I have, and shuffle inside. The cot is sound, although fist-sized cancers of white mildew grow on the canvas. It will not last long.

It is still in the airless tomb, and more silent than the silent wire. The damp prison of sand consumes the sound of my shuffles and bumps, the creak of the cot as I lie down. Lying in the bunker is like premature burial. Only the glowing coal of the Chesterfield pierces the darkness, floating in the dank air like the single red eye of a beast. I sweep it through the blackness, tracing letters in my best Palmer penmanship. The ember glows brightly, imparting the memory of its passage on my retina. "G," "R," . . .

too many letters to burn into the night at one time, but the hope in my heart captures what my eye cannot preserve: "A," "C," "E." Grace. I wonder if I'll ever see her again. I wonder if she'll wait for me. I wonder.

The butt burns low. I feel the heat on my fingers. I flick it against the bunker wall, tiny rockets shooting from the impact, burning themselves out in their fall to earth. Pretty. It falls into the sand, the single red eye still glowing, but less, and less, and then less. . . .

I wake abruptly to Bickman's chanting, "It's oh-one hundred, it's oh-one hundred." I shake myself up from the cot; my mouth is hot and fuzzy, and bits of Spam and bread clot my teeth. I grope for my canteen, swill the tepid water back and forth, and spit it into the darkness at my feet. Splatters dampen the canvas of my boots and soak through into my socks. Christ! More water into my hand, wiped across face and neck. Every muscle stiff from the damned cot. I stand up, stretching, my spine popping like castanets. A wave of nausea rises from my stomach. Think I need to puke.

I shuffle to the door of the bunker. The night air is so heavy with dampness that it's like trying to breath through a wet sock shoved up your nose, and settling on my shoulders and on the wetness on my face and uniform, it chills me. I need to piss. I walk to the wing of the bunker—not outside, never outside—unbutton, and groan relief into the night, taking pleasure from the flat spatter of urine crashing against a sandbag. "Ah. . . ." Bickman is behind me, standing at the barricade, as I cut loose a good one. "Fire one round, HEAT." Big, gassy, and satisfying. The aches and pressures melt away in the warm afterglow of flatulence.

I join Bickman at the barricade. There is nothing to report. No calls from the CP, no action on the wire. He says, "Good night," but I stop him as he moves to the bunker's interior. It won't be nearly as much fun sticking Bickman as Give-A-Fuck or Frenchy or one of the assholes who's stuck me so often. But what the hell.

"Uh, Bob?"

"Yes?"

"I forgot to tell you, I got early flight ops. Have to be on the line at 0400. I'll be waking you in three hours."

"OK."

"OK?"

"OK."

"What the fuck do you mean, 'OK'? I screwed you out of an hour."

"You're a crewchief, Zack. You need more rest than me." I can tell he means it: Bickman never kids around. The satisfaction of pulling a fast one melts, and I feel cheated, then angry with him for making me feel like the conniving shit I am. Why doesn't he get angry, show some balls? But mostly, why does he have to be so goddamned decent? Damn! He turns once again to the bunker.

"Good night, Zack."

I'm glad the darkness conceals my chagrin.

"Good night, Bob. I hope you, uh, sleep OK." The phrase lacks the passion of quoted Thoreau. What a shit I am.

He vanishes into the murk of the bunker. The cot creaks once, twice— a few soft moans and then silence. I turn toward the perimeter and begin scanning for movement, putting the anger and the guilt behind me. It isn't easy, but the night commands everything I have. There is room for nothing else, on the wire.

Night, on the wire, is unlike all other night. The darkness is not the mere absence of light but a living and hostile thing that enshrouds the killing field and smothers me. The hated darkness is friend to my enemy, obscuring his approach through the mines and the concertina. But no matter how well I am concealed, I lie naked under his gaze. And defenseless. In the darkness, the friendly whine of the generator, the beat of the rotors no longer embrace me. The distant sounds emphasize my aloneness, and magnify my fear.

Sound is the enemy too. All sound. In the bunker, each scrape of boot through sand, each creak of webbing and fetid exhalation shouts my position, but the night sounds of the perimeter come too indistinctly, with too little data to be pinpointed, deciphered, and discarded if not meaningful. Each sound from beyond the wire builds upon the apprehension generated by its predecessors. Each rush of wind or movement of brush, the muted scurry of a rat or scrape of settling wire adds new weight to the fears of the night. They build and build, and none are stripped away until the too distant return of the sun.

Phantom shapes join phantom sounds and play merciless games with my senses. Infantry training taught me to stare at nothing, to look for nothing, but to watch for everything. I sweep the landscape unceasingly with unfocused eyes and allow sensitive peripheral vision to detect motion and suspicious objects. Stare at something too long and it will become anything you fear. Stare at it long enough and it will begin to move, then disappear, then reappear! But always moving toward you and never away. Listen for it long enough and you will hear it as well. You'll hear whatever it terrifies you most to hear, but always muted, never clear enough to know. The snag and tear of cloth on barbed wire; the fall of a grenade, the swish of a knife, muted voices whispering behind you—always behind. All the myriad, tell-tale sounds that mean infiltration.

Time passes slowly, unhastened by my peering into the phosphorescent dial of my wristwatch. Every few minutes I cup the barely glowing dial with a nervous hand and raise it to my eyes. The dull green glow hangs wraithlike in midair. The second hand moves imperceptibly and becomes harder to read as the radiation imparted from the dead sun dissipates. My mind wills speed into the crawling sweep of the slender hand, and I pray soundlessly to hasten the coming of the sun. But no one hears. On the darkest of nights or under the ceaseless hush of the monsoon mists it is easy to imagine the bunker is an island, apart from all humanity. It is lonely on the wire, as bitterly lonely as a soul can endure and more. There may be no atheists in foxholes, but there is no God on the wire.

There is sound behind me.

Or is it to the right? No, the left. No, no, the right, it's to the right. I hear it dimly, through the shrouding night, and step silently to the barricade's angle to scan the darkness. Something out there is alive, coming for me.

I peer over the barricade and find a black figure, blacker than the night. It's meters away, beyond the stink of the bunker, and it's coming for me. It's probably Rock, but maybe . . . what if. . . . I can't be sure. I can never be sure, on the wire. I squat low, waiting for it to approach the bunker. I'll let it come to me! That's what I'll do. Christ no! Can't do that, can't let it come in close. I have to stop it out there. I peer over the wall, it comes slowly, slowly through the night. The M-14 is in my hands, round chambered. I push the safety to "fire." The armed weapon frightens me more than it assures.

I give the challenge, "Halt! Who goes there?" hissing it through the still air.

No answer. Did he hear? What if he heard? Why doesn't he answer? Maybe he didn't hear. Maybe he forgot the password. Maybe I should challenge again. But maybe he heard, and now he knows my position. Christ, he's got to know my position, he can see the bunker! He draws closer, cat feet on dry sand.

"Halt! Who goes there?" I rasp, louder. The words cut me, spoken with sandpaper instead of a voice.

The figure stops. A low, clear voice whispers through the night, "Baltimore." The password is good.

But wait a minute! What is it they can't say—the gooks! Is it "l" or "r"? Shit, Germans can't say "w." I remember it from the movies, Paul Douglas kidney-punching his Nazi guard. "Wee Willie Winkie, Wee Willie

Winkie," Douglas shouts. "Vee Villie Vinkie," pleads the guard. Douglas stomps his kidneys. I can see it in my mind. But those were Germans! These are gooks. What do gooks say? "Rarraparooza" and "Rirriputian," or is it "Maline, you die!" Or is it both? Shit, those were Japs, anyway. Did Japs talk the same as gooks? Christ, Chinese can't say "v" either, they say "w" instead.

What do gooks say?

We pick passwords that gooks can't pronounce. Baltimore has an "l," that means they can't say "l"! But it has an "r," too. Damn, I could hardly hear him. Did he say Baltimore or Berrimoe? He said the right password, but did he say the password right? I stare into the darkness, seeing nothing in the black-cloaked form to give me even a false sense of security. He is crouched over, thin like a gook, neither sparkle of brass nor sheen of worn green to mark the uniform of a grunt.

I point my weapon at the dark figure, wondering if the damn thing will fire. "Advance," I order, running out of things to try.

The figure speaks. "Give the goddamned countersign first, dickhead!"

Shit.

"Clipper," I squeak back.

"Dickhead."

The gook-like figure miraculously gains bulk as it nears, and the Sergeant of the Guard detaches himself from the darkness. He peers into the bunker and nods once. I am at least alert, if none too bright, and he is pleased, as pleased as he'll ever show. "Use the fucking countersign, dickhead. Next time I'll take you for a gook."

I say nothing. I could not speak if I willed it. My throat aches, its wrinkles still seized upon that last, croaking "Clipper" forced into the night. He glances at the orderly row of grenades and the -60 pointed out across the wire. He asks to see my personal weapon. I place it on "safe" and hand him the M-14, slick with moisture from the night air. He retracts the bolt and turns the weapon toward the dim light escaping the base. The dull brass of 7.62-mm twinkles. Others have challenged him on an empty chamber, and bought themselves additional nights on the wire as reward.

He lifts the field telephone from its cradle and tests the line to the command bunker. "Phone check, Bunker Two," I hear him say.

A last glance at me, another "Dickhead," and Rock melts from the bunker and blends into the living darkness of the wire.

I lean against the sandbags of the barricade. A gripping swathe of pain grips my sphincter, fanning upward along my spine and out through the small of my back. I feel as if my vertebra are being torn through my asshole, bone by lumpy bone.

The cramping seeps slowly away. Thank God it was Rock, just Rock, and nothing worse.

As the watch wears on I peer with aching eyes, trying to recall the terrain memorized hours before. Night shrinks the perimeter. I can almost reach out and touch it, fearing to, wondering what demons lie hidden on the other side of the barricade. Fantasies of fear assail me. The mind fabricates what eyes fear to see: crouching shapes with oblique weapons, moving toward the wire, shifting beyond the wire. They fade only to reappear elsewhere when I try to focus. I blink and shake my head to clear them, to see if the images will converge into awful reality. There is nothing. Was there something? God, how can I know?

The gloom seems inches beyond the M-60 atop the barricade, as if the world itself ends beyond the bunker. I force my eyes to their widest; my forehead aches from the strain. I pray for light. The flares are ready in their case behind the barricade, and one is always next to the row of grenades.

The shifting shapes of black converge and hold.

Something is there!

What was that? Movement in the wire.

No.

Yes!

No.

Over there!

Light, I need light!

I fumble in the darkness for the smooth cylinder of the pop-up flare and slide the cap off with trembling hands. It falls invisibly and rolls away on the dank sand.

I need that cap!

The cap contains the firing pin. The flare is fired by sliding the body into the inverted cap, and driving the pin up into the flare primer. I get down on my hands and knees. The smell of old urine is powerful. My hands flutter across the sand, searching.

Oh fuck. . . .

I fixate on locating the cap, rather than simply grabbing a new flare from the case. My hands stumble across the cylinder coated with damp sand. I scramble to my feet and insert the flare end into place, trying desperately not to set the thing off in my face. I aim it upward, across the perimeter. One blow and the pyrotechnic explodes away. *Whoosh!* The blast frightens me—everything frightens me. Sound and light, even the flare I desperately need exposes me and has become the enemy.

The flare hurdles aloft, then bursts in blinding suddenness with a distant *pop*. Shadows leap from the low rolling dunes and grave mounds, crisp-edged fingers of black reaching in from the wire. They race toward the bunker with the speed of painful light. Bunker One tosses a flare aloft, then Bunkers Three and Four. All down the perimeter flares blast from their launchers. *Whoosh! Whoosh! Pop pop pop!* Bunker Five's goes streamer: the tiny chute collapses atop the white-hot ball of fire and crashes to earth far beyond the wire, burning brightly in the sand. Five hurls another into space: *Whoosh—pop*. The perimeter is awash in light, night turned to hell-ish day under mega-candlepowers of illumination.

Precious seconds die as my pupils retreat from the assault of light. Magnesium clouds roil angrily aloft as the flares drift slowly, silently in the night wind. The bunkers and concertina lie evil and forbidding under the blinding glare. In the distance, light pulses against the tree line, like steady waves upon a shore that cannot be breached. The world is painted in ghastly gray and white and black, devoid of color and life, like the unholy plain at the entrance of hell: "All hope abandon, ye who enter here. . . ." The flesh of my hands displays the death pallor. I would shrink from those burning suns if I could. I shield my eyes, straining them even more painfully to penetrate the glare. Slowly, vision returns, and I scan the barren terrain.

The wire is stark against the white sand—coiling black lines of evil barbs seem to climb above the bunker, growing toward the light. I search for Charlie in the wire. As the chemical light pulses, shadows flicker, impenetrable and alive in the lee of dunes and graves, and merge: the terrain is painted anew with each pulse of the flares. I scan each new hiding place for signs of motion, but everything looks different now. The perimeter, so close in the threatening darkness, leaped away with the explosion of light. And I cannot find Charlie, but I know he lives out there somewhere, in some place that vanishes when I look into it, like the gray dots that live at the corners of squares on a checkerboard. The sour bile of fear rises; childhood bogeymen invade my thoughts, uncontrolled, like the Carkers, with leaf-crinkly dead skins and sharpened teeth, who vanish before you can look at them but who stalk you until they rend your flesh. God! I'm on the wire, and the bogeymen are real, and I'll miss them lurking in some flickering shadow. Clear my mind, Oh God! Clear my mind and let me see what's real.

I need control!

The flares settle slowly to earth; the shadow fingers lengthen slowly, reaching like dying claws toward the bunker. Toward me. I can see the concertina clearly now. It seems close, so much closer than I recall. In the waning light I search for breaks in the wire, comparing it to the images memorized at twilight. There are no changes, but I doubt that I've remembered the scene correctly, and wonder what lies brooding in the pulsing shadows. The Carker-images give way to new fears as the flares extinguish one by one. The last is close to earth, and night will soon claim the perimeter. Oh God, the night is returning.

The flare extinguishes with a silent shock. Darkness floods the land with a physical blow, and I am blind.

I force my eyes shut, willing my night vision to return. Speckles of green and red and yellow dance upon my stressed retina.

The hollowness of fear rises high, higher in my chest. A tensing spasm chokes my throat. My temples ache. Dizziness and disorientation rock me. Time melts into terror, and the few seconds before night vision returns are an eternity. The sense of unknown menace becomes overpowering. Fear crowds into the bunker with me, smothering me in an embrace of mindless dread. It rushes and roars like surf against my ears, and I struggle frantically to quell it.

Control!

God, hurry, hurry. I have to see!

I open my eyes prematurely, and the dancing speckles look like fireflies cavorting in the bunker. It was fun to do when I was a kid, a game, rubbing my eyes, then watching the dancing patterns. But I will them away. I have to see the wire! Slowly they fade. My pupils grow wide and penetrate the gloom once again. The rush in my ears diminishes, and melts into the night.

Slowly I regain control.

There's nothing on the wire.

I feel a little foolish—more than just a little—embarrassed by my private panic.

The field telephone's soft ring startles me.

"Bunker Two," I murmur, trying to sound in control.

"You send up that fuckin' flare, asshole?" It is the Rock, and he is unhappy.

"What flare?"

"The one I saw leave your bunker, dipshit!"

"Uh, well, I thought I saw something on the wire."

"You see anything or not?"

"Uh, I thought I did."

"You see anything or not, asshole?"

"N . . . no."

"You keep yur fuckin' hans off those fuckin' flares unless you contact the CP! You got that, asshole?"

"Yes, sir."

I slip the handset back into its canvas case. Down the line of bunkers, I hear the muted ring of field telephones. One by one by one. Ass chewings for all, no waiting.

I scan the wire, and the slow moments drag by. And then. . . .

What was that?

The watch crawls by. Tired. At 0230 the late moon rises behind the ARVN compound north of the base. So tired. Big moon tonight, nearly full and blood red on the horizon. It creeps skyward, silhouetting the barracks roofs and the wire, the graves and the dunes, shedding its blood on the land. Fading to pink and dirty gray, it grows in brilliance as if it draws power from its ascent. It shines over all, shrinking the ground shadows blacker than black. They become pools of darkness linked by the flow of the land, bathed in an unwholesome pallor. I shrink within the bunker. The sheen of the death moon glows upon the dirty canvas of the barricade. It glints from the -60's muzzle and flashes cool from my flesh shining cadaverous in the night. I would draw farther within the shadow of the barricade, but I cannot leave the firing step. Can they see me? I scan the pools of darkness in the lee of dunes and graves, and imagine black figures darting toward the wire. But I am so very tired. A cloud passes, lonely and black against the shining face. The moon glow struggles against the cloud, casting living shadows on the land. They flicker and flit like spirits in the night's breeze. I watch quietly, no flare launched in panic to aid the moon, or to challenge it. So tired. The other bunkers are quiet as well. Even fear cannot be sustained without limit, on the wire.

I fall silent, staring into that darkness of my soul that holds the fear of the wire. I see the dim shapes of my dream, tumbled from the nesting boxes.

"Do you feel anything that you recognize?" Tom asks.

"I feel the dream. A piece of the dream, but not all."

"What piece of it?"

"I don't know."

I want a simple answer. Let me say that the moon and the flares and the darkness are the bones of my dream, the only bones, and then it will go away. But there is more. I don't know what, but I feel it, and the dread in the feeling, of not knowing, hangs over me like the moon and the shadows of the wire.

"The moon is there. I know it. I hated the moon. It exposed me, more than it illuminated the terrain. I felt naked. It must be a piece of the dream, and the flares too. But not all. I know there's more."

I feel myself sinking into the dream, becoming lost among the rubble of nesting boxes. Searching, fearing to find.

"Go on, finish. The rest will come." But Tom is a distant voice, echoing down the tunnel of the dream. I've gone too far. I am leaving him behind, chasing the terror, drawn down by the terror. He does not know.

I try to pull away, but the wire grips me and does not yield. Like a living thing, it seizes me with grasping fingers. I can't get free. My God, I'm awake and the dream is upon me. I can't get free!

Frozen in the Center, I see myself frozen in the dream. My mind screams to run, run, run, but I cannot. The fear is large, close to my face, too close to see. My breath comes in pants, stolen gasps that cannot fill my lungs. I . . . can't . . . breathe. Run, run. I . . . can't. . . .

A voice down the tunnel of fear, a hand upon my arm: "Take . . . deep . . . breaths."

Run. . . .

"Take deep breaths. . . ."

I try to stare into the fear, but it's too close, too complex, too frightening. It has many colors, each blinding and alive, each one screaming in its own voice, a rushing wail that assaults me like the demons of the wire. Run!

I seize on Tom's words, concentrate on breathing. Slow, deep breaths, and each one drains color from the fear. It fades and fades, and the rushing voices with it, until there is nothing.

"What did you see?"

"I don't know!" Anger and frustration. The dream was there, but I could not stand against it. I could not! If only I could have forced myself to stand, I might have seen it, driven it away. Will I ever be free?

But I feel relief, as well, that it's gone.

All I can say is "I don't know," over and over again. My fear shames me; why can't I be strong?

I turn back to the memory. It is time to leave the wire.

I bend low in the bunker, letting the moon's dead light illuminate my wrist. Fifteen minutes more, then I get Bickman out of the sack.

0350. I wonder how pissed he'll be if I get him up now, ten minutes early. Even Bickman would get pissed.

The second hand moves so slowly. 0353. What about now? I ought to be able to get him up now. I can argue that it's almost five to, or just set the time ahead. Tough shit if he's pissed. It's going to take some time to get him up, anyway. I push into the Stygian gloom of the bunker and stumble against the cot.

"Bickman! Get your butt up, buddy!"

Shaking the cot. I find an extremity, a leg I think, and pull it.

"Bickman, time for your watch."

"Uh, uh, uh, huh."

"Bickman, get up.

"Uh, uh. . . ."

"Get up."

"Wha, wha, wha. . . ?"

"Up!"

I find his shoulders and haul him erect, his breath close to my face. Christ, trench mouth. Moonlight cascades through the door slit, turning the mold sickly gray, the sandbag's weeping sores even more leprous. Tiny Bickman, round-shouldered and shaking in the night chill, bathed in the ashen sheen of the moon, more Marine than human, but more pitiful than Marine.

"C'mon, Bickman, get up and piss, the world's on fire."

"Uh, uh, uh, uh, uh. . . ."

"Get up, Bob."

It takes a while for him to rejoin the living. "Anything happen?" he asks at last.

"The Rock came out here, but I saw him. Forgot to give him the countersign, though. He was pissed. Big deal."

"What's the countersign?"

"I forget. You awake now?"

"Yeah. Anything else?"

"No. Time for me to go to the line." I want to leave the bunker and the shy boy. "You going to be all right?"

"Yeah."

"I'll call the CP to secure—you know, there's no cure like see-cure." I should restrain the gaiety in my voice, but cannot. I am leaving the wire! The handset is damp in its cradle, chill against my ear, but my voice is the warmest I've spoken all evening. "Bunker Two, Corporal Zaczek. I'm securing to the flight line. Corporal Bickman has the guard."

I relish the slow walk to the flight line, passing into the embrace of the wakening hooches and shops. I begin my preflight, and the familiar feel of the aircraft is an anodyne to the stress of the night. This is my element. The bunker gloom falls as we lift into the red hue of the damp morning, outbound on the first mission of the day. It feels good to breathe again in the sharpened wind stream under the sun, to feel its warmth, to be alive—like being reborn.

It feels good to be in control.

It feels good, again, to not be afraid.

"You know, nothing ever happened out there. In all the nights I spent on the wire, nothing ever happened."

"Do you call that nothing?" Tom asks.

"In thirteen months, I was never shot at, never actually saw an infiltrator, never saw anything that was really there. I was just scared shitless, that's all. I never got used to it. I guess I never got over it."

"The moon brings it back."

"Yeah," I muse.

The dream's connection to the wire seems so obvious in the haven of the Center. But for so long, all I've known is that it gives me the creeps to be outside when the moon is full. I joke about it, saying it's the werewolf in me coming to the surface, or that it's my time of the month. But it isn't funny, and the flashes of temper and anger I vent on my family aren't funny either. I avoid going outdoors when the moon is high. It threatens me. I've even arranged my bed so that its light can't cross me as I sleep; it seems to bring on the dreams.

"Do they only come when there's a moon?"

"No, that's what makes it so complicated. I don't think the wire is the only cause. If it were, I think the dream would go away. Hell, I can't even know if Vietnam is the only cause."

I am angry. The nights of terror are pieces to the puzzle. This I know. But how many are the nesting boxes that hide the other pieces? How large is the puzzle? I am husband, father, engineer, and Vietnam was long ago. The flesh that showed cadaverous pale in the bunker's moonlight now bears the unwelcome lines of middle age. I sit in the quiet of the Center, so many years distant from those days of fear. Yet the tale of the wire fills the room with the bunker's fear-stench, fresh and awful as those distant nights. How much longer until I find all the pieces to the dream, and then will it go away? Or will I live out my life visited by the night horror, never knowing? The thought frightens and obsesses me. I want to rip through the nesting boxes, and tear from them their hidden secrets. They are mine! I have the right to know! It is so unfair. Why must I fabricate dreams from the elements of my own fear and my own life, and not know what they mean? Why must I dissect my mind in pain, under the microscope of the Center, to search for their meaning? Why can't sterile logic, the cold scrutiny of my mind's own eye, guide me to the causes? How many are the cancers of my fears, and what will it take to rid myself of them?

Tom can see the anguish that rises from my thoughts. He is consoling. "You've found some pieces to the puzzle today. It was hard, but we'll work through them. We'll talk about these again. I want you to think about something, though."

"What?"

"Were there other times when you were out in the moon?"

"Do you think there were others?"

"Perhaps."

"Have I told you about others?"

"Perhaps."

"Give me a hint."

Tom smiles. Answering such questions is not his way in this game, but I must try. Trying is my role. I know he will not share his vision, but it is enough to know that he sees something, what I cannot yet see. It means there is something else, and it gives me hope. I trust him. I must. He will guide me toward the fears and through them when the time is right for the healing to come from within, as it must.

"There's something else I want you to think about, too," he says.

"Shoot."

"Think about how it felt being in the bunker, and how you felt here today. You walked into the dream a little while ago. It frightened you. Think about how that felt."

"You think there's a connection?"

"Perhaps."

I leave the Center, tired. The walk to the car is slow. I turn, and Tom waves from the Center's doorway, softly lit in the fading glow of evening. So like looking in from the wire, but without the fear. This is home.

A few days pass, and the dream comes again. There is no moon. Flashes-light-darkness. Crying, "Wake me up wake me up!" in the chill terror of the night.

Yet something is missing, exactly what I cannot tell, as if a portrait too darkly seen and too complex to be understood has been altered when I glanced away. When I look again, I am but dimly aware that some face in the portrait is changed, but I cannot tell which or how. I sit, shivering in the night, trying to remember the dream as it was, trying to compare it to the dream as it is now. I cannot. The images fly away, flitting as always beyond my mind's grasp.

But I know the dream has changed. It is an atom less terrible. So small a loss cannot escape me, so long have I borne its weight. I do not understand, but I am grateful, even for an atom's less terror.

It is an atom's greater hope.

SESSION: INCOMING

"I've been thinking about other nights, when there was a moon."

My speech is rapid, charged with rare energy. I am eager to begin the session; there is much to say.

I'd carried Tom's directive from the Center, driven to find an answer. But in a day, perhaps two, his instruction began to elude me. What was it that I was supposed to think about? His words danced just beyond my reach like the faces of the dream, as if they didn't want to be remembered. At last I called him. "What was it you wanted me to think about?" I asked. I felt foolish and expected him to laugh, or at least chide me for forgetfulness. But he didn't. He repeated the questions firmly, and firmly restaked the path to follow in dealing with the dream. I wrote the questions down, and cheated the subconscious mind that would live with buried pain rather than endure the anguish of putting it to rest.

I carried his charge with me, and perseverance bore fruit. I made discoveries while wandering among the nesting boxes.

I'd seen another face of the dream.

"I think I discovered something!" I feel like a child bursting to tell his story.

Tom settles in his chair, silent. There is no need for him to speak. Seldom do I take the reins of a session.

I will bring forth the truth I have found.

There were no rear areas In Country, in the true sense of the term. No safe havens where we could rest secure and untouched by the war. From dawn until twilight we carried the war in our aircraft across the canopy and the rice fields, over the hamlets and villes that passed beneath our skids. "Search and Destroy" missions; "Snoop and Poop" we termed them. We hunted the enemy by day; they came for us at night. We were far easier to find.

The air base was as brilliantly illuminated as a small city. Crewchiefs and ground crews toiled long into the night to ready aircraft for daytime operations. Spotlights on mobile generators dotted the flight line, bathing the green flying machines in stark, mosquito-dense islands of white. It was a gunner's dream. Phu Bai sat in a broad, sandy plain, too flat for individual aircraft to be pinpointed from afar. But the VC had our range, and the spotlighted smears of white established the azimuth for rockets or mortar if they were after metal, or they could shift fire to the squadron's living areas if they preferred meat.

They would come while the base slept. They could come without warning, or their coming might be rumored for days. There were always rumors, always alerts, and enough sleepless nights to dull anyone's apprehension, so the rumors often went ignored. No matter: when they came it was always a surprise. They would come at their will and in their good time, and there was nothing we could do about it. There was no place to run and only the bunkers in which to hide.

I remember the nights they came, not each and individually but all of them merged into one, like a mural dashed off in panic with a brutal brush. The colors are those of darkness and fire and pale bodies glistening bunker-sweat under the ghostly pallor of the gunners' moon. The canvas screams with sound, a cacophony of sirens and curses and prayers and shuddering blasts, all cut torturously into the canvas with a fine-edged scalpel of fear. I remember the nights they came.

I am asleep.

It is an empty, dreamless sleep, the kind I seek most, untormented by the realities of the present or by aching memories. It is a warm summer night, and the moon shines full over the perimeter. It is quiet along the line of huts, save for the soothing whine of the generators. There is a breeze, rare and welcome, of dank and humid air blended of tilled earth and urine and decay—the hallmark scent of Vietnam. Any breeze is welcome after the staggering heat of the summer monsoon.

Thirteen cots fill the hut, and on each lies a thin figure clad in faded green skivvies. It is stifling under the corrugated steel roofs that turn to ovens beneath the long day's sun. The mosquito netting over some cots has been drawn away to capture the caress of the wayward breeze. But the figures in these cots jerk haphazardly, like puppets, slapping at mosquitoes. On other cots, the netting is drawn securely, tucked under the spongy air mattresses. It blocks the mosquitoes, and the breeze. The lean figures toss fitfully, bathed in sweat, soaking oily puddles into the rancid bags. The sleep grants little rest, but it is sleep, and sleep is escape.

The siren's wail pierces the night. It shrieks as the first rocket crashes into the dunes between the hooches and the flight line.

"Incoming!"

The cry echoes through the long row of huts. Nylon netting tangles in outflung arms and scrabbling legs. Cots overturn, letters, decks of cards, chess and checker pieces spill to the floor, kicked along by naked feet running, running.

"Incoming!" is the cry, "Incoming!" is the fear that drives us to the bunkers. Half the hut heads for one bunker, half for the other. Stick figures in green-dyed skivvies leap through pools of moonlight, skinny stick legs and skinny stick arms pumping, pushing. Those in the center struggle against sleep-stunned minds to pick the shortest path to survival. Right or left! Fuck! "Move! Move! Get the fuck outa my way!"

I leap from my cot, cursing the net tangled about my leg. It tears free as I head down the steps, driven by those behind me, driving those ahead. I flash through the naked moonlight quickly, as if the distant gunners can target me alone within the mob packing through the doorway. "Move!" Stubbed toes, splintered feet trip down the hooch steps. Hands, elbows, knees, feet, pushing, jabbing, poking, fear swallowing the pain.

"Move!"

The bunker entrance is the bottleneck. We shove our way in, turn left into the darkness, then right, threading around the blast maze.

The siren wails. It fills the darkness and floods my mind with its banshee shriek of terror.

"Move it, motherfucker! Get the fuck in!"

The siren's scream. The wet slap of flesh on flesh. The crush and fear of the mob. Don't fall! Propelling me into the bunker. Crouching. Grinding my shoulder against the rusted roof. Stumbling, bruised against canvas and sand. I feel my way forward, hunched down, mole blind in the gloom, seeking an empty spot far from the blast maze. I step upon limbs, am stepped upon.

"Get the fuck off me!"

"Fuck you!"

Useless obscenities, dying as the bunker fills. The last bodies pack in. Arms, legs untangle. We huddle along the damp walls of sand as far as possible from the entrance, staring back at it with night-stunned eyes. Sickly moonlight navigates the maze, turning us to dusky ghosts. Shattering echoes of rolling thunder rebound beyond the light as the crash of the first rockets sweeps through the night. The light seems to pulse with the echoes in shimmering moon glow. Then silence swallows the echoes. The light grows still inside my mind. We wait, huddled with our backs to the canvas and sand.

In the bunker, carved by the labor of our hands, we wait.

Unseeing eyes strain upward. Waiting, in silence. The initial barrage has fallen, the siren no longer wails. The generators are still, the base plunged into darkness but for the gunner's moon. The enemy waits, thirty seconds, forty. Time enough to encourage the foolhardy to run for better cover. Time enough to catch the foolish in the open when the gunners fire for effect. We gaze upward, waiting.

Incoming.

Fear erupts in waves of sound on the distant flight line. The shuddering crash rolls across the dunes and grave mounds, the hooches and the bunker like breakers to the shore. Labored breaths come as one, tuned to the rhythm of rocket and mortar. The 120s crash in earnest. Cold sweat streams from hollow armpits, snakes in thin rivulets, pooling in creases of flesh, filling, overflowing, running down the crack of my ass, trickling through pubic hair, itching, oozing through skivvies sodden on the reeking sand, dripping, dripping. The 120s crash closer, walking toward the hooches. Rotted canvas and flakes of rust fall upon upturned faces, into parched throats and slack mouths. Yellow flashes flicker weakly, then brilliantly beyond the blast maze. Hollow eyes and sunken cheeks illuminate in stroboscopic nightmare. Concussive blasts flood the bunker, pushing the air before them like an invisible smothering beast. Ground shocks surge under the piled bags and grind the sand up into aching knees. Lungs scream for air in an ocean of air. The mortars walk in, crash upon crash, lashing the bunker with slivers of steel. Sledgehammer thuds. The roof matting sings, a shriek of terror, metal rending metal, point and counterpoint to the roar of the 120s, and in the bunker the welling tide of fear.

"Oh God, don't. . . ."

The litany screams in my mind.

"Oh God, don't. . . ."

Pinioned, held immobile for slaughter. Any awareness tuned to the holocaust of the 120s.

Oh God, don't let me die!

In an instant the barrage rolls across, then onward, then elsewhere.

Stillness follows, like a slow, running tide, an eddy of peace. Aching lungs relax. Heart rates quiesce. Limbs sag into the wet sand, cool against my thighs. Strained eyes relax, relax. Shattered ears are caressed by the hush of running sand. It buries my hand on the ground behind me, fills the spaces between my fingers, covers my wrist in cool softness. I am floating, dizzily floating. Foreheads, knotted and creased and aching with stress, relax. Fear oozes out through the blast maze, into the moonlit night, sucked from the bunker by the receding echoes of the rockets.

Seven skivvy-clad assholes unpucker.

"Fuck!"

"Wonder if they got any of the birds? Christ, I thought I'd shit my pants in that last one!"

"I did."

"Me too."

"You always smell that way."

"Fuck you."

"Wonder if they hit the hooch? I think that last one hit the hooch."

"Hope my tape deck is OK."

"I hope your goddamned deck got fragged. I hate that country shit you play."

"I'll bet the hangar took it good again."

"I hope they got my ship. That pig is a piece of shit."

"Oops, safeties!"

"Goddamn you, Frankie!"

"Ah, Chanel '67. A very good year. Clear up the air in this place."

"Jesus Christ, I'm going to beat the shit out of you, Frankie."

"He already shit."

"Gawd, I'm gonna puke."

"Slugs!"

"Safeties, I called safeties!"

"No farting in the bunker! Gawd-dayum." *Punch, punch, punch.*

"Ow! Get offa me. I called safeties!" *Punch, punch.*

"I wonder how long this'll last. I got flight ops tomorrow."

"Today."

"Maybe they got all the planes and we won't have to fly."

"No such luck."

"Well, maybe at least they got the Gunny."

"No such luck."

"Fuck."

"Move, I gotta piss."

"Go outside."

"Your ass. Move or I'll do it on you."

"You pig."

"Ya goes where ya gotta go. Ah!"

"I know, let's have a farting contest."

"No fair. I can't fart on demand."

"Too bad. Safeties off." *Brrrrrrppppppp.* "Ah!"

"Fire one round, HEAT!" *Rippppppppppppppshhhhsssss.*

"Lock and load!" *Thhhhhwwwwppppshhhhhssssss.*

"Fire in the hole!" *Pluuuuuuggggghp!p!p!p!p!eeeEEEEEE.*

"That was a squid fart."

"Who you calling a squid? It was just soprano."

BLLllaaAAAATTTTtttttPSSSsssss . . . P!P!P!p!p!p!p!uh!uh! "Baritone! Oh shit! I think I shit!"

"That was a Gunny fart."

"Khaki."

"Oh, yech. P.U. Gimme your T-shirt!"

"Hey, anybody got a butt?"

Silence. Seven butts squirm on sodden sand. Seven parched throats and aching sets of lungs yearn.

"God, I need a smoke."

"Me too."

"Me too."

"Well, someone go into the hut and get some."

Silence.

"Who?"

"You, asshole, you mentioned it first."

"Fuck you, man! You hear that shit out there?"

"Pussy."

"God, I need that too!"

"Let's toss."

"Yeah, I got a nickel right here up my ass, you dipshit."

"Rocks, scissors, paper."

"OK."

Seven men, four games to choose who will go for the smokes. It is difficult to see, but the moon wash through the blast maze is enough. Shake, shake. Rock and scissors.

"Rocks break scissors!"

Shake, shake. Paper and scissors.

"Paper covers scissors!"

"Nah, dickhead, scissors cut paper. Paper covers rocks. You lose."

"Two out of three?"

"Fuck."

"C'mon."

"Fuck."

"Jeez, I went last time. Cut me some slack."

"Fuck."

"Hey, see if my tape deck is OK."

"Fuck."

I remember the night I lost. Crouching in the blast maze, waiting for a lull in the barrage, then onto the trampled sand, up the steps, through the screen door, tripping over mosquito netting and into the hut, into my cubicle, tearing through a pile of mildewed socks, moldy salami, and skin magazines to find a full pack.

"Got 'em." Out the door, down the steps, almost to the bunker. Almost safe. Then, "Matches! Shit! No matches." A book of C-rat matches is in my flightsuit under the pile of flight equipment, toilet paper, and rat traps littering the floor. The barrage is closing. *Whammmm.* Up the steps. *Whamm!* The crashes are louder, briefer, getting closer. Back into the hut, itself shuddering as the blast waves close in. *WHAMM!* I find the suit, burst back into the night, shrinking under the glow of the moon. *WHAM, WHAM!*

"'Bout time you got here, asshole."

"What took you so long?"

"Gimme a butt!"

"Is my tape deck OK?"

"Butts on that butt."

I huddle in a corner of the bunker, hoarding the Camels like gold, turning to each of my hoochmates in the gloom.

"Fuck you, and fuck you and fuck you everyone."

"Gimme a butt!"

"Butt your ass," I answer.

I pull a weed from the crushed pack, tap the tobacco down, flare a match, and inhale the raw smoke. It all seems worth it. Unseen fingers seize the pack and pass it swiftly through the darkness. There may be no atheists in foxholes, but there are no nonsmokers during rocket attacks. Seven red pinpoints trace slow arcs in the night. The butts reflect the intensity of the attack. The farther away the rockets land, the more leisurely the cigarettes traverse up and down. But when the barrage closes they glow furiously, transfixed in space between clenched teeth.

Ten minutes pass. Twenty, thirty. *Whammmm. Whammmm.* The gunners concentrate on the flight line. Sometimes long minutes pass between explosions, and we wonder if the siren took a hit and cannot sound all clear. No one leaves to check it out. Time begins to hang heavy; even rocket attacks get old. We grow impatient to return to the hooch and salvage the remaining hours of night.

"When is this shit gonna let up?"

"That goddamn asshole Gunny is probably gonna make us fix the birds tonight."

"Bitch, bitch, bitch."

"Up your ass, man, you don't have to fly in three hours."

"Tough titties."

Finally, silence, or at least the normal bitching and complaining that pass for silence among a group of pissed-off jarheads. "All clear" wails briefly, and seven sweaty bodies crawl from their tomb into the damp night air. Up and down the line of hooches the question echoes.

"Anyone hit?"

Going unanswered, it answers itself.

"Any of the hooches get hit?"

"Nah, but guess what?"

"What?"

"The officers' shitter got it."

"What!"

"The officers' shitter took a hit! There ain't nothing left but a crater!"

"Ha ha ha ha ha ha ha ha ha ha ha. . . ."

"What a ballbuster."

"I think there is a God."

Friends and hoochmates fence with black humor to mask their relief at finding one another unharmed. "Hey, asshole, I thought I'd gotten rid of you."

"No way, GI."

"Badder luck next time."

Someone remembers the aircraft. "Did any of the birds get it?"

No one knows. No one wants to know. Crewchiefs love their birds when they're well, or only a little sick. But no one wants to put pieces of a shattered aircraft back together at 3 A.M. I pray fervently for an undamaged ship, or one that's completely blown away. Nothing in between. No one with any time In Country volunteers to go to the flight line. Going there simply makes it easier for Gunny to grab you for the clean-up detail. A few FNGs will go to see what shredded Huey looks like. There's always a few who want to see the war. Gunny will get them first, and then he'll be too busy to come and get us while he's showing them what to do. Anyway, whatever happened will still be there when the sun comes up, and Gunny knows where to look if he wants us. He moves slow and needs the exercise.

I shuffle into the hooch and stretch out on the narrow cot. The netting is a shambles. To hell with it; flight ops are less than three hours away.

In the last moment before unconsciousness claims me, someone softly swears, "Jesus, shit! I hope they don't hit us again tonight."

"There!"

I punctuate the word triumphantly, proud of my discovery.

"'There' what?" Tom asks.

"'There' there! The moon, the bunker, the explosions. Flashes-light-darkness. The dream, y'know. There!"

"So?"

Tom can be unaccountably dense at times.

"Look, what more do you want? You asked me to think about other times when there was a moon, and I thought about them, and this is what I found. OK? What the fuck more do you want?"

I am angry, but also worried. He doesn't play twenty questions unless I'm jumping to conclusions, or making assumptions that will lead me off track. "Incoming" fits the dream so well, and there is the moon. It has to be a piece of the puzzle. I want it to be.

"Look," Tom says, "when you spoke about being out on the wire, you felt exposed by the moon. Right?"

"Yeah."

"But in the bunker during incoming, how much effect did the moon really have on you?"

"It lit the base up for the VC gunners." I wonder where this is leading.

"But how much effect did it have on you when you were inside the bunker?"

I don't like this. I worked hard all week, burrowing through the nesting boxes, looking for answers. The moon was the clue. We even used to call it the gunner's moon. Tom gave me the clue himself, and I followed it. What more does he want?

"Look, I don't know what the hell you're getting at."

Tom glances away, searching for the right words so that I might discover the path without being forced onto it. It is a delicate balance. The healing must come from within.

He speaks carefully. "There's a connection between being on the perimeter, being in the hooch's bunker during incoming, and what you've told me of the dream."

"How do you know?"

"It's you who know. Otherwise, you wouldn't have brought them up together."

"I wouldn't?"

"You wouldn't."

"And just because I brought them up means they're connected?" It is a novel thought, and I consider it with surprise and suspicion.

"You're the one who creates the connections. You control the connection. You are the connection. It's all inside of you. We have to deal with what you know."

I try to fit the pieces together, wanting the connections to become obvious, wanting to deal with them so that the dream will go away. I take refuge in logic; the cool analysis I use to quell the turmoil gives me control.

"Look, being on the wire, flares, the moon—they're pieces of the dream, right? The hooch bunker, incoming, the moon—it happened. I was there. It has to be another piece of the dream. It feels like it, and besides, it works. QED."

The logic holds me on course, like a sea-anchor running before a storm. But Tom isn't buying it.

"Look, you can't apply that Aristotelian stuff here. Syllogisms only work with absolutes."

I scratch my head and try to remember high-school history.

"What's a 'syllogism'?" I ask, giving up.

"All men have two legs; John is a man; John has two legs. Major premise, minor premise, conclusion." Tom smiles.

"Sounds logical to me."

"What if John loses a leg? Aristotelian logic deals with absolute premises. There's nothing in this that's absolute. Let's put it another way. Stuff happens. It'd be nice if it was all black and white; it isn't."

I ruminate on his words. Stuff happens. I can buy that. Still, I am pissed, and impatient. I spent an awful lot of time chasing that goddamn moon. I want it to mean something.

"Are you telling me the moon isn't the connection?"

"It's a connection because it led you to this point. I don't know if it has much to do with how you feel about being in the bunker during a rocket attack."

"Then why the hell did you ask me to think about it?" I am confused, exasperated, and angry, and becoming angrier. Running in circles angers me. This is too important to be run around. And I really want that damn moon to be the answer, because then it might go away.

Tom's voice is calming, and the patience in it infinite. "You had already thought about it. You made the connection between the dream and being on the wire, and you told me about the moon. I didn't put the moon there.

You did. You make the connections. I asked you to carry it forward to see where we'd end up. We ended up here. We'll deal with what we have. We have to start somewhere."

The explanation barely mollifies me. "OK. No goddamned moon. But you better give me something else to follow because I'm fresh out of shit."

"There's a deeper thread that runs through everything we've talked about in the last session and in this one."

"Care to give me a hint?'"

"I already have."

I roll the events of the last session, and this one, over in my mind, and come up empty. "I must have missed that one. Did I step outside to take a leak or something?"

Tom ignores the jibe.

"Last session, we talked about being out on the perimeter. You fell into the dream, and it frightened you. I asked you to think about how it felt being on the perimeter, how you felt in here, and the dream. Did you discover anything?"

"No. I tried, but I couldn't stay focused. It was like trying to put my finger on a spot that keeps jumping away, so I concentrated on other times when there was a moon. Look, dammit, one out of two ain't bad. This is tough shit."

Tom sighs, cleans his glasses, and perches them atop his nose, pressing them into place with a final poke of his forefinger. He weaves his fingers together, flexes them in and out, cracking his knuckles, then locks them in place behind his head and stares into the northwest corner of the room, elbows pointing skyward. The motion seems invented for him, and he uses it whenever he's about to pose a summary question. He searches the ceiling for the right words, brief and to the point, sharpened to hit the mark.

"How important is it for you to be in control?"

"Very."

"Why?"

"I don't know. I've always been that way. It's the way I survive. I've been wrapped so tight for so long I don't know how to be any other way."

"Were you that way before Vietnam?"

"I was no way before Vietnam. I was born in Vietnam." I snort the cliché, but there is too much sad and angry truth in it for me to find humor. It describes too much of my life.

"Think about the perimeter and about incoming and then about the dream."

It didn't take long. I may be dense, but not stupid.

"I wasn't in control."

"Bingo."

I recall the hysteria that drove me to the Center. I remember the first words I spoke to the large man, watching me with fierce eyes. "I'm not in control, anymore," I said. Why is it so difficult for me to discover these things on my own?

Tom reads my mind.

"Remember that first day. I said that the one thing we could do for you in the Center was to give you back the ability to make choices, and that there would be some things that affect you that would just go away, and others that you would learn to control."

I remember his words, and the hope they gave me. It was what I needed to live.

Tom continues, "I know that you want to tie the dream to some object, like the moon. If it could be tied to something obvious, perhaps it would go away. Sometimes it's as simple as that; sometimes not. What's obvious is that 'control' is integral to your personality.

"Didn't you choose aviation?" he asks, wrinkling his brows.

"Yep! All I ever wanted to do was fly. When I learned that helicopter crewchiefs fly all the time, I did everything I could to become one. Some petty officer in aviation school wanted to place me in avionics. I had high grades in the basic schools, and they wanted to turn you into a twidget if they thought you could pass twenty without using fingers and toes. I told him I wasn't interested in that fucking fairy stuff—I especially emphasized the Fs. It worked; he stamped 'ADJ, BHC' on my specialty chit—that's Jet Aircraft, and Basic Helicopter schools—and said, 'OK, smart-ass, you're going to be a raggy, and you're going to die!'"

I smile to remember that day in the naval air station at Memphis. That squid motherfucker was very nearly right.

"It's pretty unusual to get the job you want in the military," Tom observes.

"You just have to apply the right persuasion," I chuckle.

"Control," Tom smiles in return. "Becoming a crewchief only honed that trait; the war amplified the need. You were the king of the hill in the air, but when you were exposed on the perimeter or helpless in a bunker you weren't in control. It's just as important to you now as it was back then because having the dream is losing control, and that threatens your survival."

Suddenly, I remember the dream, changed only so slightly, an atom less terrible. I haven't told him yet.

"I forgot to tell you. The dream changed after the last session!"

"You . . . you. . . ," he sputters, "Why the heck didn't you say so in the first place?" His exasperation is plain, and humorous, and so rare that the moment is more enjoyable for the rarity. Springing something on a shrink, even accidentally, is a kick in the ass.

"Well, it didn't change much, and I'm not certain how, and you wanted me to think about the moon, so I forgot," I ramble, like a mischievous child.

"But it did change. I think some piece of it just went away. I don't know which piece. It doesn't seem to matter, because it's just gone. It doesn't seem important to me to know what it was. I guess I never thought that it would happen.

"But I still think I uncovered another piece of the dream. Maybe I followed the wrong trail. I think it fits."

"It probably does," he agrees. "Like I said, you make all the connections."

"Do you think the dream will go away?"

"I don't know. We came a long way today, but there are more pieces to it. We need to find them."

And so I begin to see. A fine line separates hope and resolution. In the Center I'd found the hope I need to survive, but so long have I lived on hope alone, it has become an end in itself. Going through the endless sessions, walking the twisting paths in my mind, I lost sight of the fact that hope must end in resolution, else it is not hope but delusion. "An atom's greater hope," I thought of the dream-change, unable to see it for what it is. But now I see clearly. I look at Tom, for the first time feeling not just hope, but resolution. The sense of it is warm, invigorating, addictive. I feel it surge through my veins, and want more.

We stand and shake hands. It feels good to understand. In understanding there is control.

I laugh. "You're right about the 'control' business. And crewchiefing. We are a tight-assed breed. Who knows, maybe I'll learn to control the need to be in control."

Tom grips my hand, and laughs as well, and there is no mistaking the hope in his words.

"Or maybe you won't have to."

Chapter 5

It is well that war is so terrible, or we should grow
too fond of it.

—Robert E. Lee, on seeing a Union charge repulsed
at Fredericksburg, December 1862

War Story

"I think we need to talk about the team."

"What team?"

"The team."

"I've already told you about it. How many times have I told you about it? There's nothing left in there to tell."

Tom is taking me back to the Story again. I feel like a child, hiding beneath the covers, knowing that all it takes is a peek into the darkened room to bring out the bogeyman. If I can hold my breath, and keep real still, maybe it'll go away.

"It's a place we still need to visit," he persuades.

"Look, I can go over it again, but I won't talk about the part where. . . ."

The sentence hangs like an ominous cloud. Within its shifting mists the events of that day play like a home movie of someone who is me.

"What part?"

"You know goddamned well what part. We've been all through it. How long have I been coming here? I've told it once already, and you know goddamned well what happened, so I don't have to say the words again. It doesn't serve any purpose to say them again. You just fill in the blanks."

The words I need to tell the tale—the entire tale—are bound too closely to the memory of that day. Uttering them opens hidden closets, unseals the pain. The words cut through my memory like a knife, and the sense of sound—the sound of my own voice—is a catalyst that triggers the old anguish. When I try to "say the words," my voice strangles into stuttering gibberish, then becomes mute. Bile rises, and the urge to vomit becomes overpowering. It is as if the mind must shut down the sense of speech to protect itself against the pain buried in the spoken words.

It is 1984. Three years of coming to the Center, sitting in the twilight stillness, reliving the old tales, working through the pain. Sometimes I think that if I'd known I'd be three years dealing with my year of the war, I'd have pulled the plug. When will it be done?

Yet life has been good. I have learned much of myself, and am happier for it. The old rages fill me less often, and with less fire. I am not comfortable with anger, but it is a thing I control. The dreams come less often, still with faces I cannot see; the fear seems no less for the learning.

The Story, he demands. I hate the Story. Voiced once, its blood still spatters the walls. Choking, stuttering through the pain, I told all I could, yet do not know if I've told it all. If I knew, if I really, really knew, I would never tell it again.

"Where do you want me to begin?"

When I was a child, I liked to coax stories from my father about the Good War, WWII. He never spoke freely about his conflict, but I loved war stories and pestered him relentlessly. Saturday afternoons, I watched *Victory at Sea*, and sometimes he'd spare some time from washing the Pontiac to join me. He would lie sideways on the floor, head propped up by one hand in front of the Zenith. I'd bury my elbows in the familiar rolls that framed his belly, but he never seemed to mind. Together we'd listen to the deep, bitter voice of the narrator as he spat out heinous details of Jap perfidy. The scene would cut from little yellow men on tin cans to sweat-soaked submariners with haunted eyes inside their iron coffins, rolling under the oily seas. Our Guys! I thrilled to the way the serials ended: destroyer screens laying depth charges, the carrier force launching men and planes against a distant fleet—the fuzzy, jittery flicks recorded by gun cameras as Our Guys bore in. "Scratch another one of Yamamoto's."

After each episode I'd ask my father the same question. "What did you do in the war, Dad?"

His answers were never so satisfying as the tales the deep-voiced *Victory* commentator told to Richard Rodgers's stirring theme. He became evasive, sometimes almost angry. He'd reach toward the screen and trace a meandering line over a map of New Guinea and mutter, "This is the Fly River. I was there," and nothing more. He dodged my questions about combat and tried to divert my attention to other, less-sensitive topics, the same old stories that I could recite as accurately as he. His eyes warmed as he looked into his past and selected the memories, one by one: how he used to catch the strange South Pacific crabs and steam them Maryland-style in beer; how great a liberty town Brisbane, Australia, was; how he met Una Merkle at a USO show. But never the good stuff. Never a word about the screaming Zeros I imagined boring in on his antiaircraft gun.

I have my own war now. I spin tales of liberty in Manila and R & R in Bangkok, and mourn the loss of those debauched days. Less easily, but driven by need, I tell of missions and medevacs and night flights of mercy and death. But there are tales or pieces of tales that I do not tell aloud. There are fragments of memories that need words I cannot suffer to voice, although I remember and relive the actions the words describe. Words of helplessness and fear, of pain from wounds beyond endurance, and of the awful loneliness of death in battle. The keenness in the words is undulled by the years.

Perhaps, at last, I understand my father.

There are many stories, some more difficult to speak than others. The worst are of the medevacs and the emergency extractions of wounded and dying from the zone. I have a War Story, as fine as any I've ever heard, the more so for being true, and mine.

When I remember the extracts and the zones, I think most of Khe Sanh. When I remember Khe Sanh, I recall the revetments across from Charlie Med where we parked our helicopters. Khe Sanh's six revetments were an ugly chain of blast-containment walls—UUUUUU—whose long arms embraced each aircraft to confine exploding fuel and flying shrapnel during incoming. The revetments were everything: garage, bedroom, chow hall, and pisser, and we spent more time in them than in the sack when we pulled duty at Khe Sanh.

Most of all, I remember the rags.

There were shortages of just about everything at Khe Sanh. Northwestern Quang Tri Province was Indian Country, a jumble of hills that dissolved from horizon to horizon like a lumpy green blanket. Most supplies had to be flown in when the short strip was operational, or airdropped when it wasn't. Highway 9 up the Ba Long Valley was often cut or mined and closed to convoys. There were shortages of food, ammunition, beer, and toilet paper, but most of all I remember that I was always short of rags. I used a lot of rags; there never seemed to be enough of them. One of the first things I did when I was operating out of Khe Sanh and landed at another base was to appropriate a supply of rags and lock them in my tool box to safeguard them against the fucking thieves from other squadrons.

I needed rags to clean the blood from my medevac. In flight, the shrieking wind stream drove the ooze in rivulets across the deck plating. A UH-1E's flight deck was diamond-textured, painted with a gritty non-slip surface, and there were countless grooves where blood could collect. It trickled through the deck seams into the control linkages below. Sometimes I dampened a finger and rubbed it off the wiring tags to read color codes as I traced a cable. Blood dried quickly in the sun, but would linger forever as a semireconstituted paste during the monsoon and left ochre stains on my ass when I sat in it.

Wet or dry, it was tough to remove. After a mission, when the stretcher-bearers hauled their loads away, we'd shut down in the revetment and clean the cabin for the next sortie. It was important to remove the blood: I didn't want the VIPs we flew to think that VMO-3 kept a messy house.

I soaked the rags in water and rinsed the deck thoroughly, slopping the pink froth out onto Khe Sanh's red earth. I labored like a housewife with scrub brush and toothbrush to remove the gore. It was like scrubbing sandpaper, and I didn't know how badly I'd scraped my knuckles until I finished and washed the old blood from my hands to reveal the fresh.

The lock-down points molded into the deck were the hardest to clean. Ten or so secured the crews' seat legs to the aircraft. The blood pooled and clotted in them. I'd soften the clots with water from my canteen, then wrap a rag around my index finger to push out the gelled stringers of mucous and gore. If I was lucky, they would lift out whole, like limp strawberry donuts. When the ship was clean, I tossed the rags into a corner of the revetment with the stiffening memories of other missions. I had respect and did not toss them into the corner we pissed in. But I could never bring myself to rinse and reuse them, so when I ran out I did like the other crew-chiefs and stole.

There were many missions, and their stories have merged over the years, all tired and worn, indistinguishable, like the rag pile at Khe Sanh.

One stands above the rest. I remember it, not as if it were yesterday but as if it is happening today.

May 10, 1967 has been a quiet day. I am flying left seat (copilot position) with Major Powers, ferrying observers and chauffeuring VIPs to sightsee the Imperial tombs southwest of Hue. Powers is one of the old men of the outfit—must be over thirty. Neither friendly nor unfriendly, he is meticulous and professional. Powers acts with computer precision even in combat, perhaps especially in combat. No wasted effort, no second guesses. On time, on target, every time. Nothing seems to rattle him, and we call him "The Iceman" behind his back. You'd never say anything but "Sir" to his front.

The day is warm and clear. The greenhouse heat in the Huey slick and the constant drone of rotors lull me nearly to sleep. Powers and I have been up since 0430.

Ferrying VIPs is the dregs of flight duty. No fucking action at all. Up and down, fuel and refuel. Kissing colonels' and generals' asses and sweeping out the dirt they track into my cabin. The grunt variety of Birds are the worst, especially when a Star is around. They turn themselves into complete assholes, trying to impress the Star so that they can get one for themselves. Light birds, majors, captains—none of them are as pricky-shit as full Birds, except for brown-bar looeys, but looeys are frag-bait and are easily handled.

We've been flying Birds all morning, one more pricky-shit than most. It seems to infuriate the Bird that in flightsuits, crewchiefs and gunners are indistinguishable from officers.

"Marine!"

"Yes, Colonel, sir, (your lordship)."

"Where is your insignia of rank?"

"On my cover, sir."

"Why isn't your cover on your head, Marine?"

"Because I had my flight helmet on, sir. I have to wear it in flight. I just took it off."

"A Marine is always covered. Put your cover on, Marine."

"Aye-aye, sir."

Birds love it when you aye-aye them, like in boot camp. I salute, too. Birds love being saluted and always take longer to return it than most. They inspect you to make certain you're doing it correctly, afraid that if they miss something the Star will notice and give them a demerit. Half an inch above the right eyebrow, 45 degrees, upper arm parallel to the deck, "Keep that fucking pinky in place, Marine!" I don't mind saluting Birds in the zone. It can get them killed.

"Corporal, I want this aircraft *cleaned* before the General and I return, and start turning the *minute* you see me! The General's in a hurry, and he *doesn't* like to be kept waiting!"

Pricky-shit Bird. On the other hand, generals always smile as they leave the aircraft. They've made it to the top and can afford to smile, and it's good for morale and any press that might be hanging around. They salute casually and call me "son" with a friendly wink, as if they are sharing a secret joke about the Bird that they can enjoy and I can't. The general waves his salute, tosses off a "Good work, son," and leaves as the Bird goes schizo, sprinkling rose petals and sweeping for mines in his path.

Birds suck. VIPs suck. Flying left seat is fun—if only there were some action.

We've dropped our last passenger at a forward base, and I am enjoying a little stick time as Powers pretends to doze. I'm trying to make sense of a contour map to find our way back to Phu Bai. Following Phu Bai's TACAN or ADF radio beacons would be easier, but I need the map drill, and anyway, it takes longer.[1] As soon as we land, they'll just give us another fucking Bird.

I have us lost in no time, aimlessly following the rills west of Phu Bai. Or maybe northwest. Fucking contours all look alike. The map is all over the cockpit, and Powers is no help at all, until Landshark Bravo calls. The ground controller for Phu Bai's Tactical Area of Responsibility has orders.

"Scarface Four-One, this is Landshark Bravo."

Powers responds, "Landshark, this is Scarface Four-One. Go."

"Scarface, proceed directly to Channel Fifty-seven. We have cleared Sav-a-Planes in your path. Report immediately to Money-Six on landing."

"We copy, Landshark, proceeding Channel Fifty-seven, direct."

The channel is Khe Sanh's TACAN identifier, and Powers has been ordered to report immediately to the commanding officer upon arrival. Another goddamned Bird. It has to be something hot, not just another lousy observation hop. Sav-a-Planes seldom cease fire to let aircraft pass. Landshark reports that a second Scarface slick has been ordered north and will join us on the deck at Khe Sanh. Powers takes the controls and pegs the engine to red line as we vector north, wondering what's up.

I've just finished a week at Khe Sanh, flying left seat with Cap'n Brucie. Take a red Jaguar, beat it into a tank, and lose none of the style—that's Cap'n Brucie. He's an easygoing tank—a big-man-on-campus, football-hero type who's lost his temper only once, when he ground Captain

---

1. Tactical Air Navigation (system) and the Automatic Direction Finder were two means of locating your position relative to a broadcast radio signal. TACAN indicated direction and distance, ADF direction only.

Danko's face into the concrete deck of the officers' club for insulting our CO's good name. Brucie worships Major Townsend, the same as all of us who followed him In Country as the first commanding officer of VMO-3.

Danko was an FNG, but couldn't take the bullshit that is standard FNG-ration and stood at the bar bad-mouthing the CO's methods and means of running the outfit. Old Blankenshit was picking up a few bucks by bartending in the O' club and told us about it when he came off duty. Brucie was drunk and decided that erasing a few of Danko's features was the best way to teach him manners. Brucie stopped almost immediately— well, pretty soon—after Danko managed to shriek an intelligible "I'm sorry!" through his split lips. It was hard to understand him, according to Blankenshit, what with Brucie yelling and the other officers of VMO-3 cheering, so he probably didn't hear Danko the first or third or maybe even the fourth time. Brucie apologized as soon as he sobered up. That's the kind of guy he is. We all forgave him, and Danko did develop manners that are nearly permanent, at least when Brucie is around.

We were flying mostly VIP hops, then the hill fights near the combat base heated up, and we were commandeered into hauling medevacs off of Hill 881, a few klicks west of Khe Sanh's single airstrip. It was tricky with a reduced crew. Brucie flew alone at the controls, with me in the rear to take on the wounded. The hills were too steep to land on, so he'd nose us into the slope, anchor the ship to the hillside by the tips of our skids, and fly our tail while I leaned out the cabin door and hoisted up the medevacs. It was a short hop to Charlie Med, and all I had time to do was to keep the grunts from falling out while Brucie darted away under the cover of fixed-wing. We flew back and forth while the fighters worked the hills and valleys with napalm and aerial cannon.

A spit-shined slick from VMO-2 (call sign Deadlock) was working the hill as well. It was the shiniest bird and the worst duty a Huey crew could draw in Vietnam, for its prize passenger was Lieutenant General Lewis Walt, Commanding General, III Marine Amphibious Force. Walt was a Marine legend before Vietnam, and a legendary hard-ass during it: spit and polish and Olde Corps all the way, so the Deadlock crews slaved like oarsmen chained to the benches of a trireme, pulling to the taskmaster's drumbeat to avoid Walt's ire. Walt demanded his ships spotless and his crews instantaneously obedient. He'd come to Khe Sanh to check out the hill fights and sent his personal ship to retrieve a load of crispy critters for interrogation. They'd been taken by the grunts in the wake of a napalm run and had the bad taste to ooze onto the Deadlock's pristine flight deck. The hapless crewchief swabbed the critters' goo even as the doctors processed them through triage. Their wounds proved too severe for the facilities at Charlie Med and required the attention of the burn unit at Da Nang, but deep-fried gooks were too offensive to share a cabin with Walt. Brucie and I were ordered to ferry them instead.

I railed at Brucie as much as I dared, corporal to captain. "Captain, do you know what they're going to do to my ship? Do you know how hard it is to get that shit out?"

Brucie was resigned. "Look, Zack, don't make trouble. Just get the ship ready for stretchers."

I rumbled to myself as I stowed flight gear and cleared the deck for the stretchers. "Jesus Fucking Christ. Fucking goddamned zips, why the hell can't that goddamn Deadlock ship take them? They picked 'em up. Can't offend Walt. Oh no, too fucking good to smell a load of crispy fucking critters. Damn, leak shit all over my goddamn deck. Then who's got to clean it up? Me, goddamn it, me! I'm not going to get any help, fucking Brucie's going to boogie. Jesus Christ."

Three critters arrived, naked on their stretchers, about eighty pounds apiece of raw meat connected by octopi of plastic tubing to dextrose sacks held high by the orderlies. They were human, after a fashion, in that you could tell which end was up. After that, the similarities were few. The burns across their backs were leathery, strangely white or pale cream, not the dun-yellow of normal Vietnamese flesh. Here and there the flesh was charred, and as I pulled them aboard, it crumbled and bounced onto the deck like blackened bits of burned bacon. Their faces, chests, and forelegs were blistered, and clear fluid wept slowly like heavy sweat. The stench of burned hair rose from them in a thick cloud. Every now and then, I'd suck a fresh breath from outside the aircraft and mutter curses at Walt's spit-shined slick and its spit-shined crew, parked close by on the VIP pad. Once, I thought I saw that Deadlock motherfucker smile. Bastard. My ship could hold only two stretchers. We strapped one across the aft crew's seat and laid the second on the deck, arranging two of the critters on the lower stretcher, foot to head to save space. At least the canvas would absorb the fluids.

The critters seemed in shock; neither whimper nor moan escaped their seared lips. One tried to get up, and pulled against the IV of another. I pushed him gently onto the stretcher, and he grimaced—but made no sound. I didn't want to hurt him, but there seemed to be no place to touch him that was whole and wouldn't come off on my flight gloves. It was a new pair of gloves, too. Damn. I could see the trip to Da Nang was going to be bad shit, trying to keep them from tangling their limbs and pulling out one another's tubes. Why the hell didn't the grunts croak them in the zone?

I'd hung three sacks of dextrose and was getting ready for takeoff when a gravelly voice cut through the wind.

"Hey, Marine, that hook won't hold that sack. You have to hold it all the way to Da Nang. I want these prisoners delivered alive!"

It had to be a lifer. Just had to be. Old one, too. It was the way he said "Ma-reeeeene," with the "Ma" dripping sarcasm, and the lingering sneer that carried all the way out to the "ne." Anyone in the Crotch was "boot" to anyone else with more time, and lifers said it that way to first-enlistment boots to let us know how poorly they thought the appellation fit.

I had the last critter nearly secured and didn't bother looking up as I snapped, "Look, you fuckin' asshole, this fuckin' thing is called a plasma-bottle holder." I jerked my thumb at the dextrose sack swaying from its metal hook. "That means it was designed to hold fuckin' plasma bottles." I paused for effect. "And I don't give a big rat's ass if these motherfuckers live or die, so get off my goddamned case."

And then I looked up.

Of all the lifers it might have been, it had to be . . . Him. He filled the cabin door, one of the largest men I've ever seen. I couldn't see his face; his head towered over the aircraft's door frame. No matter, my eyes were riveted on his collars. Immaculate, rice-starched, razor-edged collars. Biggest goddamned collars In Country, with six of the largest silver stars I'd ever seen. They sparkled in Khe Sanh's morning sun, and for a moment all I could see was my life passing miserably in their mirrored gleam.

Stenciled crisply atop the eagle, globe, and anchor on his left breast pocket was the name. WALT. The Old Breed had descended from Marine Valhalla to visit a mortal in an oily flightsuit, squatting over a crispy critter in an aircraft encrusted with Khe Sanh mud and Khe Sanh blood. This was The Walt of the 'Canal and Peleliu. Silent Lew, the Marine's Marine, who had been fighting battles and winning wars before I was a wayward sperm.

Holy shit.

I had just called him an asshole. A fuckin' asshole, even. Holy Shit!

He stood a moment. Without moving. Not a twitch. A silent boulder or graven image showed more life. I prayed he hadn't heard me above the wind rush of the rotors. But then the chest lowered, the collars groaning under the weight of the stars. The massive neck followed. Big fucking neck! I swallowed, waiting for the head to enter, to look upon me and with a word crush the ant who'd insulted him. A Bird hovered at his shoulder, smirking malevolently. I could guess his thoughts: "Hooo boy, you're gonna get yours, smart shit." Captain Brucie shot me a nervous glance of pity, then buried himself in adjusting the altimeter. Jelly balls—I'd get no cover from that position. My asshole flared, then puckered, then sucked up and vanished. I wished I could have stuck my head inside and crawled in before it closed.

"Fini Phu Bai, hello Portsmouth Naval Prison," I thought, and waited for the axe to fall. Walt glared through the cabin door. An enormous, exquisitely starched utility cover shaded his head. It supported three more stars, and the radiant energy of so many suns shed their heat on my face. Eyes like slivers of steel fixed upon me, unsmiling, set deep in crinkled folds of muscle that had carried them and the man into Marine Corps history. They darted around the aircraft. Up to the plasma-bottle holder meekly holding its dextrose sack. Down to the crispy critter meekly oozing bodily fluids into his stretcher. And finally returning to me, meekly trying to will myself into invisibility. They settled on my chest, on the gold-lettered rectangle of brown leather that bore my name and the rank I used to be. I hoped he couldn't read the faded letters from the doorway and fought back an impulse to cover it with my gloved hand. I wished I had borrowed someone else's flightsuit, with someone else's name. I wished I were someone else, somewhere else. I wished I were dead.

Walt grunted, a deliberate and ominous rumble. I stared dully, then realized my mouth was open and shut it with a click. He grunted again, nodded brusquely at the bottle in its holder and the critter on the deck, then turned and marched to his slick, his entourage trailing. I watched the Deadlock crew flinch and spring to action as the Master approached. Faster, faster they scurried as he neared, but I took no satisfaction. I turned to the critters, thanking the gods that favor fools, and said, "Goddamn, you motherfuckers, you're going to live to Da Nang if it's the last thing I do."

And they did, too.

At 1130 hours Khe Sanh tower clears us straight into Charlie Med. Tango 2, crewed by Lieutenant Dave Stevens and Corporal Jack Amato is already there.

Stevens has captured the essence of a high-school senior, preserved unrefined through college and brought to the cockpit of a Huey, which he drives like a '58 Chevy jalopy, fuzzy dice and all. Piloting a Huey is the best thing that ever happened to Stevens. Once, after dropping off a VIP at Hue Citadel, Stevens took us at zero-altitude up the Perfume River, nosing the skids into the water to throw a cockscomb of spray along the tailcone. "Going through the car wash," he called it. Rounding a bend in the river, Stevens found a gook canoe dead in his path. The dink went round-eyed,

Stevens went, "Fuck!" and up-sticked, and all I could see was the river swallowing the yellow soles of the dink's feet as we grazed the canoe's gunwales.

Jack and I go back together about a year before Vietnam. We had a running competition through aviation school. Fundamentals, Mechanics, Jets, Helicopters, then Jet Helicopters; we traded places for first and second in each class and never missed a chance to stick it to whichever of us came in number two. Jack is tall and San Francisco cool, and I hung around him when we posted to the Huey training unit in Pendleton, hoping to pick up scraps while cruising for broads in Oceanside. We shared a cubicle—God, Jack has the raunchiest goddamned feet you ever inhaled. Gag a vulture. He is a good crewchief; he keeps his aircraft up and his guns clean, and he covers you in the zone. That's about the most you can expect from anyone.

Stevens is already in the CP; Powers tells me to stay close and goes to join him. Jack walks over from his aircraft.

"What's up, Jack?"

"There's an emergency extract north of here. Stevens and I were flying C & C out of Scat, and we got the word to shag ass. Some -46s already tried the extract. I hear they got their shit blown away."[2]

"How many -46s went in?"

"I don't know. More than one flight."

"What we got for AirCap?"

"Some of our gunships. Fixed-wing. I hear they're all taking hits. You are not going to believe what they want us to do," Jack warns.

I chew my thumbnail as he continues, then clean each fingernail with my teeth, spit out the dirt, and pull at my nose. It is what I do when I concentrate on something unpleasant.

"The CO wants one of our slicks to hover over the zone, throw out CS, and draw fire. He wants the other crew to land, wearing gas masks, and extract the grunts."

I stop chewing.

"You are fucking shitting me."

"I would not shit you. You are my favorite turd."

The sight transfixes me. An unarmed Huey, hovering at a hundred feet, a single pilot at the controls, absorbing hundreds of rounds of machine-gun fire, plummeting out of the sky while a stupid fucking crewchief throws stupid fucking tear and vomit gas into a hot zone; a second pilot and crew-

2. "C & C" was short for command and control. "Scat" was a forward artillery position.

chief getting their shit blown away, wearing gas masks, dragging puking wounded men to a second riddled aircraft as the first ship falls on them. Everybody dies.

"I don't believe you. No one is that stupid," I say, knowing that this is the Marines, and in all probability someone is that stupid.

A mob of grunts wearing Sparrowhawk patches approaches. They are heavily armed, and the grenades festooning their green flak jackets make them look like lethal Christmas trees.

"You the new medevac ship?" the squad leader asks.

"Uh, huh," I answer absently. I can't shake the image of one Huey crashing atop the second, and wonder which one I will be in.

The grunt looks inside my slick. It is spic and span from flying Birds all morning. The emergency M-60 is under the crew's seat, zipped in its protective bag. It is part of the slick's survival equipment, and we carry only four hundred rounds of 7.62-mm ball and tracer to defend the ship if shot down. An M-14 lay behind the -60, so rusted it has value only as a club. The pilots carry .38 Smith and Wessons, but neither Jack nor I have sidearms.[3] The Sea Knight -46s repulsed this morning carry .50-caliber MG. We have little more than our dicks in our hands, in comparison.

"You're going in with a fucking slick? They shot three -46s out of the zone this morning." The squad leader turns to one of his men. "Get these guys some more rounds for their -60s." Two men run off in the direction of the ammo dump. "You want some of these?" he asks, pointing to the frags hanging from his flak vest.

"No. Too dangerous to carry in the aircraft, but can you get us some grease guns?"

I am afraid of frags. I've seen the remains of a UH-34D after a grunt

---

3. Aircrew in nearly every helicopter squadron except VMO-3 carried .38 or .45 pistols for personal defense. Infrequently, we carried a .45 "grease gun" for close work in the zone. The squadron's staff confiscated sidearms from crew members ranked sergeant and below as "punishment," after several were stolen from parked aircraft during our first naive weeks In Country. (There was little or no theft within the squadron, but another outfit's untended equipment was fair game.) Judged too careless to be entrusted with pistols, we were ordered to carry M-14 rifles, although they were far too awkward to use inside the aircraft or while trying to carry wounded men. As a result, there were many times we forayed into a hot zone armed only with a combat knife. Crewmen petitioned for the sidearms to be returned throughout 1967 but did not regain the "right" until squadron leadership changed late in the year. Pilots and staff NCOs whose sidearms were similarly liberated were quietly reissued new weapons. Rank hath its privileges. Reflecting, it is difficult to say who was less mature, those who confiscated the weapons of men at arms, or we who bowed to a discipline that jeopardized our survival. Sometimes I wonder just how brain dead we were.

dropped a frag as the aircraft landed. The ship was Swiss cheese. The recovery team puked when they saw the crewchief. He'd loaded his pistol rounds in his cartridge belt with the projectile up; it was easier to slide bullets down to get them out, since the flak vest hung waist length and interfered with the ammo belt. The grenade detonated some of the rounds, firing them upward inside his flak vest, and they ricocheted between the front and back plates through his body. No frags, no way. We can use greasers, however. The .45-caliber weapons are excellent at close range. If you suspend one from a sling around your neck, you can spread slugs with one hand as you carry a body through a zone.

"Get over to the armory. Get two greasers and as many magazines as you can carry," the squad leader orders.

Powers and Stevens emerge from the CP, running. They close the distance to the aircraft quickly.

"Corporal, which bird is hotter?" Powers shouts.

"Mine is, Major, 101.2 percent."[4]

Jack's ship is around 99.5, and I shoot him a smug look that says, "Eat shit, second place."

The pilots toss their knee boards and maps onto the dash of my aircraft.

"Transfer the gun, ammo, and first-aid kit from VT-2 into here. Dump anything we don't need for an emergency extract."

We've already stripped our ships; Jack runs to his aircraft to retrieve his weapon as the grunts return with four cans of ammo.

"What about the gas drop?" I ask as we arrange Jack's gear.

"They scratched it when they saw we were short crews. Thank God. We'd be dead meat," Stevens answers.

The pilots start the rotors turning. Powers barks, "Move it, crewchiefs. We're lifting."

"Sir, we've got grease guns coming."

"No time. Lifting."

Jack and I wrestle into gunner's belts amidst the jumble of loose equipment as we clear the taxiway. In the distance, I see the grunts laden with guns and magazines. Those lovely greasers. All we'll have in the zone are our knives, and dicks, in our hands.

Stevens calls map coordinates X-ray Delta 756530, and Powers sets course for the zone, but as we clear the hills north of the base, our destina-

---

4. The power rating of a Huey's Lycoming T53L-11 engine was measured against a nominal of 100 percent. Tango 12 had a hot engine and could develop 1.2 percent more power than considered normal for a Huey.

tion is evident: we can see the smoke marking our target in the distance. Powers switches to UHF to contact the other aircraft in the area while Stevens briefs us.

Just past midnight, a seven-man reconnaissance team stumbled into the middle of a reinforced North Vietnamese company. The enemy was well entrenched in a complex of bunkers and spider holes that riddle the hill. All approaches are covered by heavy bunkers on the surrounding ridges. The team was quickly cut to pieces, and four of their number died during the long night. Too badly wounded, and too few to fight their way out, they formed a tiny circle in the elephant grass, like settlers surrounded by Indians. The grass provided little concealment and was soon beaten flat in the fighting. From midnight until dawn, the team held out. They repelled the NVA's assaults; at times the enemy crawled to within ten meters of their position to throw grenades. In the early morning hours, the Marines ran out of ammunition, and the fighting went hand to hand. There was nothing anyone could do until dawn. Nothing.

Two Scarface gunships scrambled from Phu Bai during the night: Victor Tango 9 piloted by Major Bob Diaz and Tango 15 by Captain Bill Geiger. They, and a passing Deadlock gun that had overheard the call for assistance, arrived over the zone at first light. The first of many sections of fixed-wing rendezvoused with the gunships. Three wounded Marines huddled behind a low barricade they'd built with the bodies of their enemy, and the bodies of their own.

The fixed-wing rolled in, directed by the wounded radio operator against bunkers less than fifty meters from his position, and against adjacent hills that swept the zone with crossfire. When the jets cleared, he directed the gunships to attack infiltrators coming through the grass with grenades. The gunships rocked in, taking turns blasting the spider holes and bunkers surrounding the team. The gunships made several passes, firing to within feet of the Marines, and stopped the NVA assaults.

Such has been the pattern for most of the morning. The grunt operator maintained radio contact, calling the fixed-wing in against bunkers and spider holes and coordinating the gunship attacks whenever the NVA approached. As the fighters ran ammo-minus and low on fuel, fresh sections arrived from Da Nang, Chu Lai, and the off-shore carrier. The gunships also established a pattern of leaving one aircraft on station while the other two dashed for the fuel pits and ammo dump at Khe Sanh, seven miles away. When the lone covering aircraft itself ran low on ammo, it flew dummy passes, firing a few rounds every second or third run to keep heads down until the replenished ships returned.

The fighting wore on throughout the morning. The enemy is so close to the team that it's been impossible to silence the holes and bunkers from the air, even though the constant assault of rockets and frag has blasted the cover from the hill, and enemy positions are plainly visible. In Khe Sanh, three armed CH-46 medevacs waited on the mat as the gunships fought to bring the zone under control.

Three hours past dawn the first of the -46s attempted to extract the team. Point-blank fire from the bunkers and spider holes riddled the aircraft and holed the hydraulic system as the ship flared to land. It barely made it into the welcoming arms of the crash crew at Khe Sanh. The fighters and gunships hammered the enemy positions, and a second -46 attempted to land, but its pilot was killed in the assault. The third medevac was beaten off as well, the gunner and crewchief wounded as they fought behind their .50-calibers. There are no more rescue ships to feed into the carnage, and after seven hours of air-to-ground combat, the team is no closer to freedom.

Jack and I sort our equipment as Stevens speaks. Our M-60s, locked and loaded, rest on their bipods behind the pilots' seats, leaving room to drag in the medevacs. There are no hard points to mount them in the slick, so we'll have to fire John Wayne style, from the hip. The thought amuses me. In training, they taught us that John Wayne-ing was only for the movies. We arrange the boxes of 7.62 for quick reload and throw the empty cans into the airstream. Months before, that felt like littering, but dumping trash at several hundred feet is the normal thing to do. I never even look down anymore. We break out the spare barrels for the -60s in case one jams, then open the first-aid kits and survival packs and spread out the bandages. I grope beneath the crew's seat to find my rifle, as rusted and clogged as I expect, snap in a magazine in hope of getting off at least one shot, then place it and my Ka-Bar on the cluttered seat as well. I reserve a small spot for my ass, but it is more important to keep the necessities organized; the back cabin will clutter quickly when the fight begins, and no one will be sitting.

Our busywork done, we wait as Powers joins the air traffic over the zone. I wish that we would get on with it. After too much time, doubts begin to rise.

Will I freeze? What if I let them down? The fact that I never have before means little. Will I kill? As aircrew, you can become inured to slaughter meted out at 600 rounds per minute from a thousand feet, but we'll be landing on this mission, and wet work is something the air wing rarely experiences. I do not envy the grunts their closer familiarity with death. During every assault, especially when we leave the nominal safety of the aircraft, the doubts plague me. Can I pull the trigger when it's two feet from a face? Can I use my knife? I remember the knife-fighting instructor's exhortations: "Stick 'em, you animals. Twist it. Don't pull it out the same hole it went in." And I'd scream, "Arugah Kill!" and try to insert a tent-peg into my opponent's belly button. Will the programming serve at the moment of need? Will I be a Marine?

Powers contacts the team for a situation report. The radioman's voice is amazingly clear, and stands out amidst staccato bursts of MG and small arms that bring the zone into the tiny space of the cabin.

"Breaker One-Four, Breaker One-Four, this is Scarface Four-One. Do you copy?"

"Copy five-by-five, Scarface."

"Breaker, say your sit-rep."

"Scarface, we are three Whiskey India Alphas. Request immediate extract. We have fire in the zone, Scarface. Say again, the zone is burning."

The broadcast, in the clear, makes it obvious that we aren't keeping secrets from the enemy. Fixed-wing, dropping napalm on bunkers in a ravine beneath the hill, started the fires that are slowly burning up the slopes. The crest is shrouded in a roiling gray cloud as the grass steams, then flashes into flame. Funeral pyres fed by the jellied gasoline mark enemy positions incinerated on the lower slopes, but several are too close to the team's position to risk further drops. The team shifts in and out of view as the wind swirls the smoke across the ridge. Whites and grays, blacks and reds blend into a landscape ripped from hell. A circlet of flame drives the cloud higher, consuming the hill as it advances, but many NVA, safe underground, have survived the flames and continue firing on the zone.

Powers studies the approaches. There are few options. The team is encircled, and the -46s have drawn heavy fire from all directions. The safest avenue appears to be out of the north, but the wind drives northeast, then swirls to smear the cloud rising from the napalm fires across nearly 120 degrees, effectively cutting off any approach from that sector. Once in the zone, the smoke will make it difficult to find the team no matter what direction we choose, but we have to find the ridge first. Terrain and visibility dictate an assault from the south or southwest despite the obvious dangers.

Three spider holes controlling the crest seem impossible to take out. The holes are located in a narrow ravine, tapered upward toward the ridge, and are only fifteen meters from the team. They've proven deadly to the -46s, and we will have to fly across them as well. Two ridges guard the southern approach, and log bunkers in the tree lines that crown them deliver covering fire across the valley. Fixed-wing have flown against the bunkers all morning, some taking hits from 12.7-mm–gun emplacements. We watch a section of Skyhawks roar in low, their wings waggling and stars of 20–mike-mike cannon twinking from the leading edges. The shells explode in white bursts on the bunker's logs. Snake-eyes shred the timber on the ridge lines, but still the bunkers fight back. It seems so futile, the might of American firepower cheated by logs and earth.

As Powers circles, I know the bunkers and holes have to be silenced or we'll suffer the same fate as the -46s. We mark time in the air, flying a tight pattern clear of the jets' pullout, and wait for the gunships to call us in. From the safety of altitude, we watch the flames close on the zone, and then hear the voice rising from the burning hill.

The tale runs in my mind, like scenes from a play. The Center becomes the zone; the memory is tangible, not like Walt and the critters. I sense the familiar space of the slick. The smell of old sweat and fresh hydraulic fluid. WD-40 coating the MG. Cordite and smoke filtering through the cabin vents. The cloud penetrates the aircraft, the Center. I am inside the story, both as me and as me watching me, and I am strangely comfortable there. It is my element. But I am frightened.

I hear the voice, the grunt operator on Fox-Mike. His despair haunts me. Twelve hours within that ring of dead. Wounded, he controlled the fixed-wing and the gunships in their futile attacks. The fear in his voice rises with the flames and the choking smoke. It spills into the aircraft, and the Center, and floods my mind with its anguish. I fought it back, then. I was crewchief! I was in control. Fear had no place in the slick.

But this is not "then." The defenses that gave me strength are gone. Training, the camaraderie, the demanding responsibility of the slick and the zone are all gone. My mind parses the moments, and I cannot stop it from feeding on the fear. Remembering is too much to bear.

Tom speaks gently. He is another voice. Closer, but only slightly more real than the voice.

"It's all right. You can say it. You're not there anymore. Tell me what he said."

"Oh God, I can't," I answer. "Oh God, do you know what he sounded like?"

The tears begin, but give no solace in the shedding. The pain of them sears my flesh. I try to speak, but bile washes my throat in acid. My head throbs, and I shield my aching eyes with my hands, but the memory is clearer in darkness. My breath comes in hiccups, my lungs scream for air.

Tom takes my hand, an anchor penetrating the memory.

I try to shut the voice out. *Go away. Go away. Go away.*

I stutter out the horrible words. "Scar . . . Scar . . . Scar. . . ."

Tom is there. "You can say them. You're not there anymore."

I can see the words, branded across the vision of myself in the slick, but my tongue refuses to comply. It fights against the words that make the zone real, here in the silence of the Center.

"Scarface, Scarface. . . ." *I can't say the words!*

"You can do it. You have to do it." Tom's voice is a lifeline. I grasp it— it is the only way I can escape.

Then the words tumble, tumble, tumble. I cannot stop them "Scarface, Scarface, this is . . . this is Breaker. We're burning. You gotta get us out. Scarface, you gotta get us out. We're burning up."

Static crackles on the FM. The choking coughs of the team. Small arms. The zone is in the cabin with us. I see Jack's face, ashen—the voice possesses him as well—I see my fear mirrored in his face. He hears the voice, here in the cabin, in the Center. The voice is a plea rising from the flames, begging. It never goes away. It never goes away!

"Oh God, do you know what it was like to hear them burn?"

There is no time left, no chance that the bunkers and holes can be silenced. There is only the voice from the zone.

Powers thumbs the ICS, speaking to the crew. "We're going in."

His voice is firm as he keys the link to the team. "Breaker, this is Scarface. We are coming in *now*, Breaker. We are going to get you out *now*. You have to hold, Breaker. We are coming for you *now*."

The gunships form around us. The Deadlock will lead the assault, attacking the ridge bunkers closest to the team. The Scarface guns trail aft, fanned out to port and starboard to cover our approach and landing. One will concentrate on the holes in the ravine that troubled the -46s. The other will cover the tree line and ridge. Fixed-wing queue to run daisy chains against the hills that flank our approach. We fall into pattern and start our long run to earth.

A half mile out, the first flight of fixed-wing attacks in tandem, pouring rockets and cannon into the bunkers on our flanks. I watch in fascination as explosive heads shatter trees and walk across the log roofs. Ricochets flash wildly from the faces of boulders. Geysers of dirt erupt and splash back into the scrub, marking the path of the 20s. The bunkers fight back, 12.7 and small arms winking from slitted ports. Phantoms lead, and Skyhawks follow, sleek and swift. A pall of dust rises from the hills and grows thicker as each flight chains through their assault. At 120 knots, our section of Hueys seems to lumber toward the zone, barely moving compared with the fixed-wings' elegant speed.

At a third of a mile, the Deadlock gun opens up with rockets, then MG, concentrating on the bunkers fighting on the burned ridge. The Scarface flight thunders in behind. Close to the zone, the Deadlock breaks to port. The crewchief pours 7.62 into a bunker as the gunship clears the zone. The Deadlock begins its go-round, to get into position to cover our landing. Behind us, the Scarface guns pick targets in the trees and ravine. The pilots report ground fire picking up, but so far nothing has come our way, and Powers holds his course. The hill and its pall of smoke are large in the windshield. I can no longer see the team. Thick rolls of smoke engulf them and jeopardize the approach, but we are committed. Stevens's hand rests on the radios to switch the pilot's to Fox-Mike to talk to the grunt operator or gunships, or to the fixed-wings' frequency on UHF. He's unsnapped his .38 and lodged it between the console and copilot's seat. As we close, Jack and I watch the airspeed indicator. A Huey's cabin doors can rip themselves from the aircraft if opened while exceeding 90 knots. We wait for Powers to break into a flare and drop airspeed before popping the doors for the extract.

No one speaks. All around us, the fixed-wing and gunships are under fire, but we seem to be in the calm eye of the storm. I can hear the distant pound of 20–mike-mike, the thunder and crash of the snakes and frags. Closer in, the gunships' -60s chatter, and the *whoosh-crump* of 2.75 rockets reverberates from the hill. Why aren't they shooting at us? Three -46s bought it. At least one didn't make it as far as we are. I wonder if they are going to let us walk in. Maybe they've had enough. What the hell is going on? Jack stares out his door. Stevens sits frozen at the radios, his fingers ready to respond to Powers's need for a channel. Powers's visor is down, and the tightness in his jaw betrays nothing but concentration as he holds the aircraft on course.

The zone fills our view, and Powers suddenly back-sticks to pop a landing flare. The rotors smash at the air, *whop whop whop whop,* and we high-G into the hill. Forty pounds of flak vest, multiplied by the force of the flare, pull me down. I duck-walk forward and squat like an Indian to pop the door. The Gs eat at my calves, the flak vest at my kidneys. Christ, I need to piss. I shoot a look forward, but the grunts are out of sight. Powers must have their fix. Spider holes pass beneath our skids, and Geiger's tracers chew the earth around them. The holes are still sealed, no enemy in sight. My mind races, gathering all the data, sorting it, discarding nonessentials, keeping only what it needs to live. Nothing exists but the zone. *We're going to make it, we're fucking going to make it.* I risk a glance aft, down the line of the tailcone. Fireflies of tracers from the gunships race along it, toward me, close enough to touch, flying past then onward into the zone, holding us in a cone of fire. Still nothing from the hill. Powers pulls the flare tighter, coming in hot. *We're going to fucking make it!*

The hill erupts; a living wave of sound leaps from its blasted slopes, smashes the aircraft, consumes us like some living, hungry thing.

Powers to the gunships: "Taking fire. Heavy fire in the zone." The deep-throated chug of large caliber. Grenade. Explosions ahead in our flight path, explosions aft. RPG somewhere . . . somewhere aft. I can't see, I can't see! Geiger and Diaz rocking in from four and eight o'clock. "Scar-face One's in hot. . . . Two's in hot." HEAT bursting on the ridge. Fixed-wing against the bunkers. "Condor Zero-Five-Seven is in hot. 20-mike! 20-mike!" Where's the fucking Deadlock? Go-round, go-round. "Deadlock's in hot." Tracers from above, consumed by the burning hill. Tracers rise from the winking slits of bunkers and race like angry hornets toward the gunships and fighters. "Five-Seven: I'm Hit! I'm Hit!" Dirty smoke trailing from the Phantom. "Condor Five-Eight to Five-Seven, got you covered on your six. Break on my mark. . . . Mark, Mark, Mark!" A flash of flame and black smoke ten yards to port, low, the concussive blast of 2.75 HEAT rises like a black dome beneath us, racing to engulf us, driving the dust and smoke before it with a killing wind. Spending itself, lifting the aircraft softly, softly, with an unseen hand. Small-caliber MG. Someone: "AK! AK at ten o'clock!" *Brrrrrrppppp, brrrrppp.* . . . The *pop . . . pop pop* of individual weapons. A flash of white and flame a yard beneath us, a pale cylinder of HEAT streaks beneath the slick, gray smoke and flame lick up at our skids. *Jesus Fucking Christ!* Geiger and Diaz pull the cone in tighter around us. Earth explodes, lifting the aircraft once again on a wave of noise. Had the hill voice, it would scream.

I stare into the zone. Nothing, nothing, could live through that. Nothing. I stare, and my eyes fill with green and smoke and flame, and yet I see nothing, register . . . nothing. A blur, and becoming more blurred. Indistinct. Gray smoke, not choking, but it should be. Should be . . . something. Not breathing, not anything. I should feel . . . something. It is cold, black. Sightlessness that is not darkness. My eyes are open, accepting light without form or meaning. Two dimensional. I see . . . me. What is . . . what is happening? The firing is distant, the pound and surge of the rotors fading. Finally my voice, my own voice, telling me I am about to die.

This isn't possible. It only happens in the movies—yeah, the movies. But where's the music? Where the fuck is the Music? The 1812 Overture is nice. I heard it once. Zips jammed our freq near Hue Citadel.

The mists close and grow colder, and I struggle within them.

This can't be happening to me. *Not to me!* I rage against the fear. *It isn't fair!* I scream, but there is no sound. Me looking at me looking at me, mouth open. The scream gives no sound! *Oh God, it isn't fair! I'm nineteen fucking years old and I've never been anybody! I haven't done anything! I'm going to die on this worthless fucking hill, and it isn't fucking fair! I don't care about this fucking hill! It isn't fair!*

Not even an echo. I am alone, denied the sound of my own fear.

Finally, a voice. *Please, God, please. Let me live. It isn't fair. Let me live. I'm . . . I'm. . . . Bless me, Father, for I have sinned. I'm. . . .*

Bless me, Father, for I have sinned. I lied eight times. I. . . .

Bless me, Father, for I have sinned. . . .

The prayer rolls over and over in my mind, like a needle unable to escape its groove. The nuns told me, if you confess and get an Act of Contrition out, you'll go to heaven. I have to get it out, but I can't remember the words. Oh God, Oh God, I'm going to hell because I can't remember the words.

I feel the tears coming. *Oh God, they'll see me. I'm a big boy. Don't make me cry. Big boys don't cry. Oh God, let me live. If you let me live I'll . . . I'll . . . I'll do. . . .*

Laughter. It is myself I laugh at through the mists, and the echo is briefly warm. *Fucking liar!* I hiss, mocking. *Asshole! You are a shit, you've been a shit. You're gonna die. Die like a man, you shit.* I laugh again, and the sound is warm until . . .

The mists close. So much sorrow. So much that could have been, but will never be. Silent darkness, spiraling down, down in tight circles. The guns are a distant echo, almost gone, their chattering overwhelmed by waves of self-pity.

I see my death. It is there, and there, and there. It is many places, not just one. Wherever I look, without escape.

Such empty yearning. To live—another minute, another second—to live and hold off the end of all things. I know I will not. The end is here, in the zone.

The mists are suddenly calm, no longer cold. There should be shock, but there is none. Confusion! The fear fades to . . . what is it? It is not fear. I do not know. There is a . . . quietness. Not silence. In silence there is only loneliness, and I am not alone. It comforts me. I am as I should be. Death is there. Over there. One place, only. It is just another thing. There is an order, it is just another thing. What will it be like? No matter. Death waits, unsmiling, unthreatening, neither warm nor cold, patient.

I am not afraid.

The zone floods in. Rotors beating, beating, beating. A wave of roaring dizziness sheds itself quickly. I am back in control. Jack's door is already popped, and Powers is screaming at me to open mine. How long was I out? I glance forward. Powers has eased the flare, and the hill fills the nose bubble. I grab the latch and heave the door back. The zone is lost in clouds of swirling smoke, whipped and roiled by our rotors. An acrid gale roars through the aircraft, choking, bearing within itself the burning terror of the hill. The maelstrom screams with the explosion of the gunships' HEAT, the return fire of small arms, and the pound of aerial cannon against the land. We are low, only yards from the hilltop. Visibility is fifteen, twenty feet. Powers taxies over the crown of flame, deeper into the smoke. The hill rises slowly to meet us, and the flame of burning grass brushes our skids.

We are in. Powers taxies low, inches off the deck, feeling his way forward, then hovers in the swirling eye of the storm.

There are no grunts. God, God, God. *Where are the grunts?*

All eyes search frantically. Nothing. We lost visual contact during the approach. Powers tries the radio but can't raise the team. Five seconds pass. Ten. Powers barks to the gunship flight, "We've lost the grunts, give us a direction." But the Deadlock and Scarfaces, maneuvering at high speed against the holes and bunkers, have lost the team in the shifting smoke as well. Fifteen seconds. *Get out, get out.* I look at the pilot and see doubt cloud his face. The Iceman. We've come so close. They have to be in the smoke. Port? Starboard? We'd never find them. We won't live long enough to find them. The firing seems to grow, or perhaps only my fear of it. We've been lucky so far. I can't believe we've survived, but we have to get out. I feel the sour taste of panic rise within me, but I shove it down, willing myself into control, willing my eyes to penetrate the smoke. I lost it, and fate did not punish me, but I have it back now and will keep it. Eighteen seconds, twenty. Useless—we'll never find them in the smoke. Wounded, unable to move, they'll never reach us, although I know they hear, even feel, our rotor wash. So terribly beyond reach.

I look aft; the AirCap's tracers spear the smoke like tiny comets, chewing the dirt around the aircraft. Another yard and they'll take off the fucking tail rotor. The curtain of fire holds, and the punished earth spews itself skyward. Just like the movies. *Lift . . . lift. . . .* I stare at Powers and will the thought into his head. *Lift.* Stevens and Jack—their eyes are on the pilot as well. *Lift!*

The aircraft moves, tentatively, as if in a dream. We nose over, then taxi more quickly through the smoke.

Lift. The word is a prayer, rising with the funeral pyre of the zone.

The thrust of our rotors smashes back the circling flames, but they grow quickly under the shadow of the aircraft as we pass. Crimson tongues lick the nose, then the open door. The cabin fills with smoke, and we carry the captured cloud with us as Powers pushes the cyclic forward.

Lift.

The hill falls away, but we follow it down, down, putting it between the aircraft and the NVA guns. We punch through the smoke, still following the burned and blackened slope, gaining speed, leaving the cry of the zone in our wake. I watch the collective come up, and we rise, Gs building blessedly as the rotors pull us to safety in a clean, crystal sky. We claw upward, shedding the flame and the smoke and the fear. Gray tendrils trail from the cabin, clinging to us like ghost fingers trying to pull us back. Powers dodges and twists upward, faster and faster as the rushing wind tears the last smoke from the flight deck. We surge into the clean, clear sky.

Heart pounding, chest heaving as I gulp in enormous drafts of the sweet, sweet air. *God, I'm alive! The sky is clear. Clean! Blue! So blue! Crystal blue, and the smell is so . . . alive.* Every sense is alert. Ears hear as they've never heard before. Sounds feel as they've never felt before. *Color! So much color. Taste, smell, color!* Limbs, chest, head are light and filled with power. Every nerve tingling, bursting, burning with life. Mind racing at light speed, missing nothing, experiencing everything. The aircraft, the crew—electric with the burning energy of life.

Then tired. Oh, so tired. But tired bursting with life!

Dizzy now, really dizzy. Oh, Christ! *Thank you, God. Thank you. Thank you, God. Thank you. Have to calm down. Calm. Alive! Calm, get calm. Control. Away from the zone and get control.*

Powers slows our ascent and banks slowly to port. The vitality of the climb seems to ebb as the craft loses speed. The hill comes into view once again. Burning. The winds have shifted, and we can see the team.

Nothing has changed. Nothing. Powers raises his visor and gazes without speaking at the Marines' tiny circle, then turns toward us. Stevens, Jack, and I meet the pilot's eyes atop the jumble of weapons in the cabin.

"I guess we have do it again."

Stevens nods. Jack nods. I nod, and we begin to clear the flight deck for a new approach. The Code is the Code: Marines take care of their own.

The helicopters and fixed-wing are scattered across square miles of sky when Powers reestablishes radio contact. Our missed approach and uncoordinated escape wrecked our formations, and it takes precious minutes to assemble for another assault. As the gunships and fighters check in, we know this will be the last time. The covering force is low on fuel and has ammo to cover one more strike. There is no time to refuel or rearm—no time for anything. On the hill, yellow flame threatens the wall of the dead.

The second attack mirrors the first, except, except . . . the terror is gone. I feel no emotion save purpose as we close on the zone. If purpose and determination are emotions, if they can be distilled, made into chilling fluid, and injected into a body to sustain life when all that usually sustains life is lost, then purpose and determination are all that fill me as we close on the hill. We will get them out. We will. Marines take care of their own.

The grunts come into view on the aircraft's starboard side. Powers holds our course and watches the hill. Jack fixes his eyes upon them from the cabin. Stevens peers over the instrument console. From the far side of the cabin, I peek through the pilot's nose bubble. There will be no missed approach.

The NVA holds their fire until we flare for landing, just as before. Each man holds his position, all eyes on the team. The gunships drive in their last precious HEATs, then break off to cover our escape with MG. Amidships, the fixed-wing pound the unyielding bunkers. The hill rushes up at us, and Powers barks an order to pop the doors.

I hurl the port door aft, then move to kneel behind Jack, struggling with the starboard door's latch. It has always been sticky and has picked this moment to fail. I shove past him and heave.

"Fuckinggoddamnedthing,   fuckinggoddamnedthing,   fuckinggoddamnedthing. . . ."

I can see the grunts outside—we have to get it open. It would add twenty feet to cross the aircraft to use the port door. Too far.

"Fuckinggoddamnedthing. . . ."

My tool box lies in the jumble of gear forward in the cabin, and I pull it open. Rags, rags, fucking rags. I dump it and seize a large screwdriver from the heap. Jack strains at the door; it slides a fraction in its tracks, just enough for me to jab the blade in up to its hilt. I heave once.

"FUCKINGGODDAMNEDTHING!"

The latch rips out, popping rivets and shredding metal, and the door hurls aft, pushed by the wind stream. I fly into Jack and end up on my ass.

Jack kicks at me and screams above the wind, "Jesus Fucking Christ, get the fuck offa me, Zack."

I hurl the screwdriver at the box and kick it forward, then stare into the smoking zone. Powers taxies starboard, crabbing sideways across the crushed grass. The burned opening of a spider hole passes beneath our skids. It is empty, and I hope it will stay that way. We fly quickly toward the grunts.

The wall of dead is two high, black and white and brown and yellow. Marines as well as North Vietnamese are stacked heel to heel, head to toe, head to head, to shield the living against the firing and shrapnel that graze the ridge. A smoke-blackened face peeks over the wall, and an arm clad in green waves weakly, but the wall holds me transfixed and I do not wave back. The barrier has fulfilled its purpose. Their uniforms are punctured by bullets, shredded by shrapnel, but the pink flesh and the black flesh and the yellow flesh can no longer bleed. A young grunt lies atop it. Our rotor wash tears at his clothes. Embers lash his face, but he does not flinch against the storm. He is blond, and his hair ripples like a field of grain driven by the wind. His eyes are closed, and the blondness of his lashes stands out crisply on a tanned and muscular face, a young face, unwasted. Probably an FNG, who will never become a salt. His lips are parted, as if breathing in a quiet sleep. I can see no wound, but beneath him lies another, whose crusted neck and matted hair tell me what the quiet lips cannot. That face is turned from me, and thankfully, I cannot know him. I do not carry his image with me, as I do the young Marine, and I am glad for it.

Three men live behind the wall. Jack waves frantically as we fly the final feet, and we scream through the swirling smoke, "Get up! Get up!" hoping they can make the aircraft on their own. Two men, one with trousers soaked in blood, and another bearing chest wounds struggle to raise a third Marine, conscious but crippled. Most of his uniform is wet with blood, his blouse and trousers torn by shrapnel. They'll never make the ship by themselves. Our skids ground into the beaten grass, scarcely a foot from the wall.

Jack jumps from the aircraft, grabs the crippled Marine, and propels him to me. I catch him under the arms, his face close to mine. Jack heaves, I pull, and the wounded man shoots into the cabin. I drag him heavily across the deck. His lips are tight in pain, his breath hot and dank in my face, but not a sound escapes him. I roll him forward and push him against the guns to make room for the others.

Jack is struggling with the second man. I leave the aircraft, grab the grunt's arm, and together we heave him inside. He rolls to the rear of the flight deck and lies like a thing dead.

We turn to the last Marine, stumbling over the bodies. It is the radio operator, shot in the chest and legs. His eyeglasses lie crookedly across the bridge of his nose, a star fracture in both lenses. *Christ, the dude is still wearing his fucking glasses. Fucking shit!*

I yell at him, screaming over the roar of the rotors, "Any more alive?"

His swollen lips form the word "No," but the sound is swept away in the wind. He crawls onto the flight deck and sits heavily behind the pilot's seat. Jack hauls himself in as I turn one last time to scan the zone for life. The dead lie at my feet, their circle now empty. I look a last time at the blond grunt, for some sign of life, for anything. His young face shines as if asleep, not dead in that terrible place. Powers shouts something, and I pull myself into the cabin. We are ready to lift.

The aircraft is a tangle of limbs and weapons. Had more than three lived, we'd have to stack them like the dead wall beyond our skids. The radioman, obeying his Marine training, still clutches his M-16—empty, caked with dirt. The first man lies face down, jammed against my M-60 and the instrument console. I need to check his wounds, but there isn't enough room to roll him. I can't move aft without grinding my boots into the second grunt. I heave him, my head down, trying to rotate him in place when Powers reaches back and raps sharply on my helmet, startling me. I look up, and he taps his headset. He wants to talk. *Oh Jesus Christ, what does he want to talk about here?* I've unplugged from the comm system; the spiral cord is snagged under the second man behind me. I turn, and lift him, and it flies free, then dances away from my fingers as I try to grasp it. *Jesus Christ! Fucking Powers. Why doesn't he lift?* I fight with the pigtail at the rear of my helmet. It is the worst designed piece of shit in the aircraft. Only two inches long, the pigtail terminates in a friction connection that requires two hands to manipulate, and the bloody fingers of my flight gloves are making it all the more difficult to connect.

"Did you get them all?" he spits into his mike. I keep my thought to myself: *No, dickhead, I left them to fucking die.* I scream back at him, "Yeah . . . yes, sir. We got them all, we got them all. We're ready to lift."

"Did you get them all? Are there any left alive in the zone? Ask the grunts!"

Jesus Fucking Christ, Jesus Fucking Christ. The last grunt in, the radioman, had told Powers during our approach that only three were left alive. *Jesus Fucking Christ, what more does he want, a fucking roll call?*

"Ask them!" he orders, seeing me hesitate.

I turn to the operator. His body is rigid, every muscle contracted in pain. I have to scream to be heard over the firing. We've been on the ground for perhaps forty seconds, no more than a minute, and there's been no let up. The AirCap keeps a constant stream of fire throughout the zone, and tracers continue to pierce the cloud like fiery hail. "Any more alive out there?" I shout. His eyes open and stare through me, through the deck plating, and into somewhere that I cannot see. I draw closer, and shout into his ear.

"Any more alive?"

"No. . . ," the answer comes in a gasp, as if he's had to steal the air from some place that desperately needs it.

I key the mike a last time, "They're all dead. The operator says they're all dead. We're ready to lift." Powers nods, finally, and I turn to work on the first Marine.

I seize the folds of his blouse and roll him in position, belly up. A grunt of pain escapes his lips, but I have to get to his wounds. His trousers and web gear are saturated with fresh blood. I fear he won't make Khe Sanh if I can't stop the bleeding. I fumble with his belt and buckle, oily and slick with blood. *Fucking goddamned gloves!* I pull them off with my teeth and spit gloves and blood into the mess on the deck.

I work at his trousers. *Fucking buttons! Fucking Crotch can't use zippers like normal people.* I can't manipulate them, and think about my knife on the crew's seat. *I have to get to the wound. I have to get these fucking trousers off and. . . .*

Bam! Bam! Bam! Bam! Pistol, range zero. My left ear rings from the explosions in spite of the flight helmet. *Jesus Christ, we're being rushed!*

I move for my weapon, assuming the pilots are lost and I am next. The grunt blocks my -60. I grab his blouse and heave, no longer heeding his wounds, and jerk the weapon from beneath him. He wheezes in pain and rolls, face down, into the empty space. I stare at the gun. Fresh blood, bright red, fills the receiver and coats the cartridges that snake out of the ammo box. Clamped below the feed tray, the yellow stencil of "VT-12" is barely visible through streaks of red. *Jesus Christ, it's gonna jam. Maybe one burst, or two. The gore will cake in the barrel and the bolt will seize.* I thrust my finger into the feed tray and try to wipe it clean, cursing the helpless Marine, "Fucking grunt, you fucked up my weapon. You fucking goddamned stupid fucking grunt!"

I will die now, I think. I will die here, in the charnel cabin of the slick. I hate the aircraft, fear the aircraft. Not the hill, not dying, not death. Just . . . not here! Not like this! I will die there, outside, quickly, with my feet on the earth. I will choose the place! And maybe the -60 will work. Maybe I can save. . . . Shit! The ship is finished. I'll be saving nothing, only forcing a quick end. That's enough. I know what they do to prisoners; I will not be taken alive.

I push the safety to "fire" and rise from my knees.

But there is Powers, turning toward me. Alive! Holy Jesus! I follow his gaze out the port door and see Stevens as well. Alive! A lone NVA lies face down in the grass, two yards from the slick. A grenade beside him, pin in place. Stevens's arm is rigid in the gook's direction. Fresh cordite burns my nostrils, and the smoke of his .38 melts into the gray fog within the cabin. I throw the weapon on the crew's seat, feeling relieved, foolish, cheated, and return to the wounded man.

Stevens guards us with his pistol as Jack and I work. The bastard looks like he enjoys it, riding shotgun on the Overland Stage. Powers begins building turns for takeoff. Jack looks perplexed as he labors over the radio operator, having as much trouble with his metal jacket as I am having with buttons and trousers. The second grunt lies quietly in the rear of the cabin, ignored as we focus on the bodies in our laps.

We've been on the ground far longer than I expected, and the firing holds at a constant pitch. How much fucking ammo do they have? The smoke has become our friend, providing cover from the enemy's weapons, but the gunships' tracers penetrate the cloud less frequently, and the protective cone around us is growing thin. The ships are going dry, probably making dummy runs to conserve ammo. They still have to cover us as we rise and draw the concentrated fire of the holes and bunkers. I can hear Geiger on FM, demanding Powers's report.

"Four-One, advise when you're ready to lift."

No response. I watch the pilot concentrate on the controls, building turns. Our rotors whistle as they churn the smoke of the zone. Vibrations mount, and the aircraft strains and bumps against the earth.

"Four-One, do you copy?"

No answer. What the fuck's wrong? Geiger breaks in a third time, plainly worried.

"Four-One, we are going ammo-minus. Fixed-wing almost out of gas. Can't keep this up. You better lift, Four-One."

Stevens keys his mike. "Uh, Five-One, we're having some trouble down here. Stand by one."

The slick dances on its skids. They flex and spread like a bow but cannot shoot us into the air. We bounce a half-foot starboard, toward the circle of bodies, threatening to crush them. The gale raised by the rotors tears at their ragged clothes and web gear. I imagine them flinching in the wind. We bounce, but can't lift from the zone.

Too fucking heavy! I rail at God. *Jesus Christ! Look, this isn't fair. This is god-damned unfair! We've hit the zone. Twice! Jesus Christ.* It seems so trivial, having our asses waxed by weight. Every instant on the ground our chances of living diminish. We've been in the zone a little over a . . . I can't remember! It seems forever, and every second is borrowed.

Powers bounces the ship, tantalizingly close to becoming airborne. We rise on a cushion, then fall heavily and bounce higher. Our oscillations become more rhythmic, building on the natural harmonics of the slick. Jack and I kneel in the back, working on the grunts as Powers gooses the throttle governor to pull down max turns. We bounce and shuffle our way across the hilltop, leaving the circle behind, then simply slip over the edge of the hill. One bounce, two. Skids sliding atop the burning grass, picking up speed. The smoke and flame of the burning zone enveloping us. Like sledding on a toboggan with rotors and a tail. Borderline translational lift. Holy hell! Bounce, bounce, then one last, and lifting!

Pulling our way into that clean, clear sky.

Jack and I work on the wounded as Powers races for Charlie Med. The radio operator's shattered glasses are still perched atop his nose. His trousers are torn and dark with blood, but as Jack probes his injuries, it appears the bleeding had stanched itself. He is dazed, exhausted by twelve hours of battle, and seems close to losing consciousness. As Jack packs battle dressings over his wounds, sudden awareness flashes into his eyes. He fumbles in the chest pocket of his flak vest and produces the remains of an Instamatic camera. The entry and exit holes from a single small-caliber round appear on each side of the pocket. The grazing shot destroyed the camera but missed the film cartridge. Bits of Kodak fall in tiny pieces as he waves the cartridge under Jack's nose. "Get this to Intelligence," he orders, and Jack takes the film as he passes out.

"Wow," I shake my head. "Fucking John Wayne!"

The first grunt lies as I dropped him after extracting my -60. He is conscious, and I roll him toward me as gently as I can. He looks up at me with unaccountably clear eyes. They are brown. Not dark like my own, but brighter. Hazel, you'd probably call them. And his hair is brown, like mine, but it is hard to tell under the dust of the zone. He is about nineteen and has probably been In Country longer than I. He looks older. I, too, am nineteen, but you age badly in Vietnam. Every month seems to leave its mark. I offer a weak smile, meaning it to be reassuring, but he doesn't smile back. His lips are split and caked with dust, and not a sound escapes them nor a breath. The bright eyes seem to look deeply into my own but tell me nothing. I wonder if he can see me.

I work his belt free, and fumble again with the buttons of his torn trousers. They are peppered with ugly holes, as though savaged by a squadron of moths. The cloth is stiff and sodden with old blood below the waist, but fresh dots of red weep through the fabric here and there, like leprous sores that will not stay hidden. I manage to untangle a button, and then another, and then, and then. . . .

His . . . his . . . insides . . . his insides . . . they come . . . they come right. . . . And I . . . I try to redo the buttons, but I can't. Oh God. . . . They slip, and they're warm, and they won't go back! I look down, and there is so much. . . . Oh, Oh God. And he looks down and he sees . . . himself. Oh God, and he knows. He looks up at me, and I see it in his eyes. They are still clear, and he doesn't cry out, but he knows.

I can do nothing. I wasn't trained for this. I'm not a corpsman, and even a corpsman couldn't handle that wound with our puny battle dressings. I stare into him. There is nothing I can do.

The rags I dumped from the toolbox are everywhere. I cover his wound and hold him in my arms. And rock him. I rock him. And sometimes he looks up at me with those clear eyes. And I look back, trying to fight down the pain welling in my throat. "There, there." It is all I can think to say. "There, there. . . . There, there. . . ."

Khe Sanh tower clears us for an emergency straight-in to Charlie Med. After a while, I stop looking at the wounded grunt. Outside the slick, the gunships fly in formation, escorting us home. The gunner in one bird gives me a thumbs-up. Gungy shit! Up your ass. I hold the grunt and stare into the hills.

Crash trucks meet us as we clear the concertina on the end of the runway and escort us to the Med. Stretcher bearers swarm to the slick as we touch down. My arms feel suddenly empty as they take him from me, as when you unwillingly give up a child. The swirling wind chills the . . . wetness in my flightsuit, and I am suddenly cold. So cold. The bearers wrap him in a blanket to keep his . . . to keep him together. They take him away, into C-Med, with his eyes gazing clearly into the sky.

Powers shuts down, and I stand outside the aircraft as the big blades swing to a halt. The Sparrowhawk grunts gather and peer inside. A congealing pool spreads across the deck, overfilling the seat locks, clotting into the corrugated decking. It will be a bitch to clean. I fish around for my utility cover. It fell from the knee pocket of my flightsuit during the mission. I find it in the forward cabin, where I shoved the grunt. It is sodden with his . . . with. . . . I can't put it on.

A tall, lean Marine wags a pack of Camels at me, and I nod. He puts one between my lips, flashes a Zippo, and lights it. Eagle, globe, and anchor on the Zippo, the golden emblem twinking flares of fire in the noon sun. Wow, fucking John Wayne. I inhale deeply, and when I take it from my lips I see what is left of him in my hands. He is there, in my fingers. Encrusted nails, flakes of red-brown already breaking from the creases of my fingers as I flex them. Red prints smudging the cigarette paper, gray ash driven by the wind, sticking to the dampness in my palm. I pull on the weed, savoring the anodyne of cool tobacco.

The gunship crews join us. Someone from the CP says the mess hall is holding chow. I look at my ship, and for the first time as a crewchief think, "Screw it," and join the crews heading to the mess. The slick sits alone on the medevac pad, doors open, her main blades untethered, gyrating wildly in the prop wash of taxiing aircraft as flight ops return to normal. Khe Sanh's crumbling laterite, wind-driven like red hail, peppers her faded skin and fills the cabin, then melts to mud in the gore. Screw it. I am so tired. This once, just this once, I deserve to come ahead of my ship.

Chow was Spam, lukewarm and sweating shiny droplets of fat. There was mustard to kill the taste. The Kool Aid was green and warm, the bread coarse and dry. We ate quietly, and after a while spoke. In time the lethargy that cloaked us melted. Our speech turned animated as the myriad elements of the war story unfolded. Each man told his tale eagerly, and listened eagerly. None of us knew the complete tale, so we learned from one another as each told his part. In the end, the parts merged, and the Story was born.

The tale assembled at that table seems so much less than the tale I tell today. We recited tactics and logistics, and in the habit of warriors, we honored ourselves by honoring the skill and bravery of our enemy.

"How many NVA you figure on that hill?"

"The grunt said 150. Hardcore regulars."

"Jesus!"

"Did you ever see head-on fire like that before?"

"Nope. If I'd swerved left or right, they'd have had a bigger target. Easier to go right down the throat. Goddamn, they had balls, though. All that firepower hitting 'em."

"You bastard, you almost shot off my skids."

"What the hell were you doing on the ground so long?"

"Hey, we had a good view."

". . . and then Zack shoved the goddamn screwdriver in and ripped the fucking latch out. What an animal!"

Such was the story in that first telling, recounted while the residue of fear melted into bravado. So it is with war stories, when they are young and unencumbered by the baggage of feeling. In no way, at no time, did any of us ask, "How did you feel?" or, "Were you afraid?" Those questions are not asked when blood still crusts your hands and the smell of smoke is fresh in your flightsuit.

But much later, the emotions become more real than they ever seemed, perhaps more real than when the story was lived. The feelings run deep, and deeper with the passing years. I remember now all the instantaneous fears that happened and were gone, that were replaced in the heat of the moment by other instantaneous fears. They take their time now, and I relive them slowly, one by one. The fears are attached to places and events. And words. For me, the words are the gateway to the worst of them. To remember the place, to say the word, is to know the fear, and feel the sorrow. The tale is no longer simply logistics and tactics and a source of passing pride.

We sat around the mess tables and told the story for the first and second times, and probably the third, until we could think of no new ways to tell it. We finished our Spam and left. Powers and I returned to VT-12. Stevens, Jack, and the gunship crews went their own ways. The blood and clay had dried. I searched the debris for the rags I'd dumped from my toolbox. They were forward in the cabin, in the place where he had lain. Worthless.

I flew left seat to Phu Bai, but we wanted to get home in good time, so Powers took the controls. It was nearly evening when we shut down, too late to pick up any new missions. Powers said, "Good mission, Corporal," and left.

I drew buckets of water and cleaned the aircraft. Phu Bai was a main base and had no shortage of rags. Later, I carried my -60 to the weapons shed. It cleaned easily; the blood dissolved into rainbows of red and yellow and orange in the solvent bath. Pretty. At last I returned to the hooch and threw my clotted flightsuit into the corner for mamasan to wash. I was glad to be on an early flight schedule the next day so that I wouldn't be around to hear her bitch. She hated to launder blood. Naked, I walked to the showers and danced quickly under the cold water. It melted the encrustations from the creases in my knees, hands, and elbows. The water ran in red rivulets through my toes, across the wooden pallet floor, and into the sand beneath. I soaped thoroughly, and the pink froth of it grew lighter and lighter, and after a while it ran white, and I knew I was clean. But when I dried myself, back in the hooch, it was still there, under my fingernails, and there it remained for a long, long time.

I told the story once or twice more as night closed on Phu Bai, my part and all the others I'd learned around that Khe Sanh table—to those few of my hooch mates who would listen. It wasn't a war story to them, just shop talk. It didn't take long, only a fraction of the time and with a fraction of the words it needs today. Then it was done. War stories had a short life span. Today was today, the past was ignored if not forgotten, and tomorrow was not. It is doubtful that on May 11 I thought at all of the events of May 10. I'm not certain, however, because I have no idea what I did on May 11. May 10 might have been forgotten as well, despite its baggage, except for one thing.

The grunts decided the crew of VT-12 deserved medals: Powers a Silver Star, Stevens a Distinguished Flying Cross, and Jack and I, Bronze Stars each.

Jack and I received our awards at Marble Mountain in September. It was a clear day, uncommonly blue and uncommonly cool. At least that's how I recall it. Actually, the day might have been anything. But a day for medals should be a day for sun and breeze. Perhaps it was.

There are those who say there is no glory in war. They are wrong. Fleeting though it may be, dwarfed by the horror that spawns it, glory is there, in the Ceremony. The Spartan phalanx of Marine Air Group 16 surrounded us: hundreds of men in fighting green, faces tan and lean, weapons ready. The colors and battle streamers of our Corps snapped with loud report as we waited for the general. In the distance, the business of the war continued, though slowed a bit, stealing time from itself to make room for the Ceremony. It was not wasted time, for what would war be without the Ceremony? It is the method by which glory is conveyed. God knows, the boy within the shell of a man grasped for his taste of it, and glory was forthcoming in a patriotic bit of bronze and cloth bestowed in the presence of peers.

Oh, the taste of it that day, while waiting for the general. It was rich and fine. Aircraft on the distant medevac pad were turning up. The whine of power turbines, the whistle of rotors reached us on a breeze melded of sea air and engine gases from the fighting machines. Rather than intrude, they established the ceremonial stage. Four squadrons had been impressed as dramatic backdrop for the Ceremony. They held friends from boot camp, aviation school, and In Country, all present to bear us witness. They were faded and worn, but not shabby; at rest, but not at ease; and their ranks were straight and true. For a Marine, parade is ever parade, even In Country.

Front and center stood a captain, two miscellaneous lieutenants, Jack, and I. The captain stared straight ahead, through the distant aircraft and the hooches beyond, and into some place I could not see. His eyes were dead and unreadable, and if being burned-out had been uncommon, I might remember more of him. The looeys were FNGs with enough time In Country to pop a DFC or Navy Comm, but not enough to rub off all the green. They stared at us and we stared back, and we all wondered what the fuck the others were doing there. And we waited for the general.

He was long in coming. Anticipation, too, is part of the Ceremony. There must be time to think, to savor the moment. I felt many things as I waited. Above all, I felt self-conscious: more self-conscious than proud, but more proud than unworthy. Overall, it felt good, and better than good it felt right, that is, correct.

I was surprised at that feeling, the feeling of "correctness." I'd no expectation of being nominated for a medal, and when—in August—I'd been told the Star had been recommended and approved, I was amazed. Medals happen to others. They get pinned on people who look a lot better in a uniform than you do. Medals go to heroes of erect carriage with steely eyes and, well, to people who just look as if they deserve medals, whose portraits are born to rest behind museum glass. Medals don't go on skinny chests, or to mortals with weak chins and chicken-pox scars who smear zit cream over pimples and pray they'll go away. Medals don't get pinned on people who look like you. Oh, inside, where it is private and no one can laugh, you muse that some taste of glory will come to someone with your face. But you never really expect it to happen. That's why the winning is so totally unexpected. In Aviation School, in Memphis—half a world away—glass cases draped in Marine scarlet and gold memorialized the deeds of aircrew past. The bloodstained memorabilia brought the current conflict into our classrooms. Many of us, waiting for the general, had marveled at the deeds of others only months our seniors, but seemingly years ahead in accomplishment. We played games, saying, "This is where my picture will go," but we knew that the glass was a hard barrier and the price of passage beyond control, almost beyond comprehension. The enshrined images were always bigger than life, while our reflections were humbling. Yet, standing in that human quadrangle, I felt in some way that I'd made it "behind the glass." The zone and the grunts were reality, the skinny chest and the zits were memory.

The general arrived, the formation drew to attention, and a clear voice intoned our citations. One after the other, the loudspeaker amplified our deeds. The general paused at each man, staring into his eyes while his citation was read. The dead-eyed captain received a Silver Star for anchoring the starboard wheel of his -34 on a mountainside, saving the grunts even as

MG and small arms holed his aircraft, killed and wounded his crew, and shredded his legs. He could have waved off, fled the zone. But he did not. As his citation floated on the breeze, the words took form in my mind. I could see the zone, feel the trapped desperation within the aircraft. This man, one of our own kind, two paces down—he of the dead eyes—was now behind the glass. Someone, somewhere might take a piece of his aircraft or a blood-soaked flak vest from his crew or from himself and place it on display to shock and inspire those who followed. His name would rest in stencils on a yellowing pasteboard of scarlet and gold. The general pinned the Star to the captain's shirt, and spoke private words of honor that went unanswered. Did he measure himself by those he saved or by those he could not, or by those who died because of his decision? Did his distant eyes seek some panorama of truth and beauty, wherein lay justification for the memory that haunted them? Or had he simply seen too much, too soon, far more than eyes should witness at any age? The captain did not respond, and the general looked at him with something as near to sadness as I could imagine in a commander's eyes, and moved on.

The looeys didn't do much of anything. Theirs were scatter-gun medals, the beneficiaries of shit hitting a fan. Both had "held the controls" while their pilots did something important on a mission that meant something to someone else. The general pinned their medals with neither nod nor word and moved onward.

I straightened my back and grew another inch at attention as the voice turned to me; my moment had arrived.

Citations have a way of making everything sound more heroically glorious than you recall, even if whatever it was that you actually did was heroically glorious. I suppose they wouldn't be approved otherwise. Nor would the Ceremony seem so, well, ceremonial, and that would be a waste.

The announcer began:

"For heroic achievement in connection with operations against the enemy . . ."

Great words! Straight from *Victory at Sea*. Dad, you got nothing on me!

". . . under siege for twelve hours near Khe Sanh in Quang Tri Province by an estimated 150 North Vietnamese Regulars."

Paint the picture, set the stage! I was there!

"The situation on the ground appeared hopeless for the three survivors, all of whom were wounded and out of ammunition."

Goddamned right it was hopeless, and we got them out! There were 150 of those motherfuckers! Who could expect more than that? Me and Jack here, and the Iceman and Stevens, we did it. We all did it!

The general stood a foot away, two stars winking at me, but I stared at the bridge of his nose, right between the eyes, which is the part of a general you're supposed to look at. He stared back, and I caught the wrinkle of esteem that creased his eye. The left one, then the right one, then both. He was happy. The general was happy. I was happy that the general was happy. I grew another inch, and presented my chest as the general caressed my Star. I hung on the announcer's words, and followed each one as they echoed through the formation.

But . . . but then . . . the words . . . they began to go too far. What was that he was saying about me?

". . . unhesitatingly leaped from the aircraft . . ."

Hey. I know what I did, how I felt. Oh, God the fear. . . . That's not me—". . . unhesitatingly leaped. . . ." Hey! That's not me! I didn't do that.

Man, I was scared. I almost panicked. I blacked out for an instant. I would've shit my pants except I couldn't get that organized.

". . . unhesitatingly. . . ."

Well, I did that, but I didn't feel that. Well, I know you're not talking about what I felt, just what I did. I know it's what I did that counts, not how I felt. Yeah, I know. But I was scared, man, I was scared!

The citation went on: ". . . exposed himself to intense enemy fire . . ."

You make it sound like I did it on purpose, like I planned it, for Christ's sake. I crawled out of that aircraft, and all I wanted to do was pull my nuts inside my flak vest and crawl in before some NVA blew them out through my front teeth. We grabbed the ones who were left alive and crammed them in the aircraft.

". . . through his aggressiveness and determination under fire . . . saved the lives of the surviving reconnaissance team members."

Yeah, that's what we did, you son of a bitch. Extracted three, and saved . . . I don't know how many we saved, goddamn you. Look, I didn't do anything "unhesitatingly," and I didn't do anything on purpose, and I'm not a fucking hero, and don't you say those things about me.

Suddenly it didn't feel correct anymore, not correct at all. My place behind the glass shimmered, and the reflection was foolish and posturing—undeserved. I glanced at Jack, and saw the puzzled reflection of my own thoughts. Is that us they're talking about?

The words stopped. The general pinned the cloth over my heart, and I saw my hand being shaken by his. "Congratulations, Marine, you deserve this," and he said it as if he meant it, every syllable. He saluted me, and I saw my hand move up. Forearm parallel to the deck, forefinger atop the right eyebrow, pinky tucked in. Then he moved on to Jack and left me staring into the same far distance as the captain. The moment was over.

I know I deserved that medal. I did.

Didn't I?

At last the Ceremony was over, and the squadrons dissolved, shuffling back to their aircraft and shops and shit details. But twenty men came forward, crewchiefs and gunners—the peers of my youth, smiling male-smiles of approval and respect. They punched my arms, called me "asshole," thumped my back, and fingered the ribbon roughly, and as they threw my cover high, high in the air the misgivings, the guilt melted. I was crewchief! And these—these of my kind—who had eyes to see beyond the bullshit and the ceremony, honored me with their maledictions more than I have ever been honored. I cherish the memory of that moment as I do the medal itself, and when I remember it I think perhaps I really did do something extraordinary, out in the zone.

It hangs upon the wall, a five-pointed Star of Bronze, emblazoned with the "V" of Valor. The citation is framed beneath it. I read it, now and again. The feeling that washes over me is strange. There is pride, over that distant moment in the sun. There is rationalization and embarrassment still, about the glory ascribed to me on the yellowing paper. My children read it. It is a face of their father they cannot comprehend, for to them I am clearly "behind the glass." It is a good medal, not a great medal. Others have done better, but it is the best that I could give. It is a double-edged sword; it returns me equally to the horror of the grunt and the handshakes and honor of my peers on that September day. The memory is warming, despite the pain. Amidst the debris of ashes that the war became, the medal holds something of value.

The Story is over. I have run its gauntlet once again.

I face Tom, squarely, and ask in a tone half-bitter and half joking, "Are you fucking satisfied?"

He pulls his beard, Mephistophelian. An eyebrow arches, a brow furrows. He leans close to me, elbows on knees. His eyes pierce the short space between us. It is almost as if they pierce the memory of the zone itself. I wonder, What does he see that I do not?

"This is the first time you mentioned the blond Marine."

"What blond Marine?"

"The one in the wall."

"No, it isn't."

"Yes, it is. I've been here, remember? This is the first time you've ever mentioned the blond Marine in the wall."

"He's always been there."

"I know he has, but this is the first time you've looked at him. Why?"

Why, why, why? He always wants to know why. I look back into the Story. The grunt lies atop the wall, flesh lashed by the wind. I pull myself into the aircraft and look back a last time. I am the last to see him before we abandon the zone and the fixed-wing sterilize the hill. Erase it, eradicate it, pound the earth and the flesh into. . . .

I own the final memory. His name is on the Wall in Washington. Which are you? I read your names when I visit, you the Marine dead of that day. The directory near the Wall tells so little. Date of birth, hometown, rank. And the day of death that binds you to me, who lived.

Which are you? Somewhere there is someone with your name who wonders what the final scene was, the scene I alone preserve, who last saw you before the fixed-wing . . . before they came with cannon and fire and, and, and. . . . It is locked in my mind. I own the final memory. I know all this. What else can there be?

"There's nothing else in there." I will my voice to steel, unconvincing. Please let it end here.

"Look more closely. How do you feel about him? What happened later?" Tom asks.

Later. The word is a trigger. I sense a new memory forcing its way through the image of the death wall. Later. We landed at Phu Bai and . . . Stop, stop, STOP! The images shimmer. Jesus Christ, I don't want to remember any more.

"Look, I have a bad feeling about this. I don't want to talk about it anymore."

"You've come this far. What happened?"

"I said I goddamned don't want to talk about it anymore!"

But a door has opened. It cannot be resealed. The old memory returns.

"We landed at Phu Bai, and Charlie Shitslinger came out to the ship and. . . ."

"What?"

"He was my section leader. I told him about the hill, how we got three of them out alive."

I can see Charlie, standing outside my slick at Phu Bai. Good old skinny Charlie, the sergeant's insignia looking too big on his collars. Charlie was a baby lifer. He'd just shipped over and hadn't been to Asshole School yet. I remember, after all the years.

"Charlie asked me, 'What happened to the others?' and I said, 'We couldn't get them out, Charlie. We were too heavy. We had to leave them in the zone. And the fixed-wing came . . . and the fixed-wing. . . .'

"Charlie looked at me, and I could see accusation in his eyes, and it made me sick to my stomach. Sick and mad all at the same time. He said, 'You mean you left them, you didn't even try to get them out?'"

"And then he walked away. He just fucking walked away. I called after him, 'Hey, what the hell do you know? You weren't fucking there.' Charlie waved a hand over his shoulder, not even looking back. Like he didn't fucking believe me.

"I yelled after him, 'We were too goddamned heavy! You weren't there, you fucking bastard! What the hell do you know?'

"You . . . weren't . . . there."

I sit, wrapped in silent anguish. Tom sits quietly at my side. This is not a time for words. There is no logic with which I can defend myself against the accusation that diminishes our deed. My deed. The guilt lies heavy, in spite of its illogic. "We were too heavy. They were already dead." The phrases seem excuses to me, and not reasons. The mystique of the Corps is its religion; there is but one commandment, *Marines take care of their own.* The blond face, frozen in youth and death, will not leave me. We left him—them—to the fixed-wings' shredding bombs. He is no more, except in my memory. What is the weight of logic measured against that memory, and the compromises we made to save the living, and to survive? How sterile seem the cold equations of weights and balances, or max thrust as a function of humidity; how sterile measured against the need of loved ones to have more than just a name, only a name etched upon a Wall to remember him. My God, oh my God, how do I forgive myself for leaving him there?

At last I speak. "I know it doesn't make any sense. There was no way we could get the bodies out. There was no way, but Charlie was right. We didn't try. All I wanted was to get the hell out of the zone. I was so damned scared."

Tom answers, and closes the session.

"We'll work it out. You've taken another step today. These are things you haven't thought about in years, but they've always been there. Go home and think about it. We'll work it out."

Thinking is so difficult to do, but I have learned that thinking is the beginning of healing. Each time I think about that day, about the mystique and legend that judge so harshly, the only explanation that makes sense is that I was human. Perhaps forgiving myself for being human is all that matters when logic fails, and when mystique proves only to be mystique.

The session is finished. The Story is another spatter on the Center's walls. Even as I prepare to leave I know I will tell it again, search some other time for the hidden closets. I wonder, how will I know when there are none left to be found?

"Did I ever tell you about the health course I took in college?" I ask, suddenly.

"No."

"There was this prof named Leviton. He was fascinated with suicide. He had us read Elizabeth Kübler-Ross's book, *On Death and Dying*."

Tom nods, like a teacher watching his student in discovery.

"She worked with terminally ill patients, helping them adjust to the specter of their own death. You know, there are five steps you can go through, if you live long enough, after it first dawns upon you that you're about to die: denial, anger, bargaining, depression, and acceptance. I don't recall there being any time limit. Think it can take about thirty seconds?" I ask.

"How does that make you feel?"

"Strange. Very strange. You know, I've never known such a sense of peace as when I came out of that—catatonic state, I guess you'd call it—when the NVA opened up. The sense of peace lasted for a long, long time. Years. I haven't been afraid of death since. It's more friend than enemy, to me, and I don't mean that in a suicidal sense. But somehow, I don't think it will last. Right now, I'm afraid of becoming afraid of death, and someday I think I'll fear it again. Perhaps it's a product of age. I hope it doesn't happen. I never want to go through those stages again."

"Did you ever find out what happened to the grunts?"

"The radio operator got the Navy Cross for his work in the zone. He deserved it. I don't know what happened to the other two. Mine, the one I held—I don't know. I can't see how he lived. I wish I knew. I didn't realize that I would ever feel the need to know. You see, that day was nothing more to me than just another mission at the time. One of the worst missions, but it was still just a mission, and he was just a medevac. I carried dozens of medevacs. Dozens. Wounded. Dying."

I see the grunt again, feel the warm wetness and the cold wetness and the emptiness of him in my arms. The tears well hotly. There were others like him, but he seems to stand for them all.

"I tried to keep him alive. I really tried."

Tom reaches to me. "I know," he says softly. But it is so hard, the memory. I really, really tried.

"I never knew their names," I said, thinking about all the medevacs, all the zones. "I never knew what happened to them, whether I did any good. I wish I knew for sure—sometimes it hurts not to know. Sometimes I go to the Wall and look at the names, looking especially for the names of May 10. I wonder if he's there. It's almost a game, like maybe I'm psychic and I'll know his name if my eyes touch it. Instead, all the names I look at take on his face."

I feel suddenly angry. Why do I have to remember?

"Y'know, I think it's the damned medal. I think if I didn't have the damned medal, it would have stayed just another mission. I'm proud of that medal. I'm really proud of it. I look at it sometimes, when I want to remember who I used to be.

"But sometimes, I wish it wasn't there."

The author, a crewchief in training, posing in the fall of 1966 at the copilot's position of a "slick" in the UH-1E Training Unit (VMO-TU), Camp Pendleton, California. The crewchief's station was in the cabin just behind the copilot; however, crewchiefs often flew "left seat" in slicks because of the scarcity of pilots In Country. *(Author)*

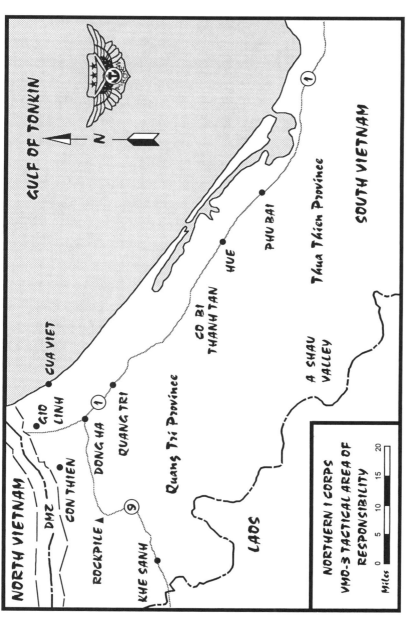

The Tactical Area of Responsibility of VMO-3, which was based at Phu Bai, ran from the DMZ to halfway between Phu Bai and Da Nang. The squadron also maintained small detachments at Dong Ha and Khe Sanh. *(Adapted from map in Jack Shulimson's U.S. Marines in Vietnam; An Expanding War—1966. Washington, D.C.: History and Museums Division, Headquarters U.S. Marine Corps; 1982.)*

VMO-3 squadron patch. Marine Huey squadrons inherited the designation "VMO," which was first assigned to spotter airplane units in 1937. The first letter, "V," signified a unit using heavier-than-air equipment; airship squadrons used the letter "Z." The "O" indicated the basic mission of the aircraft, observation, and the remaining letter stood, of course, for Marine. During Vietnam, the Marine Hueys' mission was extended to combat support, fire suppression, medical evacuation, and a limited amount of resupply. All other Marine helicopter squadrons began with the designation "HM," or Helicopter, Marine. A third letter indicated the relative "size" of the aircraft in the unit. UH-34D and CH-46 squadrons flew "medium"-sized aircraft to transport troops, carry medevacs, and resupply forward positions, and were designated "HMM." CH-53 squadrons were capable of transporting "heavy" loads and were designated "HMH." The VMO designation was changed to "HML"—Helicopter, Marine Light—in 1968.

Cpl. Anthony Zitkus—"Zeke, the boy from N.Y.C.," metalsmith and door gunner—designed VMO-3's squadron patch. The wasp, in flight helmet, is armed with the trade tools of a Marine Observation Squadron: machine guns, rockets, first-aid kit, and stretcher, with the telescope indicating the VMO's original mission of observation. *Hieu duoc* translates, "I understand . . ." and implies "I will comply." We interpreted this as "Can Do," our squadron motto. The similarity of the Vietnamese to "Huey" made it even more appropriate. *(Author)*

Semper AOCP, literally "Always an Aircraft Out of Commission for Parts," designed by Zeke Zitkus but never sewn in cloth, was a humorous and accurate depiction of the sad state to which aircraft degenerated if left too long on the deadline. Victor Tango 22, cannibalized of nearly all its vital organs, set a record for deadline occupancy.

AOCP ships found saviors in Marine bureaucracy. Wing Headquarters would "get on the maintenance chief's case" if an aircraft's flight time didn't budge for an extended period, in which case some healthy but luckless ship would be sacrificed to return the deadline cadaver to life. *(Original drawing given to the author by Zeke Zitkus on Okinawa, 1968; redrawn by Chris Zaczek)*

1 ST

MARINE AIRCRAFT

WING

GENERAL RULES OF ENGAGEMENT FOR THE
FIRST MARINE AIRCRAFT WING IN RVN

1. Visually indentify target or target marker (except TPQ-10, MSQ-77, or DIANE system).

2. Know position of friendly forces.

3. Radio contact if working with FAC/TACA.

4. Wingmen and helicopter gunners or crews are only authorized to fire on command of the flight leader or pilot in command.

5. IN EMERGENCY SITUATION with no qualified means of control, the following may designate a target:

    a. Ground unit commanders or U.S. advisors engaged with hostiles (must possess radio capabilities).

    b. Army target identifying pilot (TIP).

    c Pilot of aircraft or helicopter who is taking fire that presents an immediate threat to himself and/or members of the crew AND CAN:

        (1) Positively identify the source.

        (2) Positively orient a strike.

SPECIFIC RULES OF ENGAGEMENT FOR HELICOPTERS

FIRE ONLY WHEN

1. Under control with direct radio contact of the designated control agency.

2. You can visually identify target or target marker.

3. Friendly and civilian positions are positively identified.

4. Defending yourself against ground fire and:

    a. You can visually identify source.

    b. You can positively orient a strike against source.

    c. The fire is of such intensity to warrant counter action

5. Helicopter gunners fire only as briefed by pilot in command.

The First Marine Aircraft Wing's rules of engagement spelled out the conditions under which an aircraft could attack the enemy. In general, these rules were followed during air-to-ground combat. Many times a suspected enemy would escape because of the absence of a FAC (Forward Air Controller) or TAC(A) (Tactical Air Controller [Airborne]). *(Author)*

The author at Phu Bai with Victor Tango 12, the original *Warsaw Falcon*, in May 1967. The lightweight flightsuit was made of fire-retardant NOMEX, under which we frequently wore nothing during summer's raging heat, or full utilities and long undies during the monsoon. *(Author)*

UH-1E crewchiefs and gunners frequently cleared jams while tending the aircraft's weapons between firing passes. L.Cpl. E. M. Donaldson of VMO-2 provided a realistic simulation of clearing a jam in the port M-60s in this 1968 photo, probably taken at VMO-2's home base at Marble Mountain.

Aluminum chutes fed belted 7.62-mm ball and tracer ammunition from canisters beneath the crews' seat to two M-60 machine guns on each side of the aircraft. M-60s feed from the left and eject to the right; the gun's cocking lever is mounted on the right side of the weapon, and requires approximately an 8-inch stroke to position the bolt for firing. The Huey's starboard guns were mounted "straight-up" and were easy to service. The port -60s, shown here, were mounted on their sides to provide a smooth feed path for the ammunition, and to eject cartridges down and away from the aircraft; consequently, the gun's cocking lever was under the weapon, obscured by the bomb rack and the 2.75-inch rocket pod.

During flight, a crewchief in hard-shell flak vest and helmet, tethered by his gunner's belt to the aircraft, had to lean into a 120-knot wind stream and snake the cocking rod through the rack structure, while trying to avoid blistering-hot gun barrels, to clear jams and rearm weapons. Clearing a jam on the outer gun posed an additional dilemma. "Do I take the time (and burn my hands even through flight gloves) to remove the inboard barrel, or do I lean in front of a loaded weapon?" Jams were cleared while under fire, with the pilot doing everything he could to bring the aircraft into position to continue the attack, screaming, "Are we clear yet?" *(Department of Defense [USMC], Photo A422033)*

In November or December 1967 Marine historians visited field units to capture the dramatic and mundane aspects of life In Country. A passing historian captured the author (left), Cpl. Tom Kehoe of Imlay City, Michigan, and Cpl. Larry Zimpfer of Bloomington, Illinois, guiding a T53-L-11 Lycoming engine onto the mounts of Zimpfer's aircraft at Phu Bai. Maintenance was accomplished using a field hoist erected on the aircraft and a lot of brute force, there being no hoist-equipped hangars or "cherry-pickers" available. The author, near the end of his tour, is down to 110 pounds, suffers from malaria, and has chosen the "safest" position to perform this task. If the cable snaps—as was known to happen—the engine will land on Kehoe. If the hoist mounts snap—also known to happen—Zimpfer will be taken out. *(Department of Defense [USMC], Photo A422004)*

This scene, taken on 4 September 1967 after a major attack upon the Marine base at Dong Ha, shows how poorly constructed plywood hooches fared under 122-mm rockets and 82-mm mortars. The sandbag bunker between the hooches afforded protection from shrapnel, and could stop a mortar if sturdily built, but a rocket would erupt into a crater ten feet deep. In this area, two flaming hooches had collapsed atop a bunkered hole, roasting the occupants until they dug their way to freedom. The author had lived in one of the demolished hooches in the foreground while on temporary assignment to Dong Ha, and left the day before this attack to report to Marble Mountain for his Bronze Star ceremony. *(Department of Defense [USMC], Photo A355562)*

Unlike the Army's Air Cavalry, which distinguished units with colorful insignia on an aircraft's nose or fuselage, Marine squadrons did not employ unit markings. The squadron code, VT for VMO-3, WB for VMO-6, and VS for VMO-2, was painted on the cabin door; however, gunships did not use doors, and displayed only the aircraft's number on the tail fin, door post, and nose. Marine VMOs flew several variants of the basic UH-1, model E. VMO-6, first based in Ky Ha near Chu Lai, then at Quang Tri, flew model 204s, distinguished by large "potato masher" counterweights above the main rotor, a nose-mounted pitot tube, and a single white running light on the tail pylon above the stinger. VMO-3 flew model 540s and 540As, with no counterweights, a roof-mounted pitot tube, and one white light on each side of the tail boom, just forward of the stinger. VMO-2 flew all UH-1E variants. Each VMO painted white band(s) on the upper surfaces of the main rotor blades to enable a high-altitude observer to spot the aircraft against the jungle. VMO-3 had a single white band, while VMO-2 and VMO-6 had two white stripes of varying widths. The fuel cell of this VMO-3 gunship was ruptured when she was caught on the deck at Khe Sanh in January 1968, probably by mortar. (A rocket or other heavy-caliber artillery wouldn't have left this much of the aircraft intact.) A CH-46 Sea Knight turns in the background. *(Department of Defense [USMC], Photo A190246)*

This March 1968 photograph of a landing UH-1E gunship (from VMO-2 or VMO-6) demonstrates the aircraft's vulnerability when approaching hilltop positions. This is approximately the view NVA gunners had of VT-12 (an unarmed slick) when we extracted the "Breaker" team near Khe Sanh. Aircraft approached defended positions as fast as possible, then "flared" (raising the nose to build air cushion) to shed speed. *(Department of Defense [USMC], Photo A371387)*

Chapter 6

We are done with Hope and Honour,
   We are lost to love and truth,
We are dropping down the ladder, rung by rung;
   And the measure of our torment
Is the measure of our youth,
   God help us, for we knew the worst too young.

      —Rudyard Kipling, *"Gentlemen Rankers"*

Measure of Youth

I am standing at the coffee machine, pouring the ritualistic cup, ritualistically swirling in the chemicals, waiting for Tom. He emerges from the back office, holding the door for another vet. They walk down the short corridor, slowly. The shiny remnant of a tear lingers in the crow's-feet near the vet's eye. We exchange glances, no words—a brief nod of embarrassment and assurance. Tom thrusts his forefinger upward, "I'll be right back," then follows the vet to the Center's portal and opens it for him. The vet pauses, and sighs as I sigh when I leave the Center to return to the World. A pat on the shoulder, reassuring, and the vet walks into the evening. Tom waits for the vet to turn. He does. It is no surprise; I do. A final wave, then Tom returns.

"Ron! How's it going?"

"Hey, just fine. Fine. Just fine." I swirl the coffee. "Really fine."

"Oh. Let me get a cup, and we'll go back."

Tom draws the black liquid, another ritual. I walk back to his office, gibbering, "Fine. Just, fine."

I sit in the steel chair, the vinyl damp-sticky-warm from the other vet. Tom eases into place.

"You're pretty down today."

"Me? Nah. I'm fine." I suck at the coffee, bitter despite the chemicals, or because of them. Pick up the cup, put it down, adjust it a half inch forward, then back. I press my finger into the styrofoam, and watch the impression fade. Tom waits, a lopsided stare of appraisal stamped under his cocked eyebrow.

"OK. What's been going on?"

"I've been working on The Book," I say to the cup.

"Oh. . . ."

I began working on The Book on November 13, 1982, about midnight. It was the evening of the day the Wall was dedicated, an easy date to remember. I'd stood before it that noon, in the crush of the crowd under a gray sky. I'd arrived early to secure a position against the fence they'd erected to hold us back. It was too far from the Wall to read the names. No matter—in two days of remembering I'd learned the special places, and my eyes found the panels with ease. I chose my position carefully, and counted the panels on the eastern Wall twice to make certain I was exactly opposite the right one. Panel 29. I had to be opposite Panel 29. Ronald Joseph Phelps, Panel 29E, line 97. The next day, November 14, would be fifteen years since he was gone.

I was absorbed in remembering and recall little of the ceremony. There were many speakers, but the Wall dwarfed their words of honor. I recall mostly the somber flag bearers and the refrains of "Chariots of Fire" that pierced the autumn air. Blake's "Jerusalem" soared above the crowd; it sang for all the sixty thousand.

> Bring me my bow of burning gold
> Bring me my arrows of desire
> Bring me my spear, Oh! Clouds unfold
> Bring me my chariot of fire!

Forever after, the hymn recalls me to that day. It is an idyllic notion, but far better to remember those faces at the pinnacle of hopeful youth than to conjure the image of dust and blood and the shredded metal in which they died. Those last I carry with me always; the Wall lets me remember them as they were.

As I drove home, strangely calmed by the sad-happy days of remembering, I thought once again about how I might write the book. I'd wanted to write about Vietnam for years. The memories that fueled the story were vivid; I'd been struggling with them for more than a year in the Center. But how should I put them into print, and make the sterile paper speak of what I felt, not merely what I did?

The Wall had shown me. I lay awake that night, remembering the wash of emotion I felt while standing before the names of friends, their unchanged faces like shining shadows in the polished stone. There lay the answer! I would carve feelings from the unyielding granite, and make them leap from the printed page. I would wring those same feelings from others, and then they would understand. I would write of hope and pride, of rage and guilt, of alienation and loneliness and betrayal, of the elemental fears that haunted my dreams. And I would write of loss.

Thus began the book. It took form slowly, as I mined the memories and cut them into the building blocks of a new life. Hard labor it was, for the memories did not yield easily, and the feelings I carved from them were cruel. The book grew. After time, it became the *book*, and then The Book, an obsession that seemed to influence the very emotions it was intended only to record.

At last I came to write the chapter on loss. This chapter.

It is dangerous to dig too deeply into the past.

"I've been reading my old letters, doing research for The Book," I say.

Anxiety and confusion tear at me. And shame. Most visits to the Center portend some element of pain. But this time . . . I've never known such shame. I twist in the chair, aching from the anguish I feel. I'd read and reread the words—my words—not wanting to believe they are mine.

"What did you read?"

"Letters I'd written to a close friend. I wrote a lot. He saved them for me. I guess he thought he was doing me a favor."

I produce the torn envelopes and place the things carefully on Tom's desk, staring at them as though they have the power to inflict pain. My mind churns, watching the accusing, hurting things—red and blue envelopes emblazoned with a dragon under the word "Vietnam." I remember the rush of pride when scrawling "Free" in place of a stamp— the mark of an American warrior writing home.

I seize the first letter and present it like a piece of damning evidence.

"This thing, this. . . ," I shake the letter, searching for words to vent the hatred I feel for what lies inside. "Have you ever read something you've written long ago, and known that it was you who had written it, and known that now—today—you'd despise anyone who could feel such things?"

The question tears at me but demands no answer. The answer is within.

"I wrote . . . what's in here. It's my handwriting," I say bitterly. I would deny the words if I could. "I hate the person who wrote this. It disgusts me. This, this isn't me. But it is. Damn it. I have to admit this is me. Was me."

I spin the paper away, but the words, unsealed after all the years, torment me. I cannot escape them.

"What's in the letter?" Tom asks gently.

"It's . . . it's. . . . I can't."

But I know that I must read the words, and will. Even as I'd read the letter, in disbelief and revulsion, wondering what I had become, I knew that I'd have to face this part of my past in the Center.

I open the envelope.

> We got some good gun action today. Caught a bunch of zips in the open with AKs. We don't often get the chance. They had some balls; stood their ground and threw up head-on fire as we dove. We blew them all away. It felt great and I had a good pilot. You should have heard him, "Let's go down and kill the bastards. Get 'em, get 'em. Make 'em burn." Really kill crazy. "Kill that man, he's moving. Hit that tree, it's growing. Blast that plant, it's blooming. Make more smoke, I can still see the sun!" You know, it was great. It was really great. I don't know what this makes me, but there's no denying it. I enjoyed killing them today; making them pay, making someone pay for being in this lousy, stinking war and for the dead and wounded I flew in my slick. I went crazy on the guns. You should have seen me. Reloading, firing, reloading, firing. When we got back, my face was black with gunsmoke below the visor, like a race car driver. I've been in guns for several months now. Three or four. It's hard to remember. I couldn't take the slicks anymore. Couldn't take the screaming. I went to the line chief and convinced him to move me to a gunship. I've got a lot of "seniority" if that's what you'd call it. There are very few crew-chiefs left, still flying from the original squadron. It's better in the guns. I like hitting them. I like seeing my rounds tear into them and touch off secondaries when they hit the explosives in their packs. This action made up for the other day. I was pissed. We were flying in a free kill zone. That means, like, anything you see you can blow away, get it? Shouldn't be hard to comprehend. We spotted some women and kids herding water buffalo and the pilot just let them go! That goddamned bleeding heart! When the zips saw us coming, they started to beat feet to cross a river into safe territory. The fucking pilot let them cross. We should have greased them. Who cares if they were women and kids? They bring rockets and mortars down from the north on those damned buffalo and the kids sell poisoned Coke with glass chips or acid in it.

But we're "Americans"; we fight clean. We don't kill women and children. Shit. A couple of weeks ago we took a low pass over a group of so-called herders and I took a round up my goddamned tail pipe. It blew out the EGT probe and came through the exhaust diffuser casing. Another inch forward and it would have taken out the second stage turbine. We were too low to autorotate. We'd just have been a strawberry smear across the landscape at the speed we were going. And it was the same thing, fucking women and kids and old people, herding their fucking buffalo. Fucking bleeding heart pilot. It's real simple. If they're dumb enough to get caught in the zone, it makes them VC and they should die.

I fold the paper and put it away, slowly. Nineteen years old when the words were laid down. Nineteen. So young to burn with such hatred. I try to recall the boy I had been on the far side of the words. He is difficult to see. Where, in the year of the war, did he die? Was there a single moment, or was it a slow death across the countless zones and the screams and the blood? When, how did he become this callous thing who so eagerly sought slaughter and vengeance? I look back upon this bitter face of my youth and wonder. What does this make me now? At each moment we are the sum of our experience. Can we deny some moment and say, "That is not me"? For years, I had burned in anger at the epithet "Babykiller." But the words are undeniably mine. Was I what the world accused, after all? What does this make me now?

Tom begins to speak, but I stop him. There is another letter, a strange companion to the first. Another face of myself that lay hidden, or that I buried.

"And then I read this."

More gently, I open the second envelope. The flimsy paper bears the hasty scrawl of youth. Again, someone who had been me. I read aloud; the words come hard as before, but for a different reason. I pity the boy's anguish, he who wrote them. It seems not self-pity, for the boy is a stranger to me; I cannot remember him. And if it is self-pity, so be it. The boy has lost much, more than he deserved. I mourn his loss.

I'm afraid. I'm afraid that when I get back, I'll be a misfit in my own home and with my own friends. I've changed over here. I'm not like you anymore, not like any of you. I feel scarred by the animal life I've lead, chasing men down in helicopters, killing. The hatred. I won't belong anymore. How can I belong? I guess this isn't making any sense. I'm tired. You think crazy things when you're tired. 48 days left. I've got to get out. I can't take this place, anymore. I want to come home

and rest, and forget about this place. But I'm afraid that it will show.
The despair, misery, hatred; like a badge burned into my eyes. How
can I ever belong?

I return the paper gently to my pocket, then stare into my hands. So sad,
the words they'd written so long ago. I'd forgotten how deep was the loss.
I'd known it, even in that time.

"When did you write them?" Tom asks.

"I'm not certain. I never dated letters. I never knew what date it was, and
I didn't care. I only knew how many days I had left. Sometimes I mentioned
it in a letter—like the second letter mentions forty-eight days. I was scheduled to rotate home on January 7, so that would make it late November."

"Can you figure out when you might have written the first letter?"

"Let's see, I mentioned I'd been in a gunship three or four months."

I stare into space, glad to focus on the sterile computation. It means I
can forget the letters for awhile.

"I, uh, transferred out of slicks in May, after we pulled the team off that
hill. Failed my flight physical in June—malaria and malnutrition, they
said. I started flying guns on July 1. So . . ." I count the months on my fingers. ". . . that would be August, September, October, or . . . November.
No, that can't be right. It can't be November."

"Why not?"

"Because the second letter was written in November. Late November,
definitely." I add up the dates. "Forty-eight days back from January 7.
Thirty days hath November, April, June, and September. All the rest have
thirty-one, except. . . . That's seven plus thirty-one, then go back ten more
days from November 30th. November 20th! The first letter can't be
November too. Even October doesn't feel right."

"Why not?"

"Jesus Christ, look at the difference in these two letters." I wave the first.
"In this one, I'm a fucking animal. I want to butcher everything."

I tap the second, folded against my breast. "And here, all of a sudden
I've got religion? What the hell could happen in two or even six weeks that
could make such a difference?"

"That's what we have to find out, isn't it?"

The U.S. Marine Corps Historical Center is in Building 58 at the Washington (D.C.) Navy Yard. The passage of time has not dulled my indoctri-
nated distaste for things naval, and the lower reaches of my colon tighten
as I stride past buntings of blue toward the familiar banner of scarlet and

gold. The Marine Corps Museum is on the first floor; the administrative offices, writers, and oral-history section occupy the second. I travel to the third floor's archives to disinter the bones of the past.

The journey is Tom's idea, after my attempts to resurrect the deeds of November 1967 failed. The weeks are a blank to me, a seeming void in my life, and the void torments me. What happened? I turned to my letters. A frequent writer, I wrote two or three a day to family and friends. Writing was my only escape, my only tie to home, and I cherished those moments with the pen. Some have saved my correspondence, but careful examination indicated I wrote little and elaborated upon nothing between the last week of October and mid-November. What occupied me so that I could not write, or did not wish to write?

I search my service records for clues. The Air Medal certificates list the periods when mission credits were earned. I was a busy crewchief those weeks of November. I earned nineteen Air Medals in 1967. Each medal represented twenty missions, and three were awarded between mid-October and mid-November. Sixty missions in thirty days was a respectable amount of flying and indicate a considerable amount of gun time. But I have to know what happened during those missions.

The archives is an ossuary of the significant and the trivial, catalogued and cross-indexed in crinkling paper or microfilm that pains the eyes to view. The bones of history rest in the files' eternal rows, bleached white and silent of their former passion. They hearken only to those with ears to hear, but speak elegies and eulogies and indictments of Marines, simple and great. I come to exhume the past and resurrect into flesh and passion the memories to which the bones gave form. To eavesdrop as the ghosts of the command chronologies and after-action reports whisper in the stacks of paper, in hope and fear that they will call my name.

After-action reports are bureaucratic flotsam, dutifully completed by the flight leader after each mission and cast immediately into a sea of paperwork to be filed and forgotten. Those very AARs are now the jetsam of the archives. Every mission is recorded in deadening detail: coordinates, ammo expended, crew assignments—a few terse nouns and verbs stripped of the adjectives I felt or have come to feel. The faces of the old crews leap from the pages. I find May 10, 1967 and linger there, as Powers fights the controls, Stevens fires his .38, and Jack and I pull in the wounded.

> Went to XD756530 for recon extract. Made two approaches, took intense fire each approach. On second approach we landed and took three members of team out. Left four dead in zone.

The paper has so much meaning. It harbors powerful ghosts. The rattle of the crossfire and the wounded's whimpers rise from the printed page. I glance at the archives' secretary chatting at her desk. She cannot hear them; they echo in my ear alone. Yet here the paper lies, like so many others. Nondescript, indistinct. Just one more mission among the many.

I read on, and at last come to October and November. The bones take flesh rapidly, and there is no shortage of ghosts. In rushing moments I know the story of those lost days.

Oh God, I know.

10 NOVEMBER 1967—THE NORTHERN A SHAU, LAOS

November 10 dawned bright and clear, for the monsoon season. It was fitting and proper that the Marine Corps suffer no evil weather to spoil the celebration of its 192nd birthday. The Almighty is, after all, a Marine. A party was planned that evening, a poor substitute for the gala stateside affairs, but a party nevertheless. There would be Korean beer, 3.2 and green, then glorious, puke-racked migraines out of proportion to the mildness of the alcohol. And steaks! Grilled charcoal-black in the lee of the hangar.

A dry sky and a puke-drenched evening—what more could anyone wish? I knew it was going to be a good day.

I checked crew assignments. Hamm was slotted as pilot, with Knipp and Yantis copilot and gunner. We'd been assigned a SOG ship, which meant Laos. Laos was forbidden territory in 1967, and the political boundary of our war officially ended at its border. The tactical boundary was, however, not so readily delineated, and the covert missions run by Special Operations Group carried the war into Laos when the need arose. VMO-3 dedicated two gunships to SOG. Each was stripped of obvious national markings—stars, U.S. emblems—in a preposterous attempt to persuade a capturing enemy that we weren't American. "Official policy," the CO quoted ponderously, "is that you destroy your aircraft and escape on foot— no one is going to come in to save your ass." Marines are incurably romantic, and the policy, though possibly official, could not offset the higher, "mission impossible" pucker factor that came with SOG. In any case, policy was not the Code. Several SOG ships had bought the farm, and one was even torched by a gleeful crewchief who saw it as his best opportunity to get a new aircraft. The crews had all been saved by their chases or other aircraft scrambled from the nearest base. Even after so many months In Country, the Code was still all that mattered: *Marines take care of their own.*

VMO-3's SOG assignment usually meant flying gunship support for snooping and pooping Green Berets, or an occasional Air America Huey ferrying CIA spooks to remote landing strips deep inside Laos. This day was different. The day before, an Army Huey had been shot down amidst the treacherous hills of the Annamese Cordillera at the northern end of the A Shau. It was never clear whether they were snooping or pooping, but they undeniably caught some heavy shit, and the aircraft buried itself beneath the Laotian canopy too many klicks west of the border for anyone to walk out. In rapid succession, Navy and Air Force rescue helicopters joined the Huey, victim to the deadly A Shau gunners. Chase ships reported no apparent survivors, and the hills fell to darkness before rescue forces could be launched. After nightfall, someone activated a survival beacon, but it managed only an intermittent signal, and the ground commander at Phu Bai could not raise voice contact.

Early November 10, Special Forces launched an eighty-man search-and-rescue force from Phu Bai. SOG gunships, coded Cyclone 6-1 and 6-2, embarked with a covey of VNAF UH-34Ds to insert the mobile and lightly equipped Green Berets on the grassy slope of an A Shau hill several klicks from the nearest crash site. The insert went smoothly, and I watched over my door gun as the Berets fanned out into the elephant grass and vanished beneath the canopy.

The troops split into teams to search the jungle, trying to establish a fix on the beacon. We guided them from the air; the jungle roof was so thick we could see the crash sites only when we passed directly over them. There appeared to be no enemy about, but flying regular patterns in the confined valley made for nervous crews, and all hands kept their eyes skinned to the ground. The Berets entered the first crash site and found one body; the rest of the crew was missing. The site was too small to land in, and hoisting a corpse through the canopy would have made us too inviting a target, so Hamm ordered the Berets to transport the body to the nearest clearing and establish a hasty perimeter to cover us. We directed them to a steep hill, into a small area of elephant grass surrounded by heavy brush. Unable to land on the incline, Hamm taxied up next to the hill, slowed to a stop, then taxied sideways until the port skid locked into the earth. The slope fell away steeply under the hovering starboard skid. I watched the rotors scythe the grass farther up the slope, only inches from cutting into the hill itself. The Berets crouched low and approached along the longitudinal axis of the aircraft to avoid being struck by the rotors.

Kneeling in the door, I grabbed the stiffened body, heavy for so slight a frame, and hauled him into the gunship. Breathing was difficult in the humid air, and the flak vest gripping my chest didn't help as I reached out awkwardly to grab the corpse under his arms. The Beret shoved, I pulled, and we slid the stiff across the deck. They didn't call it deadweight for nothing, I thought. I flopped back in my seat, and propped my legs on the body for a moment to relieve the pressure of the vest. Christ, I needed air. I shoved the vest away, making room for my lungs to expand. Knipp cleared the Berets from the aircraft, Yantis kept his gun trained into the brush down the starboard slope, Hamm glanced back, and I thumbed-up "ready to lift," holding the corpse in place with my legs. Thirty seconds for the pickup, all in all, and we were off to Phu Bai. Not too bad.

The KIA lay in embryonic repose. He was an Army crewchief, clad in faded-green jungle fatigues. I wondered why the Army didn't wear flight-suits like normal people. At least like Marines. He was unmarked, proba-bly killed by the impact rather than by the shredding projectiles that had destroyed his aircraft. He was about nineteen, with a dark complexion turned an odd shade of green that matched his fatigues. You ripened quickly In Country. His uniform was stretched drum-taut by the bloated flesh. I leaned forward and prodded his swollen thigh with my forefinger. It yielded with elastic resistance, like poking a water-filled balloon. Here and there mounds of sickly skin swelled through holes in his trousers, encir-cled by the frayed tufts of worn fabric. Black hairs sprouted from them like coarse grass, whipped this way and that by the wind. He was a hairy bas-tard. Italian, maybe Greek. Hard to tell from the olive hue. Except for being green, he appeared quite normal. Dead, but normal. Black stubble sprouted on his face, and his hair was matted flat; the Berets must have removed his flight helmet before carrying him out. Too bad. I needed a new helmet, and the Army always had the latest issue. The whiskers both-ered me. I couldn't believe that any crewchief, even an Army crewchief, would fly looking so unkempt, and so I decided that what I'd heard must be true: beards do grow after death. At least he didn't smell, not that I recall. But after ten months of blood and urine and decay—who noticed?

We flew him to Phu Bai and gave him to the stretcher bearers. They took him away, to identify and tag, then zip snugly into a green plastic bag a few shades darker than his skin's ripe hue. They'd store him in a refriger-ated CONEX box, stacked neatly in the silent rows with other wanderers into death's green land: a corpse-sickle, ready for the long journey home.

Eventually, the Beret teams reached each of the crash sites. They found more bodies and confirmed that several crew members were missing. Our hopes faded when no visual signal, no flare or smoke grenade rewarded our efforts. We wearied of the search, but it was better flying the treetops while sitting on our asses than slogging through the growth beneath the canopy. I didn't envy the Berets. The beacon eluded us; it didn't emanate from any of the crash sites and didn't broadcast long enough to triangulate a positive fix. Was it a wounded man with a damaged set, wandering in the jungle, or the enemy leading us to an ambush? We needed readings on two headings, preferably three, to pinpoint the signal's source, but it would cut out before we could change position enough to compute a solid bearing. Several times we fixed a location beneath the canopy, but succeeded only in pissing off the search team as they cut their way through brush following our lead.

"Joker, this is Cyclone Six-One."

"Go, Cyclone."

"Pop some smoke. We have a new heading for you on the beacon."

"Roger, popping smoke."

The air was dense on the jungle floor; it took time for the smoke to filter through the green ceiling. Saffron mists, like a pollen cloud, rose slowly above the trees.

"Joker, I have yellow smoke. Confirm."

"Yellow smoke, Cyclone. Confirmed."

"New heading coming up, Joker."

The voice was tired and testy. "We hope you can do better than the last one, Cyclone. It's pretty thick down here."

Hamm passed low over the drifting smoke, lining the gunship with the beacon's suspected location. "Come to heading 275, 300 meters," Hamm ordered.

It took time for the Berets to chop their way through, and we spent it in attempting minor course corrections to give them a better heading. But each time, the teams arrived to find nothing, and the beacon seemed to leap a hundred, two hundred meters elsewhere. If someone was playing a game, they were winning.

We spent the day in chasing signals. As it passed, leaden clouds advanced upon the zone, the harbinger of the returning monsoon. At last it was time to quit; soon the valley would be consumed by cloud with the night not far behind. The Berets regrouped to their drop zone to await extraction by the VNAF, expecting to resume the search the next day. The gunships stood on station, impatient to return to Phu Bai with its promise of green beer and charred beef.

Instead of the VNAFs, the Berets received new orders. Dig in for the night. The Cyclone flight was ordered back to Phu Bai but remained on call for night operations should the team come under attack. There would be no dry, safe bunk for the Green Berets, but for God's chosen few—the Ball!

The Green Berets were pissed. Hamm offered half-hearted condolences that went unacknowledged. They shouldered their packs and headed up an A Shau mountain to dig in. Knipp mutilated the Berets' song as we lifted from the valley and found Phu Bai's TACAN radial.

> *Fighting Sollll-jurs from the skieeee,*
> *Fearless fucks, who march and DIEEEEEE.*
> *Eighty men got screwed this day,*
> *Those sorry shits, of the Green Toupee.*

"Isn't that 'those sorry *pussies*, of the Green *Bidet*'?" Hamm wondered.

"Um, that works too," Knipp agreed.

The monsoon followed on our heels. An hour after landing the ceiling fell to zero. A thick, drippy, wonderful ocean of soup choked the long flight line. Flight operations shut down in all squadrons—the best of all possible worlds. There'd be no night ops to interrupt the party. Yes, indeed, it had been a very good day.

The Ball began at 1930 hours. Day crews cleared the hangar and pushed aircraft under maintenance out into the mist. They sit there like plucked sentinels, stripped of their lethal arms, sad-tired in the rain. Icy tubs of beer and soda dot the matting deck, the ice itself a welcome novelty, and we suck on it despite the grimy tubs that encase it. Groaning boards hold deep trays of mess hall beans, lethal ammunition in the proper intestines. There is potato-ptomaine salad with funny-colored mayo, and slabs of warm bologna and Spam that sweat in the night mist. There is coarse bread with crunchy black specks, and enough mustard and ketchup to smother the taste of anything. In the hangar, drum-halves of burning charcoal wait to greet the steaks. A fine feast.

The first sergeant calls, "Attention!" and in keeping with tradition, the CO recites the proud history of the Corps. We fidget nervously and watch the beer grow warm as he speaks of other Marines' wars. Then, with an intonation proper to men at arms, he recites the deeds of VMO-3. Our deeds! The squadron grows silent. This is not the stuff of legends, learned at our DIs's knees. The story is our own. Rabble though we are, in filthy flightsuits and bush hats, we grow still and listen. He speaks of Khe Sanh, and the battles of 881 and Dong Ha Mountain, the A Shau, and other places VMO-3 has shed its blood. He speaks as if we are part of the long story. It seems that his words become one with the familiar tradition. Suddenly, in this foreign land and on this most revered of nights, we join the long line of men in green.

A cake is produced, laced in white icing with the emblem of our Corps splatted upon it. A small piece for each man. Top cuts the first slice with an M-14 bayonet, his right as the oldest in the outfit. He steps forward in rice-starched utilities, black chevrons glistening night-wet on his collars, six up and six down with the diamond of first sergeant. Top draws the blade slowly, with a reverence that carries the CO's words into the cutting of the cake. The cake is cut by the oldest, and then the youngest—thus our Corps survives.

Top turns solemnly to pass the bayonet.

To Lewish.

Holy shit. Lewish.

A groan rises from the crowd. The slimy little bastard still has the clap from his last R & R, and Top stares at him in disgust as Lewish reaches for the bayonet. The clap is still bad enough to leak pecker tracks through his utilities, and Top looks as if he'd rather carve Lewish's dick than let him touch the cake. But the Ceremony is the Ceremony, Olde Corps passing the baton to the New Breed, such as it is. Top extends the weapon reluctantly, holding it by the blade near the base of the blood groove so that Lewish won't befoul anything but the handle. There is little chance the dumb shit washed his hands. He slices once, then sets the knife on the table and grins. Someone produces a camera, a semi-tradition to capture the Olde and the New and freeze it for eternity. Top stands woodenly, refusing the photographer's urging to shake Lewish's hand.

The Ceremony over, we form a long line, snaking through the hangar as the steaks burn on the grill. Rank has no privileges: Olde Corps and New Breed, officer and enlisted, line up alike, clutching wet cans of Crown Korean, sucking the green 3.2 as if it is stateside Michelob. Satisfying clouds of smoke waft through the crowd, then out the doorless entrance into the soft mist of the monsoon. The mist and the fleeting spell of the Ball transform the hangar's harsh work lights into party lights, but beyond their warm glow chill cloud shrouds the flight line. Cool droplets of water float in the air, dot the chill metal, building, joining, glistening in the party glow, and pat, pat, patting all about us. All but the nearest aircraft are lost to the consuming fog, and the steel matting that ubiquitously frames our world vanishes as it stretches into the night. The hangar seems an island locked in gray-green swirls of cotton. I feel an insular, twilight sensation, comforting and false at once, but not unpleasant. It is good to be among my own kind.

In the hangar, the mood is light. Phelps is ahead of me, Moose behind. We suck at our brew and pass gas as we shuffle toward the food.

"Can you believe Lewish?"

"I can't believe it, he didn't even change his trousers."

"I thought the Top was going to puke."

"I thought the Top was going to cut his dick off, then puke."

"I wonder if he'll get mess duty."

"After the clap clears up, I hope."

Lewish walks by.

"Hey, Lewish, easy on the brew. You're going to be bending pipes if you don't watch it."

Lewish grimaces, and apes grappling in pain with the water pipes of a urinal, holding on while taking a leak. He toasts us with his Crown.

"Fuck it, man, the pissers here don't have pipes."

"Yeah, well, just keep your pee-ater in your drawers."

Lewish walks over to us, snuffling conspiratorially. "You want to hear something good?"

We close rank around him, in spite of the pecker tracks. Lewish is great for gossip, and there are damned few secrets where 153 enlisted men sleep, eat, and shit together.

"You know Samson's ol' lady?"

Samson is a former crewchief who fucked up so often at so many things, the line chief assigned him to permanent guard duty, which means he has most days off in the hooch. He spends them porking the hooch's mamasan, a dumpy, pleasant Vietnamese who is happier making her piasters on her back than washing the hooch's clothes.

"Yeah, yeah," we say. This is going to be good.

"Well, I've been porking her since I came back from Kuala Lumpur. You guys can start watching Samson's trousers, now."

"Snuff, snuff, snort, snort, snort," Lewish walks away. "Snuff, snuff."

I am ten men away from the grill, ten fucking men, when Timmy, the ready-room watch, runs into the hangar.

"SOG crews up!" he shouts.

Hamm and Knipp are on the far side of the hangar. Yantis is working on his steak. I glance from Timmy to the grill. One of them is mine, goddammit. I look at the cotton-mist surrounding the hangar, then back to Timmy.

"Up your ass," I say, "we can't fly in that."

Hamm passes by, peering doubtfully into the weather. "Hey, Captain. Captain! Do we have to go out in that?" I whine. I showed him my empty mess plate, like a supplicant begging for gruel. "I haven't eaten yet."

"Hold your place in line. I'll see what's going on," Hamm says.

I yell to the grill-master, hope fading. The fucking Crotch is certainly dumb enough to send us out on a night like this. "Gimme a steak. Any steak," I plead.

"How much is it worth to you?"

"It's worth your ass."

We begin to argue, but Hamm returns quickly.

"The Green Beret are being hit. They want us to pull them out. Get the aircraft unbuttoned."

I scream at the grill-master, "You better save one of them for me, you bastard," and hang my empty tin in the corner of the hangar.

We dash onto the mat, throw the engine and gun covers to the ground crew, and turn up quickly. The rotors swirl the mist about us, softly reflecting the red and green running lights against the glistening airframe. We lift to a low hover; a chain of taxi-directors with glowing wands leads us through the fog to the runway. The ARVN's small LZ is only a mile away, but Phu Bai GCA has to talk us all the way in. It is fucking ridiculous. How can we find our way to the A Shau if we need ground-controlled approach to hop over to the LZ next door?

The pilots head for the Berets' briefing room, while Yantis and I guard the aircraft. We are in enemy territory—ARVN and Green Beret—and they can strip the ship in moments. Two VNAF -34s are parked in the LZ, their ARVN crews also standing sentinel against scavengers, including us. Our pilots return quickly and brief us as we strap in. Ahead in the fog, the VNAFs start turning their rotors. A probing NVA force found the Berets astride their mountain. Under assault and running low on ammo, they called for an emergency extraction. The Berets' commander has ordered Hamm to escort the two VNAFs to the zone and cover them during the pickup.

Right, I think. A night extract in this shit, under fire in heavy canopy. Right.

The -34s lift and fill the ether with excited Vietnamese. We follow close on their tail. It doesn't sound as if they are any happier than we are.

Hamm cuts in, "Cyclone Six-Three, speak English."

The chatter continues.

"Cyclone Six-Three, this is One. Speak English!" Hamm repeats.

"Cycrone Six-One, this is Cycrone Six-Three, ceiring is too (something Vietnamese)."

The VNAF voice is agitated, pissed, and difficult to understand even in English.

"What the hell did he say?" asks Knipp.

"Something about the ceiling, I think," says Hamm. "I did not copy your last, Three, please say again."

"No fucking visibility!" yells the VNAF plainly.

Hamm laughs, "I copy you on that, Three, but we have to try."

We can barely see them in the fog ahead of us. Hamm closes on their running lights, then shifts to starboard and comes up level on their three o'clock. I can see the dark shape of the -34 across my gun, its red collision light sweeping in the mist. The close formation makes them nervous. Hell, it makes me nervous.

"Cycrone Six-One, fall back. Fall back! You are too close," the VNAF orders.

Hamm ignores the outburst. "Cyclone flight, maintain this formation so we don't get separated."

We rise slowly, in radio silence for several thousands of feet, straining to top the IFR[1] and the fog-shrouded mountains between Phu Bai and the northern A Shau. The air grows thin, and we lose power steadily as we climb. At eight or nine thousand feet, the Huey's begin to pull ahead of the older -34s. Hamm has to throttle back to maintain formation. It is plain— the VNAFs are peaked out; the -34s can climb no higher.

"We cannot go farther," radioes the VNAF leader. His voice expresses the obvious fact with obvious relief. "We return Phu Bai, now!" he exclaims and changes radio freqs before Hamm can respond. The VNAFs dive earthward and vanish into the cloud. Cyclone 6-2 holds steady on our six as Hamm takes the gunships higher, searching for clear sky while Knipp tries to raise Phu Bai to report that the extraction is aborted. Hamm switches his freq to the Beret team for a sit-rep, and hears the team's operator on the net to his commander in Phu Bai. The team is under heavy fire—the rattle of MG and small arms chatter like static in the receiver.

Hamm calls the Berets' commander. "Blackjack Six, this is Cyclone."

"Go, Cyclone," answers the Beret.

"Blackjack, the extract is no joy. VNAF is returning to base. I'm looking for the top of this weather." Hamm addresss the team. "Joker, what's your visibility?"

---

1. IFR is a region of low visibility requiring the pilot to conform to Instrument Flight Regulations.

"We've got a hole in this stuff right over the valley, Cyclone. I can see some stars. Don't know how long it's going to last. We need your guns in here. Can you make it?"

"I'm passing ten thousand now, Joker. Still pretty thick out there. I can't find your hole if I can't see it."

"Cyclone, this is Blackjack. I've got a flight of A-1s en route, but they won't arrive on station for another forty-five minutes. You're the closest we've got."

The A-1E Skyraiders—piston pushers—are slow, highly maneuverable aircraft that carry greater than their weight in high explosive. Jets have difficulty navigating the twisting valleys in the best of conditions, and these are far from the best.

"Roger, Blackjack. We'll do what we can."

I watch our airspeed drop as Hamm pushes the aircraft higher. Our normal cruising speed of 90 to 120 knots falls to 60, then 50 as the air grows thin. In a few thousand feet, we will peak out as the -34s did, unable to maintain more than a tenuous hover in spite of the screaming engine. There is no danger to the aircraft as long as the engine holds together. We'd simply dive and reestablish power at a lower altitude. But we can't locate the zone unless we escape the soup. Hamm is following a TACAN radial from Phu Bai, and we know the approximate distance to the team, but "approximate" isn't nearly good enough, and it is easy to wander off a radial. A degree or two would place us kilometers away from the team. In spite of the technology that assists us, in the end it is naked eyeballs that do the job. We have to be able to see the zone.

We break through the clouds at eleven thousand feet. Hamm calls the team.

"Joker, we got lucky. ETA your position in one-seven. I hope your hole is still there, Joker."

"Roger, Cyclone. ETA one-seven. We're being hit; NVA are massing in the valley beneath us. Bring in those guns."

It is bone chilling above the A Shau mists. There is a crescent moon, a meager cold sliver shedding little light, and the harsh stars pierce unlovely through the southern sky. Beneath us, the monsoon blankets the land in impenetrable fog. The Cyclones labor through the night. Yantis and I huddle in the center of the aircraft, as far away from the windblast shrieking past the open cabin as we can manage. Hamm and Knipp, sheltered in the forward fuselage, give us their flight jackets, and we press together beneath them, seeking warmth. The Huey's meager heating system is shut down—it siphons too much power from the engine.

It is so cold.

"You're remembering the flight," Tom says as I fall silent.

"Yes." I gaze into the chilling memory. So cold. Cold enough to penetrate the years and breathe its keening wind into the quiet of the Center. I see Yantis and me sitting in the cabin, our weapons hanging darkly in the door frame, glinting dull red in the glow of the instrument lights. Strange, I remember the warmth I drew from him, along my left side where our arms and legs pressed together, the warmth of my own breath flooding across my face buried beneath the flight jacket, the smell of sweat, the damp, soft touch of the fleece against my face. But I cannot feel the warmth. It is gone, and the loss is chilling. The fleeting warmth in that short flight is as a memory of a memory. Empty. I cannot touch it.

"What altitude was it?" Tom asks.

"Eleven thousand. It was a lot of altitude for a helicopter." I answer his question, but I am distracted. When did the warmth leave me?

"Had you been that high before?"

"Not in an open gunship."

"Is that why it was so cold?"

I breathe deeply, and the deep cold once again enters my lungs. Yes, it was high, and cold, and my threadbare flightsuit was damp with monsoon mist. But the dead chill runs deeper than altitude and weather can explain. I see the answer. It lies at the bottom of a hole, on a TACAN radial out of Channel 69, ETA one-seven and closing.

At ten minutes ETA, Hamm raises the Berets and finds good news. Landshark has located an O-2A spotter and directed him to the trapped team. The O-2 has found the hole and is on station, and has even made a firing run with Willy Peter to keep the NVAs' heads down. The spotter is circling the hole as we arrive and leads us quickly down. The firefight is distantly visible at its bottom. A fickle providence, not to mention the unpredictable winds of the northern A Shau, has carved a huge funnel of clear air open to the heavens. At one end, the harsh stars twinkle; at the other, lines of red-orange leap from nothingness, trace short-lived paths, and disappear into nothingness, like an angry spiderweb of flame cutting through the night. The O-2's dim cabin light is our only beacon as we spiral into the hole. Our own marker lights are extinguished to avoid drawing fire. Hamm studies the altimeter and levels off at 1,200 feet amidst the hills and ridge lines, blacker than black, that loom around us. The hole feels a mile or so wide—difficult to tell. It is damned little airspace for three aircraft, and collision is as great a danger as the enemy's weapons. Ground fire marks the Berets' position halfway up the mountain ridge they climbed at dusk, near one side of the hole. A narrow valley carves upward through the ridge, separating it from another that disappears inside a wall of cloud. It is from this cloud, along the ridge and into the valley, that the NVA have come.

The spotter accelerates away from us.

"Cyclone, this is Charlie Oscar. The main force is in the valley between the ridges on the west end of this hole. They've got small arms and 12.7. I will mark with Willy Pete."

Hamm rogers the O-2 and raises the Beret team.

"Joker, this is Cyclone Six-One. Charlie Oscar is starting a marking run. When I give you the word, cease fire and flash your survival strobes one time. I say again, one time, Joker, then cease fire. We will fire on the muzzle flashes. Do you copy?"

"We copy, Cyclone. One flash and cease fire."

"Joker, how close is the enemy?"

"Ten meters, Cyclone, maybe fifteen beyond our perimeter. They're downslope from us, coming up through the valley. We can hear them talking."

The spotter flashes a red beacon to signal his position. Tracers search the sky but fly by harmlessly as he banks away. The O-2 begins a steep marking run, unleashes two Willy Petes, and pulls out sharply. Muzzle flashes burst from the valley's floor, and tracers chew the night sky, some hurtling upward to challenge the harsh stars, others disappearing into the monsoon mist. The spotter is a dark ghost beneath us. He jinks to port, then starboard, twisting away from the zone and the killing fire. One rocket sprays dirty yellow flame in the valley, the second on the slope below the Berets' position. The flames flicker briefly on the sodden earth, then die, reflecting only for moments within the lethal cloud. But the flash is marker enough, and we track in on its fading glow as the O-2 pulls away.

"Cyclone Six-One in hot! Joker, mark your zone," orders Hamm.

The ridge looms at ten o'clock, blacker than the night sky. Hamm aims to bracket the O-2's rocket. Our dive steepens, but the ridge remains in darkness.

"Joker, mark your zone. Mark your zone," Hamm repeats. We need a reference point, something to fix the relative position of the friendlies for an instant.

Three brilliant bursts of light pierce the night in response.

"Roger, Joker, we have your marker. Stand by and cease fire." The strobes flash once; Hamm corrects course and launches his first HEAT. The cylinders leap from their tubes, and we follow the flaming wakes to earth. The fire searching for the O-2 turns to us. Hamm fires two more HEATs, farther upslope in the direction of the Berets. More muzzle flashes erupt from the jungle, spears of tracer light hurtle toward the gunship, sweeping parallel to our axis and beyond into the night. "Guns!" Hamm orders. Knipp switches the firing selector from ROCKETS to GUNS, and Hamm brings the M-60s into action, raking the enemy with 7.62. Knipp grips the firing handles of his TAT, and the twin -60s follow his gun sight as he sweeps long bursts across the zone.

Hamm banks to port, and I open up. I hear Yantis on the starboard gun; it is difficult for him to aim, but he sprays suppressive fire liberally along the right flank; the team is on my side of the aircraft and out of his range of fire. I turn my weapon against the muzzle flashes bursting beneath us, walking the tracers from one position to the next. The night is full of them. I chase one, then another, then still another. They seem to go out, only to appear elsewhere in the valley or up the ridge in the direction of the Berets.

Hamm pulls out at fifty feet and courses through the zone, following the valley rising with the ridge. It curves to port; the wall of cloud is not far beyond. We cannot enter, or we'll be lost. Hamm pulls the left bank tighter, straining to put the Berets' ridge between ourselves and the enemy fire. At six o'clock, Scarface Six-Two begins his run. He unleashes his HEAT, then goes to fixed guns. I watch for Two to draw off the fire, but he does not. We are too close, still the ones they want. We must clear the zone to cover him. Hamm's fixed guns go off target, and he ceases firing. We are trapped in the rising valley; safety is in the blackness behind the ridge. Death is on the hill, and in the cloud beyond. Knipp's TAT is fully depressed; he rakes swaths of tracers down, point-blank into the muzzle flashes passing in a blur of speed, then goes off target. Hamm's steep bank to port has Yantis out of action. He stops firing to conserve ammo for the next run.

I am the only gun on target. The flashes and tracers burst up. I fire almost straight down, grabbing the handhold above the cabin door, pressing my shoulder into the gun to steady myself. I am welded to the aircraft, the weapon—a piece of the machine. I feel its power. The gun fires, fires, fires, with a life of its own. Hot brass ejects into the cabin, thrown against my cheek by the wind stream, burning. Power! I feel the power of the gun. I track against the flashes, short bursts. Ten o'clock, swivel, fire! Nine o'clock, pivot, fire! New flashes, higher up the ridge in the direction of the Berets. Four positions, flanking steady fire. I track the new targets, arcing tracers in graceful, deadly curves as the bullets shed the forward speed of the aircraft. Fire low and work uphill, holding a tight pattern on the flashes. Converge, consume, annihilate! One position, then another fades to darkness. Destroyed. Fire at will. Power! Fire, fire, fire!

"Cyclone gunship, Cease Fire! Cease Fire!"

Fire, swivel, eight o'clock, fire! Position three destroyed! Lock on, pivot on four. I am the weapon! Fire, fire . . .

"Cease Fire! Cease Fire! You are firing on our position!"

What? What?

"Cease Fire Cease Fire Cease Fire!"

I fall back from the gun; it bounces in its mount, then swings aft in the wind stream, lifeless. The deck is slippery with cartridges rolling under my boot. I come down hard; they burn my ass.

Hamm rounds the ridge, skirting the monsoon cloud.

The Beret's scream echoes off the chill mist: "Cease Fire! We have wounded in the zone!"

Hamm: "Joker, say your situation."

"We have at least one wounded. Machine-gunned in the stomach and the groin. Your door gunner shot up our position."

I sit where I've fallen. What. . . ? How could. . . ? My gun. . . ? My gun! Stomach and groin. My God! Oh, my God. And we can't get him out. We can't get him out. *Oh, my God.*

We round the ridge into safety. Behind us, Cyclone Six-Two is in his run. I sense his dive and the gauntlet of fire rushing up from the zone, but it is unreal. I sit on the flight deck amidst the brass and the trash of battle, and my heart grows as cold as the dead mist outside the aircraft.

The Beret operator raises us again, urgency and accusation and hopelessness in his voice. "Cyclone, Cyclone, we have one wounded in the zone. Your left gunner raked us on the pass. We have one wounded in the stomach and groin. You shot up our supplies, Cyclone, medical supplies destroyed."

Six-Two is in the gauntlet; Hamm starts his second pass. I rise to my weapon and reload, but it is useless. I am useless. I cannot fire. I try. I try to pull the trigger. I fucking try! But the weapon is impotent, its power has vanished. All I can see is him. Stomach and groin, entrails seeping through the clenched fingers. I'm sorry . . . I'm sorry. . . .

The A-1s sweep in, led by the O-2 spotter. We don't stay to watch. The Cyclones are ammo-minus and low on fuel; we circle high to clear the mountains, pick up the radial to Phu Bai, and reenter the cloud.

Yantis and I huddle once again beneath the pilot's jackets, but there is no warmth. My mind is drawn inward, and everything is as black and chill as the monsoon.

I failed. I am crewchief, and I . . . have . . . failed. No one speaks of the Beret. The twenty-minute fuel light teases, then threatens, then glows madly red, and the pilots' faces are tense as they nurse the power settings to eke out the gallons we need to get home. GCA brings us down a glide path to the Berets' compound; it is a few klicks closer than the main base at Phu Bai. We shut down next to the ARVN -34s. Their crews are gone. Some other time, we might rifle their aircraft, out of need or out of spite. But not tonight.

The pilots leave us to report to the Berets' CO. I am glad to remain with the aircraft. My shame is deep, and I know what I have to do when we return to Phu Bai.

Yantis is there, and Six-Two's crew also. My friends, my own kind.

"Leave me alone," I say. "Just leave me alone."

"Zack, what happened up there?"

"Leave me alone."

Arms clad in green on my shoulder. Dark faces speaking, caring in the night. But they cannot know.

"It was a hot zone, man, the dinks were in tight. You couldn't help it."

"Leave me alone."

"You couldn't help it, man."

I turn to them, not in anger but shame, and I let the pleading show in my voice, faltering, begging. Stomach and groin. I am crewchief, and I have failed.

"Leave me alone!"

Hamm emerges from the darkness. I busy myself with the aircraft, avoiding him.

"Come on in, they saved some chow for us."

I search for an excuse. "I'm not hungry. I'll guard the ship, sir."

"You need some chow. Come in. That's not a request," he adds.

I follow him silently. The Berets' CO greets us as we enter the mess hall.

"You men did a terrific job. You saved the team. You held the NVA off long enough for the Skyraiders. The team wouldn't have held if you hadn't made it through the monsoon. We'll pull them out in the morning."

He passes among us, shaking hands. The Beret major is a big man with a broad grin.

It comes my turn, but I cannot take his hand, and have to force myself to meet his eyes.

"Major, I'm the one who shot your man. I did it. I don't know what happened. I don't know. I just fired on the flashes. And I'm good, Major. I'm good! I just. . . ."

The words tumble. Stomach and groin. Oh, my God.

"I did it, sir, I shot your man in the zone. I killed him. I'm sorry." I want to cry, but the need shames me too. Marines don't cry. I choke back the tears and stare at the gritty deck of the mess hall.

"You saved the team, Corporal. That's what you did out there tonight. You saved the team.

"You did your job, Marine." He phrases the word carefully, with the emphasis on the "rine" and the understated "Ma." He says it the way men do when they mean it to be a title, and an honor.

"But you couldn't accept that, could you?" asks Tom.

"Accept what?"

"He tried to let you off the hook—your own hook—you couldn't let him, could you? Your pilots knew it was an accident. The rest of the crew knew it. They were there. They knew what it was like. They were professionals, like you. They were good pilots and crewchiefs and gunners, like you. You couldn't let them let you off. You couldn't let yourself off."

His tone challenges me, and I grow angry.

"What the hell do you know? You tell me, how you can fabricate an excuse for shooting a man's . . . a man's . . .

"You tell me how it can be OK to shoot a man in the . . ."

The words carry images I cannot bear.

"What the hell can you do to fix that? What do you fucking think, you can say, 'Gee, I'm sorry—it was a mistake,' and that makes it all better? You think being sorry makes it better? I was a crewchief. A fucking crewchief! Those guys depended on me. I was the only weapon in the ship. It was all up to me. I had to be per . . . per . . . I had to . . . be. . . ."

I am railing, and know it. And I know that what I struggle to say, "I had to be perfect," is nonsense. But the zone is unforgiving. The zone does not accept excuses. What excuse can you give for destroying a life? What words or circumstance can excuse that? And to compound my shame, I had frozen, and in freezing had lost self-confidence and self-respect.

The Berets' CO makes us sit, and ushers his mess cook in. He smiles. "We heard it's your birthday."

Steaks.

I sit and eat. It is expected. I pick at my plate, and know what I have to do when we return to Phu Bai.

The ARVN dumps a few jerry cans of aviation gas into our gunships to cover the short hop home. It isn't jet fuel, but it will burn. GCA guides us back through the mist, and taxi-directors lead us to the hot pad. It is past midnight when we touch down. I secure the aircraft and straddle the engine nacelle to tie on the exhaust cover.

Hamm looks up at me. "You did a good job out there, Zack. I'll see you in the morning."

He forms the words curiously, as in a question, and waits for me to answer.

I pause a moment, then make up my mind. "You'll have a new crew-chief tomorrow, sir. I'm finished. I'm turning in my skins."

"Don't do that, Zack. You're a good crewchief. What happened tonight could happen to anyone."

The compliment is bitter, the excuse shallow. It did not happen to anyone. It happened to me. I cannot forgive.

I finish with the aircraft. Yantis and I walk past the hangar on our way up the flight line. The birthday ball is over. On the grill lie eight charred steaks, grim and wet with the night's dew. They did not forget us after all. I leave Yantis and head for Operations.

Major Burke has the duty, and the expectant look on his face tells me Hamm has already spoken to him. Crawly things churn in my stomach. A hot flush burns as I stand before him, and when I begin to speak, I can do nothing but stutter.

"S-s-sir, I-I-I want t-to t-turn in my fuh-fuh-flight s-s-skins." I am angry, and frustrated—I can't even resign flight status properly. And the shame is crawling back as well. God, the shame!

Burke leads me to a chair.

"I spoke with Captain Hamm. Do you want to talk about it?"

I pour out the story of the zone, needing to talk, searching his face for signs of revulsion, accusation, anger, disgust—anything. I find none, only understanding. He listens with patience as I stammer through the tale of darkness and fire, and the pain. Once again I hear the voice of the opera-tor—"wounded in the stomach and groin."

"And, Major, I froze on the gun, too," I say. "I couldn't bring myself to shoot. I lost my nerve."

It is like going to confession—Father, I lied three times, I stole, I forni-cated. I shot a man in the balls. I lost my nerve. Mea culpa, mea culpa, mea maxima culpa. But no Act of Contrition or ten Hail Marys can turn back the night and make things as they were. No litany of Our Fathers can drown out the radioman's accusation, "Your door gunner shot up our position. . . ." No penance can absolve the terrible sins of this night.

I finish the tale, and know in the marrow of my bones that I am finished as a crewchief as well. The evidence is damning, the sin deserving of condemnation.

I knit my fingers and press them to my lips, crushing bone and flesh against my teeth, trying to lose myself in the pain. We sit at the plywood table where the officers play cards between missions, and I stare into its harsh surface. The night assault plays over and over again in the swirls of grain, like the flowing contours of the A Shau. I see the aircraft, racing along the ridge. Eight o'clock, pivot, fire! I shake my head at the vision, press my eyes shut, then stretch them wide, but the images will not fade. I was on target. I was. *What the hell went wrong?*

Burke has listened without question, like a Father Confessor who knows and understands the sin. The stillness in the hut is complete, emphasized by the soft drip of the monsoon. Burke leans forward, and knits his fingers, like mine. Staring into them, he speaks softly. From afar, we might look like brethren of some monastic order, meditating, praying.

"I was on patrol, north of Phu Bai. Scat One-Four, you know," he begins.

"Sure. Yes, sir," I mumble. I keep my fingers pressed to my lips, staring into the table.

"Landshark scrambled us to cover a Marine unit caught in the open. They were out of ammo, and the fighting was hand-to-hand by the time we got there." Burke's voice grows thin, his eyes distant, as the ghosts of that day come alive in the room.

"'We're overrun,' the grunt operator said. 'Fire on Red Smoke.'"

I share the fear and dread in his voice. A new nightmare penetrates the room. I watch Burke remember, relive the pain.

"We attacked. I ordered the other guns to stay out of action, so that no one else had to . . . had to. . . . I made one pass. I killed three Marines. Three."

The word lingers in the chill air, and the number becomes men to me, dressed in tattered green, sprawled in terrible death on the torn earth. Friendly fire—the hopelessness of the zone—the fear. I understand the pain, and share it with him.

"The VC broke when they realized the Marines would sacrifice themselves. They ran, and we killed them. But I killed . . . three. There was nothing else I could do."

Suddenly, Burke asks, "How many missions have you flown?"

"Three hundred and nineteen," I answer. The number seems larger this night. It makes me feel old and tired.

"Did you ever shoot any of our guys before?" he asks.

"No, never!" I say with indignation. "I only shot at the muzzle flashes. I'm good on the -60, Major. I'm damned good. I never miss my target." But then, in a small voice, I add, "Not until tonight."

I remember the exploding lights. Nine o'clock, swivel, fire! I did my job, welded to the weapon and the aircraft. I did my fucking job!

Burke speaks softly, persuading, explaining. "Zack, you know it was a tight zone—night firing, no visibility. The Berets knew it too, when they called in that air strike. They knew the odds, they took their chances. So did you. Captain Hamm said you did everything right up there. What did their CO say to you?"

"He said I saved them," I answer flatly, not wanting to believe.

"You did. You all did. Look, all I know is that if it weren't for all of you, including you, Marine, there might be eighty dead men on that hill right now."

I stare into the table. I am so tired. He is trying to absolve me. Is this what I must do to gain absolution? Weigh the one against the many? How much does a life weigh, when it is my finger that tips the balance? Mine! I want to believe him. I want to let his logic soothe me. But how do you absolve yourself when you've destroyed your own kind? Has Burke gained absolution? Can he? I see the pain in his eyes, the haunting terror of his memory. I know that he did only what he could. I can absolve him, I know. I understand. I can play Father Confessor and relieve him of the blame that shows in his eyes, that I hear in his voice. Yes, I can. Three men died under his guns, and many lived. This night, one man may die because of me, though the others will live. I understand him, and forgive. Can I forgive myself?

Burke continues, persuasive. Persuading.

"You're a good crewchief, Zack. We need you. We need you to train the new guys, to show them how its done. We need you to fly. The Green Berets in the zone—you helped save their lives. They need you in the morning."

I am too tired to feed the sorrow or nourish the shame and too tired to think. Burke's words hold peace and rest and acceptance, and if I accept them, I can live with myself, at least through the night. I need the night and sleep. I need so badly to escape.

"You've done your job," he says. "Three hundred and nineteen missions is nothing to be ashamed of. If you want off, you can have it. But think about it a while. Think about it in the morning."

I thank him, and walk across the silent sand. The night is damp against my face, like the breath of a cold tear. I find my hooch and walk carefully past the sleeping forms, undress, and sit at the edge of my cot. I reach into my desk and find the rosary in its hiding place, and carry it with me into the sleeping bag. The damp cotton is chill against my flesh; it will take hours to return the warmth it steals. I finger the wooden cross with the relic. The guys don't know about my beads. It is, perhaps, the only secret I keep in this place. I try saying it each night, never finishing. Sleep comes with the second Hail Mary, or the third. At least I have the intention, and the nuns said that is enough. Sometimes I pray for myself, sometimes for my folks or my girl. Sometimes I thank God for letting me live through the day. Sometimes it is sincere, and other times I am just sucking up to Him, hoping He'll give me something I want because I am being nice enough to remember to thank Him for my life. I think about praying for the Beret, that he will live. I think about it. But I do not. Instead, I pray that it should all go away. The memory of the day, the shattered Beret. This night—I want it to all . . . go . . . away.

The night watch is shaking me. "Zack, boy, up and at 'em." It is 0500 hours.

I raise up on an elbow. "You awake?" he asks. I stare into the gloom. The face is happy to be coming off watch; he does not know about the night.

"Uh, huh."

I lie back in the sleeping bag, warm against my flesh. The memory of the night seems like a distant nightmare. The numbing process has begun, the forgetting. I recall Burke's words. They are all that seems real. I want to believe them, I want it so badly. I know what he said has truth. I believed so firmly that I fired only into the enemy. I did save the Berets. I did do my job. *That* is all that matters. The need to explain how my fire went astray is gone. His words of praise, "We need you to fly," touched me. Perhaps it is bullshit, perhaps not. I need them, and the need is all that matters.

Three hundred and nineteen missions. Burke said it is enough. What should I do? But Burke is still flying despite his tragedy. We began our tour together, and he is still flying. Can I do less? Would a Marine do less? I need to believe him. If I don't, what else do I have? Only the pain and the shame? And what about the fact that I froze off the gun? Can I trust myself? Burke trusts me. It is implicit in his words, "We need you." Shouldn't that be enough? Can I forgive myself the loss of control? No, it is too much to ask. I cannot forgive.

But I will not quit for that reason, either. The mind, like the machine, can be controlled if there is will. I steel myself. I will not fail again. I will not lose control. I will not! I cannot forgive, but I will never fail again. *I will never lose control.*

"We need you. . . ." Burke chose his words well. Are these the words that serve him, also?

I resolve to believe the words. I will myself to believe them, to accept this absolution. I dress quickly and start for the flight line. Yantis has prepared the aircraft, wondering if I'd show for the mission. Hamm has pulled the preflight, and sits watching at the controls as I walk across the mat. My quandary has caused me to be late for flight ops, which warrants the most severe of ass chewings. I run to the gunship, climb into the crewchief's position, and cinch on the gunner's belt, saying nothing. Hamm and Knipp look back at me.

"Glad to see you could make it, Zack." He says it without sarcasm. I offer no explanation. Hamm raises "thumbs up," a double meaning in the air wing. "Ready for take off?" is the jargon of our trade. "Gung ho!" usually receives an up-thrust middle finger in return. This morning, it means something of both. "Thumbs-up," I respond, and lower the smoke-tinted visor to shield my eyes. He understands. They all do.

The growing day burns the mist away as we fly into the A Shau. A VNAF -34 is already en route, and we join him to extract the wounded Beret. It is a quiet flight, the sun warm on my flesh. I feel a distant tremor of shame, but it comes only once. I cannot change the night, and do not have to look back upon it unless I choose. Those with the power to absolve granted absolution. The Berets' CO, my squadron mates. And I was in control.

The valleys and the Green Berets' mountain show little evidence of the night's battle. The forest consumed our Willy Pete and HEAT. A few small craters give testament to the Skyraiders' strike. In a few days they will fill with water and join the thousands and tens of thousands of their brothers until the jungle reclaims them. Forgotten.

The wounded Beret is alive! A team has carved an LZ out of the elephant grass. A VNAF lands and drops new supplies to replace those expended and destroyed in the night's carnage. The team leader speaks with Hamm as we orbit the zone. I can see a wounded man being carried down the slope to the waiting -34. The shame rises again, but only slightly. It is a new day, and I resolve to feel nothing. I am in control. Still, I am glad the man cannot see me, glad I don't have to face him in my own aircraft. As the wounded Beret vanishes into the dark maw of the -34, the radio chatter seems to fade. The words become distant and indistinct and are lost as I stare into the night's memory, drawn to it against my will.

"Did you get that, crewchief?" Hamm's question brings me back to the living.

I fumble with the radio switch. "What, sir?"

"The guy's going to be all right, Zack. Didn't you hear what they said?"

"No, sir. I didn't copy."

"He's going to be all right. You didn't ding him that badly." Hamm laughs, a smile mixed of encouragement and relief spreading under his smoked visor. Knipp smiles, and Yantis—all smiling at me in the aircraft.

Hamm keys his mike. "You shot him in the ass! Just in the ass. It splattered blood over his groin. Just a few flesh wounds. He's going to be OK."

"And it wasn't your fault," the pilot continues.

"What?"

"The Berets went for cover after marking their zone with that one flash, just like I ordered. The gooks were in tight, and when we returned fire on their muzzle flashes, some of the Berets got nervous. You were shooting less than ten feet from their front line. Four of them put their strobes back on to mark their perimeter, then you started tearing up their zone. They dropped the lights and ran. You shot them out. Can you believe it? A hundred knots at night and you shot out their goddamned strobe lights. That is what I call shooting. One of the Berets tripped, and you shot him in the ass before he could get away from the lights. If his buddies hadn't dragged him into the darkness, you'd have waxed him for sure. Some shooting."

It was their fault, all their fault and not mine at all.

The shock of the news rocks me, a flash of amazement like flares bursting in my flight helmet. *It was their fault.* I make myself laugh with the pilots, wanting the humor to flush the night away. We all laugh. Hamm and Knipp smile as the Cyclone flight meanders through the A Shau. Yantis leans against his gun and smiles, his flight helmet nodding in agreement. I nod too, and smile.

We remain on station throughout the day, but the beacon transmits no more. The Berets search but find no one, and Blackjack terminates the mission late that afternoon. The night was for nothing.

The story of the wounded Beret spreads through the squadron. In the sameness of the days, the black humor in it brings a few good laughs. I play at enjoying my role.

"Zack! Did you really shoot him in the ass?"

"Where else would you ever get a chance to shoot a Green Beret?"

"Tell it again."

"Well, we were covering the fucking Green Toupees up in the A Shau. Army, Navy, and Air Force birds got shot down, so naturally they sent us in. . . ."

"Ho, ho, ho, ho. . . ."

The laughter echoes through the years. Chill and hollow, unforgiving as the memory of that A Shau night.

"Ho, ho, ho."

18 NOVEMBER 1967—THE AN HOA BASIN, QUANG NAM
PROVINCE, R.V.N.

For a week Hamm, Knipp, Yantis, and I share the same aircraft, the same missions. We complete the SOG assignment and pick up a new one: fire and support for a Marine battalion in the An Hoa Basin. We are sent to the *Tripoli*, shifted south to support the grunts. It is OK duty. The helicopter carrier's bunks are clean and dry, and the chow plentiful if not exactly prepared to please the gourmet palate. It would be great duty—if the ship weren't run by the fucking-Navy.

The fucking-Navy runs things about as close to prison as you can get. Shipboard life is a gray world of whistles and loudspeakers. Although there are many places to squirrel yourself away aboard ship, it is impossible to fuck off in peace and quiet. The bosun's pipe penetrates all hiding places.

Woooooooooooooeeeeeeeeeee. Woooooooooooooeeeeeeeeeee. Wooooooooooooeee(trill)*eeeeeeeeeeeeeeee. Weeeeeeeeoooooooo* (dribble off and die miserably). Then "Sweepers, sweepers, man your brooms! Clean sweep down fore and aft!"

Some similar nonsense precedes "Garbage detail to the fantail!"

The Marine crews respond to only one call. *Wooooooooooeeeeeeee!* (Period). "Flight quarters, flight quarters! All flight quarters personnel to your flight quarters station!"

Flight quarters sounds at dawn on November 18. We lift from the carrier's deck and land on the airstrip at An Hoa a dozen minutes later. Operation Essex has just finished; it has been under way for a week or so with few results. We've flown in a number of skirmishes covering elements of the battalion as they've swept through the valleys, hunting the enemy, but the force reputed to be in the area has eluded us, and we've flown long hours, trying to establish contact.

An Hoa control scrambles us immediately to check out a report of enemy movement along the Song Vu Gia, near Razorback Ridge northwest of the base. Hamm approaches the river a mile upstream from the reported coordinates and settles into a high-speed run, skimming the water's surface. A narrow trail runs along the riverbank; behind it the forest and outcroppings of rock afford perfect cover to an ambushing force. We could scope out the zone from the safety of altitude, but advertising our presence so blatantly usually ensures we'll find nothing.

Hamm runs the gunship down river. The rotor wash ripples the quiet surface and sends a "V" wake to the shores, like waterskiing. It reminds me of the rivers off the Chesapeake where I learned to ski, only a few years before. Suddenly, the river takes a ninety-degree right hook, and Hamm banks hard to negotiate it. My -60 lifts skyward, then returns to horizontal as we complete the turn. Magically, a figure in black pajamas appears neatly atop the gun sight, eyes round in surprise. The roundest fucking eyes I've ever seen on a gook. He turns to leap off the trail, but a dirt cliff rises behind him, trapping him next to the river. He begins to run, toward us. Mistake. He shouldn't run. Or he should leap into the running tide. That might give him a chance. Hamm says he sees a weapon. That seals it for Hamm, and he seals it for me. But the gook shouldn't run. Anything that runs in the zone is dead meat. He runs like a sprinter, long-legged strides and pumping arms, his head back, sucking in lungs-full of air. I can see his forehead recede beneath the conical straw hat. Hell of a hat to wear into combat. He is young, that I can tell, and strong, and his face has the hardened features of a soldier, or a farmer. He shouldn't run. Anyone in this war should know that.

It's not easy to kill a man when he's running parallel to your line of flight. Everyone knows that you lead a moving target, but when you're moving yourself, the rounds travel with the forward velocity of the aircraft and lose only a little of their speed before impact. Aerial gunners learn to lag a target and sail the rounds in, like smashing empties out of a car into highway signs. If the target is running opposite to the direction of flight, you lag a little more; if you're moving in the same direction, a little less. It takes practice to kill at high speed.

My rounds catch him in the chest. Five or seven, more than is needed. I see the tracers go in, disappearing inside the black pajamas. I know they come out the other side—they almost always do—but I can't see them exit from my angle. He goes down hard, carried forward by the momentum of his run, skidding headfirst into the dirt with his feet arching in the air, as if his legs were seized by ghost hands driving him like a wheelbarrow. I continue to shoot as he falls, but my rounds are wasted, going over his head and into the scrub on the cliff. He slides several feet, facedown all the way. Funny, his hat stays on, held securely by the chin strap. It shields his face.

The dust whipped by the wind of his death rises about him, and is rising still as we round the river's next bend. A dirty, tan cloud of dust, settling onto the black pj's.

He shouldn't have run. You don't run from a gunship, in the zone. Everyone knows that. You just don't do it.

There are no main-force NVA north of the Razorback. Another false report. Another wasted mission. Hamm reports the shooting incident to An Hoa on the way back. I wonder about the figure on the road, and why he had to run.

Why?

"You recall that scene vividly, don't you?" asks Tom.

"Yes."

I wonder what happened to the gook on the trail? Did he just lie there and rot? Did some animal take him away? Did the brush conceal his friends, who watched him die rather than open fire and betray their position?

"Why? You'd been in search and destroys before. But you remember this in such detail. Why?"

"I remember a lot of them. I have a good memory," I say in dismissal.

Tom stares. I fidget under his gaze, and search for an answer.

"Maybe it was because I could see his eyes, if only for a moment. I seldom saw the faces of the men I killed. You know how air wing was. We just blew them away. But this guy—our eyes seemed to lock for just a moment before I opened fire. He knew who was killing him."

"Is that all?"

"Maybe it was because of the way it happened. So sudden. Round a river bend, then *Pow!*, someone is dead. Just because he ran. You know, I think if he had stayed still, Hamm would have let him go. I would have let him go, I think. I don't know how—I mean, we were out in the boonies. There were no grunts, and we couldn't land anywhere to take him prisoner. Somehow, I think he wouldn't have died if he hadn't run."

Tom leans toward me, pressing the question. "Is that all?" he asks again. "How about what you wrote?"

I think of the letters, and the days between them; the days hidden in the archives that I relive in the Center. Did I ever think of the black figure with the wide eyes, caught in fatal surprise on that river trail? Certainly not later that day, nor the next, nor any time in the ensuing years. Yet, the letters are the letters; the change in them undeniable. What did I feel, then? It is so difficult to see myself as I once was, to feel myself as I was—so difficult to separate those feelings from those I have today. What did I feel, then? I look back and read the words of a killing machine, effective in its task. Was there pride, as well? Surely pride should not be denied the warrior for mastering his task. Is it not the right of a craftsman, no matter the craft? What emotions lie in the second letter? I read regret, and guilt, loneliness and fear. Are these, too, the lot of the warrior? Every warrior? No matter the conflict? No matter that the war was Vietnam?

I wonder aloud, "Maybe I was just getting tired of all the killing. Maybe it was getting to the point that I'd just had enough. Done enough. Maybe it just gets to you after awhile. Maybe it's as simple as that."

I stare at Tom, with eyes that feel as dead as that day, long ago. "I killed him, just like a machine, not thinking, not feeling. I remember flying around the river bend, looking back. Seeing him in the dust. That first letter, as bad as it was, I guess it shows there was a time I felt something. Maybe I just got tired. He ran, Hamm said, 'Fire,' and there you go."

"There you go," Tom echoes.

"Yeah. . . ."

The Marines engaged the NVA at noon. Or rather, the NVA engaged them. It depends on your point of view. The grunts discovered a bunker complex northeast of An Hoa and attacked, advancing across an open plain that afforded the bunkers an excellent field of fire. There were only low scrub and a few rocks for cover. Suddenly, a string of 12.7-mm guns erupted from a weapons trench flanking the advance. To make matters worse, the bunkers were reinforced by the enemy's main force. The grunts were pinned in the open and quickly being chopped to hamburger.

The trap was only a few minutes old when the NVA mounted a ground assault. It came with such frenzy that the grunts expended maximum firepower to keep from being overwhelmed. The enemy withdrew to a tree line and began dropping mortars. The Marines dug in, scraping shallow holes in the rocky earth.

We arrive low on fuel and attack immediately. Hamm empties our rockets into the forest while Knipp, Yantis, and I strafe, but the cross fire that traps the grunts envelopes us as well, and we are too few to sustain the assault.

Hamm breaks contact and hurtles into An Hoa, calling an emergency straight-in to the fuel pits. We receive word that a section of carrier fixed-wing is en route and will rendezvous over the zone in a few minutes.

Hamm and Knipp refuel while Yantis and I throw rockets into their launchers, then rearm the TAT. Some grunts throw ammo cans onto the flight deck for us to rearm the fixed externals. We'll reload our own -60s in flight. The pilots finish fueling, but reloading the remaining guns will take several more minutes, so they walk a short distance from the aircraft to take a leak. At that moment, an unarmed slick from VMO-3 lands. The pilot leaves the aircraft and walks into the field to talk with Hamm.

Yantis and I finish rearming. We strap in as Hamm returns. Instead of taking off, he pauses and turns to face us. He looks awkward in his seat, the armor and shoulder straps impeding him.

"We got some bad news," he says slowly. "We lost a crew."

Who? I want to ask. But I don't. Hamm will tell. Which is worse? The pain of losing someone close, or the guilt that spreads with relief when it is someone else.

I am afraid to know.

"Two days ago. Captain Carter and Captain Kelsey were killed. Major General Hochmuth was on board. He was killed, too. They were all killed. The aircraft blew up in flight."

The words numb me. I feel the gripping emptiness. It begins high in my chest, spreads to my throat, paralyzing me. It fills my ears with rushing blood, flashes hot, aches in my eyes, burning. Roils my stomach, tightens my sphincter, and grips the base of my spine in a vice of pain, shooting, up, up, up. I am filled with its consuming. How can emptiness feel so full?

But two days ago. Oh God, no. Oh God, no! "Who was the crewchief?" I ask despite my fear. But I know, I know. Oh God, I know. The tears rush, driven by the emptiness, and a burning wave of dizziness explodes inside me.

"Phelps."

Oh God, no. Oh God, no. The pain explodes inside, doubling me over. I rock in my seat. It hurts. It hurts. The cramp wells upward, tightening every muscle. It rises higher into my chest, tightening my lungs, tensing every tissue, constricting. Impossible to breath. My lungs pulse in shallow, panting spasms, hyperventilating. Dizziness, blindness join the pain. Phelps! I want . . . to cry. I need . . . to cry. But the rush of tears runs dry and I can't. I can't. Oh, God. I . . . he. . . . My mind races in spasmodic bursts of thought. *Make it be a lie.*

Hamm returns to the controls. "Lifting." I see us, spinning inside a black dream. We rise into a hover, turn on the axis of the rotors, then cross the perimeter wire, heading for the zone. I stare down. The mud of the combat base turns to green earth. We rise higher. The green is a blur; green earth, green aircraft, green uniform. Green life. Green death. All . . . green.

Phelps. I see him, and myself. Two days before, sitting in his slick at Phu Bai. Phelps and me. "You wouldn't take a man's last cigarette?" "Fuck you, hand it here." So few of us left, still flying from the old squadron. He had—we had—a month and a half left! All those missions. All those zones. "Hey, Ronny Joe!" "Don't you call me fucking Ronny." Oh, God. "I hate that name. My mother calls me that." A month and a half left. Phelps! And now he's gone. I am alone. If it happened to him, it can happen to me. God, all the zones. I'd seen. We'd seen. Eleven months as aircrew—eleven months! Three hundred and something missions. Stevens, dead. Tommy and Mouse, and the Jew, and Tiny, wounded. Captain Bill, the CO, and Saint, and Brucie, and the rest. Wounded, all wounded. Me and Phelps left. We were going to make it! Phelps and me. I know death, administer death, cheat death. It means that I accept death, doesn't it? Doesn't it? The matter of my own mortality. "Yea, though I walk through the valley. . . ." I can handle my own mortality. "I shall fear no evil, for I am the meanest motherfucker. . . ." My mortality. ". . . in the valley."

The zone, God, the zone. Get in control. Get in fucking control!

Hamm is setting up the firing run: "We're attacking on the A-6's wing; there's not enough fixed-wing to saturate the zone. I'm on the main force. Copilot, take the tree line to port with your TAT. Gunner, cover the A-6 on pullout. Watch your fire when his playmate attacks. They're dropping nape on the bunkers, then running 20-mike. And watch the grunts, gunner, they're on your side. Crewchief, take out the weapons trench. We have no cover on that side."

I hold the black comm switch in my hand. The grunts, the killing trench, the bunkers, and the shredding metal. Phelps. It can happen to me. Push the switch. Push and scream, "Take me back, take me back!" Three hundred missions is enough. *I don't want to die!*

"Crewchief!" the pilot calls.

Closing on the zone, fifteen hundred feet. Mortar plumes erupting among the Marines, heavy caliber ricocheting from the rocks. Green uniforms like rag dolls clawing at the ground. Take out the trench. Take out the trench. A-6s closing, "Dreamhour One and Two on station, angels eleven; nape and 20-mike, Scarface, mark your target."

"Crewchief! Answer up!"

Mark your target. Closing on the zone, open before us. Mark your target. I see the trench, a linear scar in the land: black figures on black weapons, jagged spears of flame, leaping. And . . . and . . . Hatred. Mindless, vicious, savage, flooding hatred.

My . . . friend. My. . . . You killed my friend! We . . . I. . . . YOU KILLED MY FRIEND! I will make you pay. I will. Phelps and me, we'll make you pay.

I don't remember killing the first gunner in the trench, nor the second, nor the tenth. Nor the last. There was no counting, only an unbroken stream of butchery as we fell through our dive and the ground fire leaped up from the zone. My mind preserves a long, frozen image of myself as I killed them. Bent over the gun, firing, firing, firing. I remember only the long burst of rage. Four hundred rounds of ball and tracer spewing from my gun in unbroken hatred. All consuming.

The gunners turn their weapons upon us, 12.7-mm on aerial mounts. It is a long pass, and I fire continuously, raking lead up and down the trench. The gunners fall behind their weapons, and others leap up to take their places. I cut them down. I smash them into the dirt and lash their bodies with MG, for Phelps and me. My weapon jams. I throw the feed cover open and grab the burning casing with my naked fingers, then throw in a fresh belt and fire, fire, fire, loving it. Reveling in it. I would leap from the aircraft and do the job by knife if I could, not for Country or Corps—for me, me and Phelps. I make them pay. They pay. Phelps and me—we make them pay.

Then, they are all down. No figures crawl to the hand grips of the weapons; the 12.7s hang slack in their mounts, pointing skyward on tripods of steel that vanish into piles of shredded flesh and cloth. Hamm breaks low over the zone, banking left to start a new run. I rake gunfire along the trench. Up and down, churning the meat. Suddenly, the weapon stops firing. We are almost out of range, and I still need to kill. I cock it, pull the trigger, cock it, and pull again, then throw the feed tray open in fury. The ammo can is empty, and I feel angry, more than foolish, when I see I have nothing left with which to kill.

We save the grunts. The A-6s destroy the bunkers and main force with bombs and napalm. We make several more passes, firing into the tree lines while the -6s drop their heavy loads. The trench remains silent, and I pay it no further heed. At last it is over. The grunts crawl from the meat grinder, care for their wounded, and line up the dead for bagging. CH-46s evacuate the wounded, then others ferry supplies from An Hoa, trading C-rations and ammunition for the dead.

I inspect my aircraft when we refuel. There is a single large-caliber round in the nose radio compartment, in some piece of electronics we seldom use. We fly out to the *Tripoli* and spend another night. They have chicken. And fucking Navy beans, and potatoes and bitter coffee, and there is a movie in the enlisted mess.

The next day, we continue to patrol east of An Hoa. It is quiet; whatever survived of the main force has withdrawn. We fly a few feet off the deck, skimming the dull earth. It is a gray day. Everything seems gray and tired and old. Even the green seems gray. We cross a dirt road; a corrugated pipe buried beneath it tunnels a thin stream from one side to the other. I glance backward as we cross. It is my job to watch our ass as the pilots concentrate on where we are going. A North Vietnamese soldier in gray combat dress emerges from the pipe, and I can see others crouching inside behind him. He holds a weapon by its barrel, and I know it would take him too many seconds to raise it to fire. One word to Hamm and it would be a simple kill. I could take out the soldier, and Hamm would plunge a rocket through the tunnel like a pipe cleaner. The soldier freezes. I meet his eyes for a moment that seems so long. The face is hard, flat-nosed and broad with thin-slitted eyes, more Chinese than Vietnamese. They follow the receding gunship in expectation, rather than in shock. Certainly they show no fear.

Hamm flies on, unaware. I stare back, into the cruel eyes of my enemy. Unless I act soon we will be out of range, and they will go to earth in some better hiding place. It is easy for them; they are the masters of this land, not us. I finger the microphone switch; I should report the sighting; it is my job. I am the crewchief, and I am good.

I make another decision before the aircraft's speed takes the choice from me.

I will not kill, this day.

I look to Hamm and Knipp, then Yantis, each absorbed in his own task. No one has seen them but me. When I look back, I can no longer find the pipe or the men.

I return to scanning the fields. And say nothing.

Chapter 7

Raised in the land of the free America's youth has
developed a deep love, appreciation and respect for
the ideals of peace and freedom.
As they mature,
Each willingly accepts the duties of citizenship
in a democracy, because he realizes
how much he owes his country.
Many young men serve in the armed forces fulfilling
their responsibilities to preserve
the ideals of democracy and curtail the spread
of communism.
Earnestly fighting to procure
what cannot be won by treaty or pact, these men
are willing to give up their lives
for a cause in which they believe. These are
Men whose courage and loyalty have promoted them
to leave those they love
and risk their lives. They have done this in the hope
that there will someday be
Brotherhood and tranquillity among all people
and that such sacrifices will secure for citizens
of enslaved countries the God-given freedom
and personal dignity now denied them.
Every one of our boys in Viet Nam is fighting to
prevent the extension of a creed
of hate and atheism. We want to
Reassure them of our support of their presence
in Viet Nam and our continued prayers for their
safe return. We are grateful to them
for their unselfish and courageous
defense of our country and its ideals.

Therefore,
We the Senior Class, proudly dedicate
Thabor 1966
to
Our Boys in Viet Nam

—1966 Yearbook, Saint Saviour High School, Brooklyn, New York

256

DEROS

● ● ● ● ● ● ● ● ● ● ●

"How long have I been coming here?"

"Gosh—I can't remember! I think you started in '82," Tom answers.

"Nah, it was '81. Right after the Iranian hostages came home. Remember? The Ayatollah wouldn't release them until Carter was out of office. I wandered in here a few months after the Gipper took over. Remember?"

"No! It can't be that long. That makes it . . . this is '87 . . . over five years. It can't be that long!"

"Yep. Time flies."

I study Tom closely. There is more salt than pepper in his beard, and the lines that feed his smile have deepened.

"You know, I think you got a lot older than me."

"You've been a tough case."

We laugh.

Five years. The man who is my counselor has become my friend. It has been a strange relationship: usually cooperative, sometimes adversarial, never dull. Doctor-patient, student-teacher: the roles have on occasion reversed. It has been a working relationship that worked.

I know that our time together will soon end. Much is as he promised. Control, choices—I have both. I still anger easily, but seldom does it flash to rage. Circuit breakers of self-awareness kick in and dissipate the rage into understanding. Anger is a thing that I control. I've almost forgotten what it felt like to despair; it is a memory whose loss I welcome. The dreams come seldom, the flashbacks not at all. Some of the guilt is gone, some I live with. I have not completely learned to forgive. I am not certain that I want to.

We have dissected my life, then rebuilt it, atom by atom. The redefinition of my self is almost complete. It is not a new self, not even an old self. It is something in-between, but something that I understand. I understand more of my self than ever I desired, and I am grateful for the understanding.

I have survived, and shall.

"I think it was the hostage situation that set me off," I offer, remembering the early days in the Center.

"The parades touched off a lot of bad memories for Vietnam vets," Tom observes. It has never been his way to say "yes" or "no." A shrink shall ever be a shrink.

"Yeah. It still rankles me, the homecoming they got. Not the hostage's fault, of course. There was no 'fault' to it at all, though I didn't think so at the time. Those poor bastards deserved their parade. Still, I've never forgotten how it was for us. Nothing seems to make up for that. It's just another touch of bitterness for the poor fucking Vietnam veteran to live with."

Tom keys on the sarcasm. "What about when the Wall was dedicated?"

"The Wall was the best thing that ever happened to us. Jan Scruggs should get a medal for giving us back our pride, but it didn't erase the past. At least not for me."[1]

"You made it through the dedication pretty well."

Tom had insisted on meeting during the gray days in Washington. I hadn't known what to expect, searching for Phelps in the granite. His memory lived in me; how would I handle finding him in cold stone? I'd expected tears, or anger. Instead, I sat soaking my ass on the damp earth, staring at the carved letters, seeing him in them. Not even thinking, just remembering. People paused and patted my shoulder. I guess they thought I was sad. I don't think I was. I don't know that I felt anything, except that it was good to be close to him again.

Two hours I read his name, so my watch claimed. Didn't seem that long. At last I said good-bye, feeling good that I knew where to find him

---

1. Jan Scruggs, a front-line grunt wounded in action, is president of the Vietnam Veterans Memorial Fund. Scruggs conceived the idea for a memorial dedicated to all who served in Vietnam, with special tribute to the men and women who did not return. The Wall is a meeting place that makes no political statement, and serves as a healing place for veterans and the nation.

when I needed to. Tom met me at a Thai restaurant. We talked about cooking and R & R in Bangkok, and Vietnam not at all. It felt good to be with him, to be with them both.

"We spent a lot of time preparing for it. You didn't know if I could handle it, did you?"

"You hadn't been in therapy that long. I knew it would be tough, but it was something you had to do."

"I made certain I got to the parade route early," I began remembering. "I wanted to witness it all. It was important to be a witness. They'd hung a giant American flag from two construction cranes, that formed an arch under which the parade passed. It was a sight to see! All these guys in their boonie hats and uniforms marching, people applauding. 'Welcome Home' and 'Thanks' signs. Aircraft flying overhead in the missing-man formation. I wore my old camouflaged hat, with aircrew wings pinned to the flap. Didn't want to drag out my uniform. Hell, I couldn't have stretched it over this lard ass, anyway."

I pat my tummy, thinking I would hate to have to strap on a gunners' belt.

"Some grungy old jarhead walked over and said, 'Come on, brother, come march with us!' and I couldn't. I just couldn't. He hugged me. I got so choked up I couldn't speak. I just shook my head. He looked at me like he understood. Hell, I didn't understand it myself. I just couldn't join them. But I saw it. I witnessed it all, when the Vietnam veterans came home."

The memory mists my eyes, warm and cold.

I was there when the vets came home.

Tom looks surprised. "You never told me that."

"I didn't want to."

"Why not?"

"Hell, I'll never tell," I joke, wiping my eyes.

"OK. It's time to put on the doctor-patient hats."

"Yeah, yeah," I nod slowly. "I like it better when we're friends.

"You know," I begin, "we never really talked about the end. What happened at the end of the tour, coming home, et cetera. We've talked all around it. How is it that you've let me get away with it after all these years?"

"You haven't been ready," Tom responds softly, "until now."

I numbered each card.

I savored every number, and carved them into the polished paper with the relish I reserved for an infrequent steak, or an especially fulfilling shit, alone in the four-holer with a fresh roll of Charmin.

52, 51, 50, beginning with a red king and working toward a black suit, then a red, then the black. The thrill of it was at once tantalizing and excruciating—a perverse delight, like anticipating a make-out session with your girl—sitting and rocking, and rubbing your arms and hugging yourself because you knew, you just knew it was going to be so good!

34, 33, 32, 31. Resurrection! That big ol' stone rolling back slowly, oh so slowly on Easter morn, letting just a twinge of bright sunlight into the ol' sepulcher!

17, 16, 15. Every down count was an emotion becoming tangible in my hand. The numbers were real, more real than anything had seemed for a long, long time. More real than the medevacs of yesterday or the incoming of today. More real than the missions of tomorrow, or of the next fifty-two days. Feelings I hadn't known for a long, long time leaped from the pasteboard. Hope instead of despair. Living instead of surviving. The feeling that there was a "someplace" that was not "this place."

Then the last days, the special days. Only spades were fitting for the special days. Five of spades, four of spades, three, two of spades, and then the one they'd wanted to deal me all along. I'd shove the ace up their ass on the morning of that last full day before I left for the World.

The last card was the best of all: the joker, and on it I drew the hook, straight shanked and viciously barbed. This was the long-sought hook tagged to every number on my short-timer's calendar: "216 and a hook," "95 and a hook." The hook was the dawn of the first day of the rest of my life.

DEROS was at hand! In just a few days, the end could begin.

CARD 52

Card 52 fell into the sea.

I don't recall seeing it hit the water. I probably couldn't have seen it; it was windy beneath the carrier's flight deck, and getting dark. Maybe it didn't hit the sea. Maybe the bow wave's wind rush threw it back into the superstructure. Maybe it's still there, even after all these years, stuck in some gray metal girder.

No, I'm sure it's in the sea, the South China Sea. I'm certain it's there.

It was either the king of diamonds or the king of hearts. I don't remember. I know it was red, not black, because we saved spades for the end. I know it was a king, not an ace. Some guys started with an ace, but I didn't because the last full day had to have the ace, and I was consistent in everything I did.

Everything.

Fifty-two—the beginning of the end. Special. I'd wanted him to get it, I really did. He'd have been so pissed. Diamonds or hearts. Funny, now that I think of it. Either would have been appropriate. He . . . uh, he . . . he was a hell of a . . . he was a hell. . . .

We'd been together on day 55, but he was going his way, and I was going mine. Doubtful that we'd be together on 52, probably on 51 or 50. Ohhhhh. . . .

I could have just given him the card, or planted it on him before he flew north. Maybe that's what I should have done. Maybe. He would have had it with him when. . . .

Yeah.

But there was the ritual. I had to follow it. It had to be done in person. You picked a new victim each day, carefully, someone you liked or someone you hated, but definitely someone you wanted to shove it to. You planted the card in their clothing or flight equipment, then crept away to watch, sniggering, from some hiding place.

They'd find the card, and look around muttering curses as they tore it up. It was so good! There'd be a lot of short-timers around, so the victim couldn't know who the perpetrator was, until you leaped up, crowing maniacally, "Got you, fucking got you, asshole! Fifty-two Days! And a hook! Haaaaaaa haaaaaa haaaaaaa. . . ."

Oh, man.

It had to be done that way. It was no good unless they knew whose card it was. The game was cruel, and crude—exquisitely so, and all the more satisfying for it. The cruelty had purpose. Envy. You needed to see envy, to be cursed and have the card flung back in mock anger or real anger but, most of all, in envy. If you were envied you knew, you really, really knew, that DEROS was at hand. In this way, only, could leaving be for real.

Cards were supposed to be given out on the day they were numbered. It wasn't the same if you didn't do it on the proper day. Kind of like cheating. Still, it would have been so much better to stick the first one to . . . to stick it. . . .

I decided to keep it until I got back. I put it in my flightsuit pocket. I'm right-handed, so I put it into my right breast pocket, the one I didn't go into very often. That way, I wouldn't lose it, or get it dirty. Then I'd stick it to him!

Flip me the finger. You bastard! We'll see who has the last laugh.

It was the end of the day, the last night on the ship. The operation was over. Everything was over. The *Tripoli* was sailing up the coast, and we'd fly off in the morning. We double-lashed the gunships to the flight deck—leaden seas were picking up under a driving gale. A typhoon was coming;

it would be a rough night.

Everyone had gone below, even the fucking-Navy rust-pickers. I walked down the starboard ladder, forward of the gun mount, and watched the bow slice the sea. Up and down, up and down. The cold salt spray cut the nausea quivering in my stomach. Too bad I couldn't spend the night there. It would be easier to barf. I always puked on a ship. Go figure. I flew gunships all day and never got sick. Not once! Put me on a fucking bucket, and it was all she wrote.

Everything was gray, fading to black. Gray sky, gray sea, gray rust bucket, gray heart, gray soul. Gray. Whitecaps were white, though. Guess everything wasn't quite gray. Felt like it, though. Remembers like it.

Fifty-two. I held it between my thumb and forefinger where it fluttered like a butterfly struggling to free itself. I didn't know what to do with the fucking thing. Shit. I should have given it to my gunner, but I couldn't. Just . . . couldn't. It didn't seem right. I'd wanted him to have it.

He'd never have it. And today, today I didn't even shoot that lousy gook! I let him go, y'know? I let him fucking go! The bastard probably went out and killed a Marine, and it's my fault. I'm nineteen fucking years old, and I'm going to carry this all my fucking life, and I should feel . . . I should feel. . . . There are names on that Wall, and it's my fault! And I . . . don't . . . I don't even feel bad about it. I want to feel bad about it—guilty. But I can't.

God damn me!

I

Don't

Feel.

The wind—it took it from me. Like it decided to take it all on its own. I didn't let it go. I jerked my head aft and watched it sailing, tumbling. I lost sight of it against the jumble of waves.

No, that's not how it was. I loosed my grasp, just a little. I remember. I began remembering, you know. I loosed my grasp. Just enough. Deliberately, until the wind claimed it. It didn't belong to me.

I let it fall into the sea. The South China Sea. . . .

CARD 43

Dear Tom and Joe,

I may make it out yet! Wing has passed an order that anyone who left the States before 10 December last year will leave here by the 20th. It'll be tight, but I could be home for Christmas!

I have to get out of this place. Going out of my mind. Have to get out before Christmas. Wish I could explain how I feel. If I spend Christmas here, I'll snap.

We're having heavy monsoons. It's bitter cold and rains constantly. Nothing is dry anymore. The work stays the same, though. In the rain all day and half the night. You have to see it to believe it. Flying is hell. Well, when I get home I can say I've been through something a lot of people haven't, whatever that's worth.

You wanted to know what I thought about the "peace" demonstrations. I guess the best way to answer is that I asked my parents to quit stuffing newspapers into the Care packages they send me. The protests depress the hell out of me and I'd just as soon not know about them. Most of us are pretty sick of the way things are going back home. We are laying our lives on the line and no one is doing anything to back us up. We're all pretty disillusioned with our country. All I hear about is the demonstrations. It gets me that no one cares enough about us to show support.

All of us have proved that we are willing to do something for our country. Now it's time for the country to prove that it's worth the effort. None of us expect to be treated like "conquering heroes" but we do want to be shown a little respect and we don't want to be made fools of, or made to feel that what we've done is for nothing. Don't know if I'm making sense but that's how I feel. I guess its not a very "popular" opinion. Too fucking bad. I earned the right to it.

> Your old friend,
> Ron

CARD 30, THEREABOUTS

It was getting harder and harder to get into the gunship.

For three hundred mornings I'd crossed the flight line, preflighted and readied weapons, then waited for the pilots to arrive. I'd stood near the pilot's door, holding the cabin fire extinguisher as the engine ignited, returned the extinguisher bottle to its place after a safe start, then walked aft to take my station. For more than 350 missions I had hauled my skinny ass into the cabin, strapped the gunner's belt on, and sat back for takeoff with nary a thought in my mind.

I don't recall when I began thinking. That's what did it, I'm certain. Thinking. You shouldn't think too much.

It was like this. I figured God dealt you just so many chances to do some particular thing—like you could pick your toes 10,423 times or eat 5,432 hot dogs. He figured it all out when you were born, stuck you with the number, and that was that. I was certain there was only one special number for one special thing per person. It would get too complicated if there were more, and God liked things simple. If He didn't, He wouldn't have created Marines.

Most things you could do as often as you wanted. I'm certain Frankie had been given an unlimited number of farts or God would have greased him before he reached puberty. As long as you didn't exceed that hidden number, you'd be OK. But just try to pick toe 10,424 and you were screwed. You could live forever, as long as you didn't exceed God's number. God knew the number, and God picked the thing. We didn't. It was His little joke.

God wanted me to die in the air. I just knew it. It wasn't that I thought He had it in for me. I just believed it was His plan. It was a good plan. I counted missions. Why shouldn't God count missions, too? The irony appealed to me. God loves irony, or He wouldn't use it so much.

I had begun hesitating at the cabin door—only for an instant, mind you. I doubt that the rest of the crew suspected anything. God knew I wouldn't tell them. They'd have thought I was fucking crazy. Perhaps they thought I was just getting a little slow, tired.

I hesitated, thinking of all those missions when I'd come back without a scratch, when so many of my friends had been hit, or crippled. And me, without a scratch. How long could it go on?

It couldn't.

I would die in the air. Someday. God tracked the number on His celestial abacus, and every time I flew He flicked His celestial finger and moved another bead into another column. *Whap!* Move a bead. *Whap!* Another. Got shot at this mission? Two more beads, *WHAP! WHAP!* It got louder every time. Hear it?

Each morning I paused, wondering. And yes, I was afraid. How many beads were left? I didn't pray. I'd given Burke my word that I'd fly. I was not brave, and do not like to think of myself as stupid.

Only crewchief.

CARD 30, ONWARD

Someone was fucking up my orders. *When would they come?* There was so much that could fuck things up. There was normal Crotch bureaucracy— orders to be signed, orders to be countersigned, orders to be re-signed because they'd been countersigned incorrectly; orders to be lost or mistyped because I had this dumb fucking Polack name with too many Zs and no "ski."

Then there were the Raiders. Some dumb, motherfucking Remington Raider FNG REMF who did not in the least give a damn about a poor dumb bastard of a crewchief withering on the vine in Phu Bai. I knew he didn't, I just knew! The bastard was hanging out at Hill 327 or China Beach, slopping milkshakes in the USO, and my orders, my fucking orders, were lying on his desk, or maybe blowing onto the goddamned floor to be shit-canned by some slope-headed dink trash-hauling motherfucker. Oh Christ! He wouldn't even remember them when he got back; all he cared about were his own orders, and getting laid, and he'd sure as shit see he got both taken care of, and and my whole fucking life depended on this asshole!

I began harassing Crisp on Card 30, humorously at first.

"Crisp, old buddy, old pal, my orders in yet? Buddy, buddy, buddy."

Crisp checked the mail from the Wing. "Nah, Zack. Nothing in yet."

"Hey, OK. Just checking, you know. How much time you got?"

"Almost too much to count—183 and a hook."

"Yeah, well, we all can't be short. Know how short I am? I'm so short I have to use a stepladder to wipe my own ass. Heh, heh. I'm so short I can't see over the top of my boots. Heh, heh, heh. See ya', Crispy. Heh. I'm so short I plan my craps by the quarter-inch so there's not as much to wipe. . . ."

Whining began around Card 25.

"Jeez, Crispy. Can't you call Wing? They said we'd be home by Christmas. That's twelve days away, Crispy. Jeez. . . ."

"Calling doesn't help, Zack. It pisses them off."

"Yeah, yeah. But why'n't you think of something else to call them up about and just sort of say, 'Hey—what about Zaczek's orders'? Huh, Crispy, huh? Remember those nights I let you sleep late on the wire?"

Threatening took hold at Card 20.

"Look, you fucking asshole, *(smash)* the next time I come in here if you and that ass-wipe buddy of yours at Wing don't have my orders I'm going to cut an inch off your dick. *(Slam)* Then an inch after that, then another *(slam)* and another. *(Bang)* I will frag you and every motherfucker . . . goddammit, goddamn all you miserable Remington . . . goddamn . . . fucking . . . Raiders . . . horses' ass pieces of shit. . . ." (Unintelligible muttering as I exit.)

Desperation began as Christmas neared.

"I have to get out of this place. I am . . . I am. . . . If I don't. . . . Look, I'll do anything you want me to do. I have to get out. . . . I have to."

CARD 15

I helped kill my last man two days before Christmas, 1967. I can't say who actually got him; all four of us were pulling triggers.

Of course, I didn't know it would be the last at the time, and I don't remember thinking of him as a man, either. I do now, though. I think of all of them as men.

People, actually. When did they become people?

It's funny. I don't remember the first time I killed In Country, nor the last time that I killed all by myself. But I remember the last of all. It wasn't much of a mission. Hell, it wasn't much of a killing, not compared to all the others. Just a lousy lone sniper delaying a troop advance. The grunts called us in on a spider hole; we raked it with MG fire and shoved a rocket in for good measure. I don't remember who else was in the crew. It was near the Imperial Tombs, west of Hue. The jungle was green, and there was a lot of brown dirt. The grunts confirmed the kill from body parts. Mes-sy! We flew away with a good BDA and an attaboy. It was hardly worth our effort to clean guns that evening.

There was nothing remarkable about it at all.

CARD 14

Dear Grace,

Christmas Eve in Vietnam.

I'm looking at an artificial tree about 2.5 feet tall covered with tinsel and tiny lights. It has a white sheet for a background (snow?) and is surrounded by wreaths and cards. At its foot is a manger and a few wrapped presents.

I was alone in the hooch, feeling pretty bad. It stinks being here at Christmas. I miss you. Some of the guys came over to get me. As I write this, we're all sitting in semi-circle around that stupid tree. I think Jerome wired it up with instrument lights. I hope he didn't get them from my aircraft.

There aren't many of us left in the huts, anymore. The Seabees built these damned two-story wooden barracks and most of the E-trash (Enlisted men with rank from Private to Corporal) in the squadron are living in them now. You have to be an E-5 Sergeant or above to rate living in a hooch.

I wouldn't want to be in the barracks during a rocket attack—too far from the bunkers. I like living next to a bunker just fine. Fortunately they left us "short-timers" in the hooches, but its not like the old days. So many of the old guys are gone; there's only about six of us left from the advance unit—VMO-3 Alpha, remember? Used to be 20. Not too many left from the main squadron, either. 14 days.

This letter was getting pretty depressing until the party started. Your folks sent Care packages that were out of this world! A complete ham dinner, down to the mints. Everyone shared their packages from home. I sliced off hunks of ham with my K-Bar, and we passed the bone around and gnawed on it one by one. You'd probably think it was pretty disgusting.

As I write this, the guys are singing 'Jingle Bells,' though not like you've heard it.

Jingle Bells,
Mortar Shells,
VC in the grass.
You can take your Merry Christmas,
And shove it up your. . . .

I'm managing to forget a little about how lonely this Christmas is for me. Refreshments (beer!) have arrived and we're all pledging good cheer. I must rejoin the party, so I'll continue with this later, if able. This is the first brew I've seen in I-don't-know-how-long. I may crawl home tonight, but by God I'll crawl like a Marine!

Well, you needn't worry about my sobriety. About five minutes after that last sentence, I returned to my hut and I'm just sitting here. I guess I was just fooling myself, in thinking I could forget how lonely I feel. I miss you so much. But I'm better now! Some of the guys wandered by singing Carols. I stuck my head out and they serenaded, and then walked over to the Officer's four-holer and toppled it! I guess the sight of that is about as good a Christmas present as any in this place. 14 more days.

More letters to write then hit the sack. Tomorrow might be Christmas Day but it means little here.

All of My Love,
Ron

CARD 13

I walked slowly to the flight line. I don't recall whether I was scheduled to fly, or not. It made no difference; I would not fly.

I headed to Ordnance, to the little flatbed truck we used to run ammo to the aircraft. I had a novel in my hand, and a towel. I wrapped the towel around my head like a turban, and sat on the truck with the rockets and MG ammo, crossed my legs, and laid the novel in my crotch, cover up.

Some friends stopped and said, "Hey, Zack," and I said, "Hey."

A few FNGs stared, but did not stop or talk. I think I made them nervous.

No one told me to get off the truck. I probably would have, if they did. I didn't want any trouble.

But they didn't.

A couple of times, the ordnance men came out, started the truck, and drove onto the mat to rearm an aircraft. I sat and watched, letting the rotors' familiar wind wash over me, breathing the engines' perfume. Three hundred and ninety missions. I was still crewchief. It was light monsoon. The mist was in my face. That was all right.

I didn't say too much, and when I spoke it was to no one in particular. I said the same thing, to anyone who wanted to listen, to no one. Over and over.

"This is Christmas, and I'm not going to kill today."

I got tired of it after a while. I don't know how long. Long enough. After a while people stopped looking at me. I had this strange feeling, as if I were invisible.

Eventually, I put the book away and went to work.

I didn't fly that day. No one asked me to.

The novel was *Catch-22*.

CARD 12

The monsoon lifted enough for us to cover three troop insertions. Missions 391, 392, and 393. It would have been nice if we had gotten to shoot someone. Or just shoot. Didn't matter. Then it would have been 394, 395, and 396 as well. I wanted to shoot something! Shit! I needed four hundred missions for my twentieth Air Medal. Eleven days and a hook. I was pretty short, and the monsoon stood a good chance of keeping us grounded before I could pick up the additional seven missions. I liked the idea of going home with a round number.

There's always tomorrow.

CARD 11

The Christmas party began after flight operations shut down. The CO had delayed it for two days because of reports that we'd be hit. There was a great deal of enemy movement in I Corps. Khe Sanh and Quang Tri were getting plastered daily, and we'd spent several nights reinforcing the perimeter against an enemy who never came. The nights in the open sand had drained me, and I looked forward to an evening's rest.

It was a strange gathering, unlike those that came before. The men of VMO-3 were celebrating as a squadron for the last time.

We had been more fortunate than most who served in Vietnam. The majority of U.S. troops arrived and departed alone; each man's DEROS separated him from his neighbor even as they fought a common fight. VMO-3 had begun as a new unit in Pendleton in late '66. I'd been among the first to arrive In Country in an advance detachment, called "Alpha," that covered an amphibious-air operation in the Mekong before rejoining the main unit, which had departed the continental United States on ship. Together, we called ourselves the "first increment" of VMO-3. With 12 aircraft, 23 officers, and 102 men, the squadron covered all of Quang Tri and most of Thua Thien Provinces at half strength from January until May.

The arrival of the "second increment," populated by many friends who'd remained with the Training Unit in Pendleton, swelled the squadron to 26 aircraft, 33 officers, and 194 men. We'd welcomed them, at first, for entirely selfish reasons. Northern I Corps was a large territory for twelve ships to cover, and after five months we were burned out. We'd eagerly awaited the arrival of fresh meat, upon whom we would dump every shit detail and lousy mission we could connive. It was right and proper. After all, they were only FNGs.

Clothed in sheets and burnoose like a sheik, Give-A-Fuck greeted them by distributing a bullet and a Band-Aid to every man, chanting. "You're all going to die!" Drooling in the wings, the old crewchiefs sized up the new meat like slaves on the block, anxious to run them through our aircraft like pipe cleaners.

We came to resent them almost immediately. There were many reasons. The sun was the problem, and the heat. And their attitude. And ours. And DEROS, which in the end both divided and bound. There were a lot of reasons. The sun came first.

The FNGs melted. Phu Bai was like an oven at 250° by mid-May, on its way to high broil. The FNGs' stateside flab couldn't stand the heat. They dragged their duffels to the hooches, and there they parked for two weeks, languishing in the smelter beneath the tin roofs. While they lolled, we flew. And flew and flew and fucking flew. I threw murderous glances, fingers, and epithets at them as I walked off the flight line after a day's missions, and they threw them back—without right, I thought.

The clash of attitudes was as severe as the heat. The cockiness we'd brought In Country had burned out; its charred remains flashed in resentment to the FNGs' show of stateside attitude and the eagerness we had lost to the zone. There was a great deal to resent. They had moved into the huts we had furnished, walked the mat we had laid, avoided the mistakes we had suffered. And worst of all, they acted like they belonged. The months we had spent building and improving the base from which we flew, and the endless missions, had welded the original outfit together, forging bonds that could not be extended to embrace the newcomers, even these friends who'd remained behind. We were VMO-3, not them! With the perversity of men who flaunt hardship like a badge of honor, we resented them for missing the pain we could not have avoided.

Even so, as the months wore on, "first" and "second" slowly learned to work together, though we "of the first" rarely missed an opportunity to remind the others that they were second in every way. The missions and the heat had a leveling effect, I suppose, and by August we were a squadron again, though, it seemed, never in the sense of the earliest days In Country.

December's onslaught of newcomers should have been no surprise. Having arrived in two large groups, we would be leaving as we arrived. Yet it was a surprise, even a shock. The flood of new faces with names I refused to learn meant that the war, which had no beginning, and no end, was ending. Proximity to DEROS again distanced the first and second increments, and both sets of old timers avoided the new meat.

I accosted Shifty, a crewchief from the second increment, waving my can of Crown Korean under his nose, pretending the 3.2 was a hundred proof.

"How much time you got, boot?"

"Go away, Zack."

"Eleven days. And a hook."

"Go away, Zack."

Smitty wandered over. He was Alpha detachment, and we'd be leaving together.

"Hey, Smitty. Old Smitty. Smitty, Smitty, Smitty. How much time you got?"

"Gee, I don't know. Let me get a ladder and see. I'm so short I can't see the minute hand on my watch."

"Smitty, I'd like to help you, but I'm so short I can't reach your wrist."

"You first-increment assholes make me sick," Shifty tossed back.

"Ho, ho, ho. Even if you second trash weren't second, you still wouldn't be good enough to be first. You know what they did when second increment was born? They kept the afterbirth and threw away the babies."

"Zack, you know you're an asshole," Shifty answers, not angry. Shifty and I have flown together for months. In mid-summer, we'd been temporarily disabled—he with high blood pressure, and me with the same and vertigo to boot. We spent long hours working shit details when we couldn't fly. We'd become friends in spite of first/second rivalry. The banter was not insults, only the way we communicated, yet DEROS had come inescapably between us. I would be gone in a few days, and he would not.

"Yeah? Well, I'm a short asshole."

Zeke (another Alpha) passed within hailing distance, a brew in his hand. I waved mine.

"Zeke! Zeke, Zeke, Zeke. You skinny bastard. How much time we got left, Zeke?"

There was food: beans and franks, shredded green stuff. Piled all together on a paper plate, it looked as if someone had been mowing the lawn and killed a puppy. We were warned that this was dinner, and that we weren't allowed in the mess hall. Not to worry. Music poured from a loudspeaker: Mommas and Poppas, Nancy Sinatra—the same old songs. We never grew tired of them.

I grabbed some chow and looked for a place to sit. There was an empty spot next to Dr. Kyle, the squadron's flight surgeon. Old Kyle was a little strange even for fucking-Navy, but he'd been trapped among the jarheads so long he had an excuse.

"Mind if I sit here, sir?"

Kyle smiled and motioned to the seat with a plate of beans.

"You're pretty short, aren't you, Zack?"

"Yes, sir! Don't know if I have the time to finish this sumptuous meal."

I liked Kyle. A friendly guy, more doctor than officer, he was easy to talk to. "Got a girl to go home to?" he asked.

I fished into my flightsuit and pulled a creased photograph from my wallet. A lock of brunette hair tied with red ribbon fell into my lap. I replaced it sheepishly. Kyle smiled and took the picture.

"Very nice," he nodded and passed it back. "You're luckier than most that you still have one."

I stared at the picture, knowing exactly how lucky I was.

The flight surgeon was first-increment VMO-3 and would leave a few weeks after me. Kyle had grounded me when I failed my flight physical, and offered to keep me grounded until I truly recovered, but he'd understood how important it was for me to get back in the air and fed me vitamins and magic potions to turn me back into a crewchief. He'd stitched my wrist when I'd slashed it while replacing the number four jet on VT-25's transmission. It was night, and I was working by touch and feel in the hole beneath the gearbox. A wire snapped as I tightened it, and though I'd felt something sharp against my wrist, I thought the warm stuff trickling down my arm was oil. Kyle lit a cigarette and placed it in my lips as I lay in sick bay. Real John Wayne, except John never bled like a pig.

Doc fed me antibiotics when my right hand got nicked, and again when my left arm bloated, and he pumped salt into me the time I passed out in church, and he was there when Saint came back pumping through his femoral, and when Mouse lost most of his triceps, and. . . .

Kyle surveyed the party.

"How do you feel about leaving the outfit?" he asked suddenly.

"I can't wait to get out of this stinking hole," I shot back.

Kyle took a slow breath, glanced around, then lowered his eyes to his plate.

"I'm going to miss all of you. This has been the most important year of my life."

Missing. I had not separated that thought from the jumble that filled my mind. Phelps had been gone more than a month, but a month was forever, In Country. Missing. . . . you needed to be able to feel to miss someone.

"Zack, do you believe that it's OK for guys to have strong feelings about each other?" he asked suddenly.

The question made me nervous. I watched the squadron. Mouse and Moose, Saint and CC, Shifty and Tommy and Hazel Baby. Eleven days and a hook. Back to the World. Would I ever see them again? We'd never talked about getting together "after the war." We'd never spoken of doing anything that would keep Vietnam alive or remind us of it. It seemed crazy to even consider that kind of talk. I tried to envision it—not seeing them again. I thought of Phelps, who was gone forever. There had been a time when I'd known none of them, then a time when they were the most important people in my life, and in a few days I'd know none of them again. I pondered the flight surgeon's question and felt a distant pang, like an ache barely noticed, gone before it could be identified. It all felt so remote, like thinking about a thought rather than having the thought. Doc watched me. *Do I believe?* I didn't know what I believed; I only knew what drove me, and that was DEROS.

I answered, thinking about thinking about a thought. I liked Doc; neither the answer nor the thought was a lie, just so abstract that I could not feel it or touch it. It was safest that way.

"Yeah, Doc, I think it's OK."

Trouble appeared in the form of the ready-room watch, who spoke softly to the duty officer, who escorted him quickly to the CO, who summoned the maintenance chief to call the squadron to attention.

"We just received word that VMO-6 has been hit at Quang Tri. They've had a number of aircraft destroyed, and we have to send replacements in the morning," the CO said.

A groan rose from the crowd as the chief picked up where the CO left off. "You all know we don't have any replacements, so we'll patch together what we can.

"Section leaders, inventory the deadline ships and see how many you can rebuild. Strip the worst ones down to their skivvies, then take the skivvies. Do anything you have to do to get the most birds up. Everyone else report to their shops and stand by. Ladies, the dance is over."

The night was surreal.

Disembodied voices shouting, "Move the generators in—there, and there, and there. Not there, there! Dummy!" Spots of light erupting through monsoon mist, bathing the aircraft in clouds of vapor.

Inventorying the deadline: men moving purposefully through darkness, acolytes among their aircraft, silhouetted by the spotlights, bearing tools, moving heavy equipment onto the mat.

More shouts. Flashlight beams slashing the flying machines like swords, slicing across the green metal.

"What you got?"

"Engine, transmission, mast. Rotor head, no blades, no tail rotor, FM, no UHF. You?"

"No tail cone. No seats. Floor plates ripped up, linkages missing. Servos! Hey, it's still got servos. Get Hydraulics over here to rip them out. -11 needs servos. I think we can rebuild a stabilizer bar from these pieces. Anybody need a stabilizer? Look! I found a diary. It says, 'In the event that I am lost. . . .' It's signed, 'Amelia Earhart.'"

All up and down the line, pieces and parts of dead machines recited like the organs of a cadaver under autopsy.

The maintenance chief emerging through the mist like Hamlet's father, but more threatening. "Just write it the hell down and get back to the shack so we can figure out what we can piece together."

Crews swarming over the aircraft. First increment, second, newcomers, working together.

"C'mere. What's your name?"

"Hess."

"Here, Hess, take this short shaft (main engine drive shaft), clean it, and get Gunny to inspect it. See this seal here? There's a small crack. Right here. No, not there! Here. Get Gunny to look at it after it's clean. If he says it's good then repack it and install it in -25."

"What if he says it's no good?"

"Then repack it and install it in -25 after he leaves."

"Uh, I don't know how."

"Yeah, well tonight you're gonna learn. Come with me."

"What's the big deal about getting so many aircraft up to Quang Tri?"

"Fixed-wing can't fly in this shit. The grunts are going to wake up with no AirCap and no cover for the medevacs. We gotta get these ships up."

"What about the Army? I hear their squadrons are three times as big as ours."

"They're four times as big as ours."

"Well, what about them?"

"Look, if -Six had wanted the fucking-Army, they'd have called the fucking-Army. We take care of our own. That's how it is, over here."

I ran the crew rebuilding VT-11. It needed an engine, the collective servo, blades, and radios, which my crew was cannibalizing from the hulks on the deadline. There weren't enough radios to go around, but we figured VMO-6 could salvage some from their wrecks. -11 had been mine once. Not this -11, and not the one before, either. The one before that. Both were gone. When this one died, the squadron would kill the number as unlucky, like the third VT-4, Phelps' ship. Of the twelve aircraft in the original squadron, all but one were gone; only the *Warsaw Falcon* remained. She was a gunship now, refitted after taking a mortar at Khe Sanh. It felt good to see her, armed and proud on the line. It felt good to know that she survived.

Frenchy had invented a dirge when we began losing ships. It was dumb, and difficult to remember. The words changed often as ships failed to return or erupted under incoming. I hummed the most recent version.

> VT-1 your time is done.
> VT-2 is finally through.
> VT-3 is in a tree.

VT-4 is no more.
VT-5 didn't come out alive.
VT-6 is in a fix.
VT-7 went to heaven.
VT-8 is way too late.
VT-9 didn't make it this time.
VT-10 has crashed and burned.
VT-11 has followed seven.
VT-12 is going . . . to . . . hell.
Crash and burn,
Crash and burn,
See the little Huey, crash and burn.

I checked my watch. Past midnight. Ten days and a hook. -12 and I: we'd cheated Frenchy's prophecy. Suddenly, an engine levitated toward me through the darkness, and I called my crew to work.

We came off the line at dawn. Four aircraft stood ready for flight test and Quang Tri; the grunts would have morning cover. We'd relieved half our work crews at 2 A.M. to grab a few hours of sleep. They trudged in as we trudged away, zombies passing zombies. I don't remember falling asleep; I think I became unconscious first.

CARD 10

"WakethefuckupZackyourordersareinWakethefuckupyourordersarein-GoddammityoulittleshitWakethe. . . ."

"Uh . . . Uh . . . Uh. . . ," I mutter. Someone has me by the shoulders, shaking me. My eyes are glued together; I can't see. He shakes, and shakes and shakes, hurting my neck.

"Wakethefuckupyourordersarein!"

I pry an eye open. It is Saint. Fucking crazy Saint. The asshole hasn't been the same since that 12.7 sliced his leg, and has recently confirmed his insanity by extending his tour for six months. I mouth gibberish, wanting it to sound like "Fuck off and lemme sleep," but even in my daze I can tell it doesn't sound anything similar. The asshole keeps shaking me and screaming, "Wakethefuckupyourordersarein!"

I try to dissemble a meaning but it escapes.

"Wakethefuckup yourordersarein!"

"Huh, Huh, Huh?"

"Wake the fuck up your orders are in!"

"Huh, Huh?"

"Wake the fuck up you shit you're going home!"

"Huh!"

"You're going home, asshole!"

"YAAAAAAAAAAAAHHHHHHHHHHHHHHHHHHHHHHHHH-HHH!!!!!!!!"

I race from the hooch, Saint trailing in my wake. He is stronger, faster, but I am Power, Strength. It is over.

It is ended, and I've lived. I've beaten you bastards. Oh God, I'm alive!

In an instant, I am on the flight line. The remaining few of Alpha Detachment are already there, waving their sacred papers. There are Zeke and Smitty and Tommy, Bunny Morton and Sergeant Francis and Captain Bill and Major Burke. The excitement is infectious. Grins running ear-to-ear on faces unused to grinning. We are the first of VMO-3 to live the months—to take the full measure—and survive. They laugh, scream, "Hurry up, Zack, we'll leave you behind." I run to S1; my papers are there. The Remington Raider, the hated, accursed Raider, hands me my check-out sheet, and I love him for it. All is forgiven. The sheet has two columns, one of places to go, the other for signatures to prove I've been. This day I will collect signatures instead of missions, and when I am done, I'll have something more than the missions and medals to show for my effort. I'll have my life. *My life!* It begins at the bottom of the checkout sheet, beneath the column of signatures. All I have to do to begin living is collect them. I seize the precious paper and race onto the flight line.

Operations is first. I confront the officer, giggling, jumping, twitching, unable to contain myself, and shove the paper under his nose. "Sign this!" He glares. "Sir!" I add, quickly. Indian war whoops filter in from the flight line. The Ops officer is pissed: others have already shoved their measure of glee up his ass. I squirm in front of him, thinking, "Sign it sign it sign it!" Checking out of Ops means you can't fly any more! His signature guarantees my life; it stops the missions forever. He glares at the sheet, some FNG with a name I never bothered to learn, and will never learn, then pulls my mission card from the file. Nineteen Air Medals. Three hundred and ninety-three missions. Seven short of the twentieth Air Medal. No more! He signs the sheet slowly. He has to sign it. The rules of DEROS declare he must. He knows it, I know it, and I watch his unknown name fill the sacred slot. It is done!

The Guinea is outside, clad in flight armor, a gunship turning on the hot pad.

"Hot mission, Zack. Want to come along?"

"Not me!" I cackle and show him The Paper. "Not me! Fini Phu Bai, Fini Vietnam." I shake his hand and wish myself well and run off, parroting, "Fini Phu Bai, Fini Phu Bai. *Bwaaaawk, bwaaaawk!* Fini Phu Bai!"

The shops fall to me. Line Shack, Flight Equipment, Tool Room, Metal Shop. The shop heads all sign. Hydraulics, Avionics. I swear I've stolen nothing, and they take the lie, sign, and wish me well. Checking out of Supply takes longest. I return to the hooch and gather the remains of my Marine issue, then race to Supply.[2]

"Where's your helmet?" the clerk rasps.

I pull it from my seabag, make a show of polishing the scratched and dented object with the sleeve of my flightsuit, and place it delicately on his table.

"One pot, right here!"

He grunts and makes a check.

"Bayonet?"

"Here!" jabbing it into the counter.

"Haversack?"

"Lost in combat."

"Cartridge-belt suspenders?"

"Ooh, lost in combat, I think. Suribachi, it was. We had just put up the flag, and this asshole took a picture, and . . ."

"Tent poles and pegs?"

"Combat!"

Last stop is the armory—the Rock's kingdom. For nearly thirteen months the Rock has allowed only pristine weapons into his loving custody. Smitty and I enter together, seeking strength in numbers. We find him surrounded by his gleaming guns, the air heavy with the familiar perfume of oil on blue steel. A nervous pair of FNG ordnancemen toil under his gaze. Timidly, we deliver our rifles. They are pathetic: the surface rust is leprous on the pitted steel, the stocks scored and caked with dust. The Rock glances at them, then at us. He smiles. The Rock smiles! He summons an FNG and orders him to take the filthy objects and clean them on pain of death. The FNG obeys, and slinks from the hut like a beaten cur. The Rock shakes our hands and says he hopes to share a brew in the World. We exit, astonished.

2. Marine Corps property, known collectively as *782 gear*, was primarily for grunt use in the field, e.g., helmet, bayonet, packs, one half of a two-man shelter, tent poles, etc. 782 gear stayed with the outfit when a man left. Air wing used the helmet, bayonet, and shelter-half; the remaining equipment was frequently discarded or lost; however, all had to be accounted for on departure.

The Rock is human. Truly, a day of miracles.

Ordnance is at the far end of the flight line; Smitty and I walk the row of shops slowly, waving our papers, stopping all we meet.

"Want to see something?"

"I've been seeing it all fucking morning. Go away."

"Here, look. Want to touch it? Just one finger, though. You're not short enough to touch it with your whole hand. You couldn't take the shock. Your pecker might fall off. See this. This is Ops, and this is Supply, and this, this is the Rock. The Rock made some FNG clean our weapons. Can you believe that? Look. You see this? This is my paper. It says Z-a-c-z-e-k. Just read the first three letters. I don't have time for you to read them all. You know how short I am? I, I . . . I can't tell you how short I am! I am *ti ti* (Vietnamese for little) fucking short."

"You know, you are a real asshole, Zack."

"I'm a short asshole, hey? Gotta go, hey? Gotta go. Freedom bird is waiting."

A dozen feet onward, smiling, waving.

"Hey, you! Dickhead! C'mon over here and look at this. Hurry up! I don't have much time. . . ."

We march into S1 like real Marines, salute the clerk like dogface-fucking-Army and lay the papers down.[3] Tommy is already there. The clerk examines the signatures one by one, running a finger down the columns of shops and names. He selects a rubber stamp, sets the date, inks and tests it. *Smash.*

"OK! You need," he begins, "to be in . . ."

Tommy: *thump, smash!*

"Da Nang tomorrow!"

Smitty: *thump, smash!*

"Stop here first . . ."

Zaczek, Ronald J., Cpl. 2155755, USMC, Combat crewchief, retired:

". . . thing in the morning . . ."

Thump!

". . . and pick up your orders."

SMASH!

Done!

I leave the hooch, a foot tall for every month, look up the flight line, then down, then out to the mat. It is just past 1400 hours, too late for chow, too

---

3. Marines do not salute "uncovered," and only cover the head indoors when under arms.

soon to return to the hooch and pack. I wander the line, fascinated by the strangeness of having nothing to do, nothing to be responsible for, no one depending on me. Crews wave from their aircraft, envy in their eyes, shouting, "Hey, short-timer!" and, "You lucky bastard." It is all happening the way it should, except that once again I have the odd sensation of being invisible, in spite of the waves and kudos and curses. My body is still here, but I am not. I am thousands of miles away, and it is but a minor detail that my body hasn't caught up with me yet.

I enter the line shack, deserted except for Gunny, and notice my flight helmet resting in a bin. Not my bin; I'd emptied it earlier when I returned my flight equipment to the paraloft. It seems funny, seeing the vacant space beneath my name. I examine the helmet. What the hell is it doing in someone else's bin? No name on the bin, either, though I can still read, "Francis" beneath a fresh smear of white. Wonder if Sarge knows some asshole is using his bin? I turn the helmet in my hands. Crewchiefs personalized their flight equipment, and no one ever copied another man's style. I'd sprayed mine matte black, and my name had stood out crisply in stenciled white, "ZACK."

Fresh black paint covers the stencil, though I easily trace the outline of the letters with my finger. Guess this guy likes black, too. Well, what do I expect? Are they supposed to turn it into a fucking shrine? Serviceable equipment is turned around quickly. Still, the damn thing wasn't even cold before they gave it away. There's no name on it, guess the guy didn't have time to mark it. Probably some FNG. I replace the helmet, wondering who it belongs to. I glance at Gunny, behind me at his desk, but don't ask. What difference does it make?

Something is missing in the shack. It finally dawns on me; I'd seldom been in it during the day. There are no crews, just old Gunny making love to his flight schedule, juggling tomorrow's assignments. So that's what he does. I check the date scrawled above the yellow-sheet books: 28 DEC 67. How about that! DEROS is supposed to be 7 JAN 68. I wonder why I'm going home early—never thought to ask. I guess they were aiming for Christmas after all and missed. When was Christmas, anyway? Let's see. We worked all last night. That was last night, right? Right. And we had the party the night before. No, we had the party just before we worked. We had the party on the same night we worked, I mean. I mean. . . . What do I mean? We partied, and then we worked. OK, try it this way. I didn't fly on Christmas. That was (count fingers) one, two, three. . . . Wait a minute. I have 393 missions. That was the day after Christmas, and this is ten and a hook, so that makes Christmas, uh, how many days ago? It was. . . . Oh, fuck it, anyway. I wonder what day this is. Didn't make it to chow at all today—fucking Saint was in my face. They only put the sign up at breakfast, if they even put a sign up. Only two chances out of seven that there was a sign. I wonder if anyone knows what day this is?

Hey, my yellow-sheet book is in the wrong place. We arrange them by section. It's in the wrong section. It belongs over there, in Charlie's section. This is Sergeant Francis's section. No, he's going home. The new guy has it now. Who the hell is he? I open the book. There is a name I do not know in the crewchief's place. It pisses me off.

"Hey, Gunny, who's flying my bird?"

"So and so has it now, Zack. They're out on a mission."

I am not certain which face matches so and so. So and so is an FNG. Not even NATOPS qualified. Shit, the hydraulic damper on the mast has a leak. If he's any kind of a crewchief he should catch it on preflight. These new guys are sloppy, though. Well, it was full last night—leaks slowly in flight. It will take days to run dry. He ought to know enough to catch it, and the pilots check it anyway. I hope he has the sense to refill it. Goddamn aircraft will crash if it runs dry. Happened in the States. I shake the vision away.

"He's not NATOPS. What the hell is he doing in my aircraft?"

"Not enough qualified crewchiefs, Zack. You know."

"Yeah, yeah."

"You could extend. . . ," the Gunny offers, half-smiling. "Vic is staying, and Larry, and some from the shops."

I glance to the flight line, shaking my head. "No way, GI," I answer and replace the book. VT-25 is stenciled on the cover. Bureau Number 154755: the last *Falcon* of them all. The bastard probably changed the name, too. Who cares; the fucking pig is someone else's problem, now.

DEROS is two weeks away for Gunny. "You going to extend?" I ask.

"Bull-fucking-shit!" he explodes, "I did my time."

I look around the shack and suddenly realize that when I leave it will be for the last time. Everything that I do or have done, today, is for the last time! There isn't time left to do anything twice, even if I want to. One dinner left, one night's sleep, one breakfast, probably one good shit. It feels wonderful, strange, confusing, euphoric, sudden. All that time, and now I'm out of it. I guess I really am "short." I take the shack in: the coffee urn, the cluttered crew's lounge, Gunny. My helmet. My bin. At least it still bears my name. That's how I'll remember it.

"See you in the World, Gunny."

He rises and shakes my hand, and as the door closes behind me I know it is for the last time.

I walk a short distance, stop, stand, think, turn, then reenter the shack.

Gunny looks up. "Forget something, Zack?"

I march to the yellow-sheet book, pull up the current mission's page, and scrawl, "Hydraulic damper leaking. Remove and replace. Aircraft grounded." Maintenance entries have to be signed off by Quality Control before the bird can go up again.

"Nah! I didn't forget anything."

CARD 10, 1800 HOURS

I examined each item carefully, piece by piece. Each sock, each paperback, each carefully preserved artifact of life In Country. I'd rehearsed this night for so long.

"Green skivvies. Don't need green skivvies in the World. Need some skivvies? No? Once, twice. . . ."

The hooch looked like a Goodwill collection center as the skivvies joined a growing pile of fatigue uniforms, flightsuits, paperbacks, blankets, extra 782 gear, towels, and mosquito netting in the center aisle; each item considered, packed or discarded as an act of deliberate will.

Saint was sitting on a nearby cot, the leg carved by the 12.7 thrust out awkwardly and the other crooked in behind it to keep him balanced. Saint had always been off balance. He was Alpha Detachment, but he wasn't going home. We'd been friends since Pendleton, before the war. It was funny to think of something that way—before the war.

"You're crazier than shit for staying," I noted for the hundredth time.

"If I hang in here another six months, I'll be short enough to get out of the Crotch when I get back."

"There's no guarantee of that."

"They said if you go home with less than a year left, they'll let you out early."

"Shit in one hand and try to wipe it with what they say. See how much crap's under your nails. You are so fucking dense. Whaddayou think, the tooth fairy is going to write you an early out?"

"You are always so negative. You should be glad I was here to take care of you. Where would you be if it weren't for me?"

"You! Take care of me! You are too fucking much."

"I've looked after you, just like I promised Grace I would."

"You had some balls writing to my girl."

"She asked me to look out for you."

"Yeah? Remember when you got shitfaced on Guam when that bitch johnned you the week we left the World? Who was it that pulled you puking off the lawn before the Air Police showed up?"

Saint countered quickly, "Who was it that hustled you off the flight line when you pickled your brain on VO and tried to shoot Gunny? You'd be in Portsmouth if it weren't for me."

"It took six of you," I sneered, "and I only weighed 108 pounds. You were lucky you got the rifle away from me. I'd have blown you all away."

"Yeah, well who hustled around and got the other five?"

"I pulled you out of the gutter in Olongapo, you dick."

"I covered you at Cua Viet . . ."

"I covered you at Khe Sanh."

I held up my writing gear. "You want this stuff?"

"Yeah, thanks. I've been looking out for you the whole goddamned tour."

"The only place you look is in my care package for brownies, you goddamned mooch. What did you get from home? Asparagus. Canned asparagus. No one wanted your fucking canned asparagus. I have been sleeping in the cubicle next to you for almost a fucking year, smelling you sweat asparagus."

Saint shrugged, shifted his leg, and grimaced slightly.

"Does it still hurt?"

"Nah."

I hadn't been there when he was hit. They'd told me how the heavy caliber tore into the cabin as Saint stood behind his gun; how he'd kept standing, and shooting, his leg twitching in spasms as the blood spurted. They'd fixed him up in Japan. Two months he was gone while they repaired his back, stomach, and leg. I remembered the day he returned to the squadron, fatter than a pig after being out of Country, walking to the mess hall with his tray slung across his back, gibbering, "Hi, Zack!" as if he'd never left. He still walked as if his feet were cast in cement. I was so glad to see him. I never thought I'd see him again. Couldn't show it, though. We didn't do things like that. Yet, in another day I'd be gone, and he would not, and I knew I'd never see him again.

I packed the last of my gear. The cubicle was nearly empty. I left one clean uniform for the morning; if I could find a fire, I'd burn what I had on.

I moved the cabinet that had held my clothes, and from a hidden recess withdrew a long-concealed object and pondered whether or not I should pack it. I decided it would be fitting to leave it behind. A touch of me, something tangible that would remain. Saint watched as I scribed the wooden handle with my Ka-Bar.

"What are you going to do with it?"

F-i-r-s-, I carved the letters.

"Give it to you."

S-e-r-g-

"What the hell am I going to do with it? It's all rusted. How long have you had it, anyway?"

S-t-o-l-

"Almost since the beginning."

Z-a-c-k, the etching continues.

"What should I do with it?"

E-R-F-

"Give it to Top when I'm gone. Then steal it back. That's how I'd like to remember it.

"There!" I produced the finished work with a flourish and laid it in Saint's lap.

First Sergeant's Saw—Stolen By Zack

VMO-3

16-DEC-66–29-DEC-67

Semper Fi

CARD 10, 2000 HOURS

There was nothing left to pack, give, or throw away. Only good-byes remained. Alone, I walked the row of huts. It was quiet; the squadron was exhausted from the previous night's labor. I had to hurry, although my own adrenalin was giving out. The squadron would retire early, and there was little time.

Shifty: "Take it easy, old turd."

"Yeah, you too. Don't puke anymore!"

We laughed, and remembered.

"'Member when you got airsick?" I began. "I busted a gut. Your gunner said you were Marine green from the hairline down, cheeks stuffed with chunky vomit, sweat popping and bulging eyes. God, I wish I'd been there."

The gunner had warned the pilots that Shifty was about to blow his lunch. Evil bastards—they jinked the aircraft all over the sky, chanting,

"Get it out. Get it all out." Shifty blew his cookies into the pilot's helmet bag. Said he didn't want chunky smears down the tailcone.

"I fixed him," Shifty continued. "He had a meat loaf sandwich in there. Doubt that the waxed paper helped too much."

I laughed, and made a motion to leave. It was a short night. But then we laughed some more, and . . .

"Remember the nights we spent on the wire?" he asked.

Most times we snuggled together under a poncho, dug into the wet sand reinforcing the outer perimeter. God, we smelled; like a canvas-and-wet-fart sandwich. We talked a lot, those long nights on the wire. He was from Pittsburgh, and me from Baltimore; I still had a girl and he didn't; I was headed for Lejeune, and he six months later; I was going to college, and he was going to be a mechanic, and. . . .

"We thought we lucked out the night we drew the command post on the roof of the old French bunker. It was your idea to try to sleep inside that night of the monsoon."

"It was your idea," he challenged.

"No fucking way! Fucking rats! We must have looked like the two stooges pulling each other out of the way, trying to clamber back onto the roof."

"You kept pulling me back."

"You were on a first-name basis with them. I figured you could discuss family history."

I stood to say good-bye. "I'll see you in the World, you ugly bastard."

Shifty extended his hand. He was a gnomey-looking guy, with a twisted front tooth in a mouth that just always wanted to smile. I remember, his nose pointed off to one side. Christ, he was ugly. We shook, long enough that the seconds began to matter.

"I'll see you in the World, Zack."

Mouse: "Go 'way! Don' bother me!" he snapped over his shoulder.

Mouse was arguing chess with Smitty. Mouse had been arguing chess with Smitty since the first week In Country. Mouse was tired of hearing that he had two weeks more In Country than me, or Smitty or any of the rest of us who were leaving in the morning. Smitty cocked his head and gestured with a thumb.

"He has no sense of humor. What an asshole."

Smitty was packing between moves, and reached up to remove the shade from the naked bulb over the chessboard. Mouse, or Smitty, or both—they were inseparable—had appropriated Madame Nga's straw hat a year before, punched a hole through the cone, and declared it a lamp shade. Mamasan had been pissed. And then they wired her chopsticks together so that she couldn't eat her fish heads. And then. . . .

"Vhat you doing vith that?" Mouse challenged, mimicking a Russian accent. Mouse did good Russian.

"It's mine."

"Bullshit! It mine!"

"I ripped it off Mamasan."

"I hung it there," Mouse shot back, dropping the Russian, standing.

"It's mine."

"Bull-shit. Bull-fucking-shit."

"I came over to say good-bye," I interjected.

They tugged at the hat.

"It's my fucking hat."

"You're going to tear it."

"I'm not going to tear it. You're going to tear it."

Tug.

Tug.

"OK OK OK. Take the fucking hat!"

"Screw you. Take it yourself."

"I don't want the fucking thing."

"You have to take it."

"I don't have to take anything."

"You are an asshole."

Inseparable.

They sat, backs to each other. The chessboard and the hat in between.

"'Bye, Mouse." I said, leaving. Mouse was West Coast; I was East. I knew I'd never see him in the World.

"Yeah. Yeah." Mouse didn't seem to want to look at me.

I walked away, stopped, turned, then walked back to the hut. They were playing chess again, the hat nowhere to be seen.

"Zack," Mouse called, his eyes on the board.

"Yeah?"

"You take care, huh?"

Joe: "What the fuck are you coming over here for? Why don't you get out?"

Joe had extended his tour and moved to the new barracks. I'd walked the length of the building, saying good-byes quickly. Sarcasm had always been a part of Joe's humor, but the accusation in his voice seemed real. I tried to ignore it.

"I came over to say good-bye, asshole."

"Good-bye," he sneered. "You're such a baby."

"What the hell are you talking about?"

"Go ahead. Run away."

"What do you mean, 'run away'?"

"You see these new guys. Who's there to train them? All of you are just running away. 'I've got to get back to my girl,'" he sneered again.

"Who the hell do you think you are? This isn't any 'run away,'" I bristled. "The tour is over! Maybe you're too damned dumb to figure that out. Or maybe you just don't have anything to go back to."

"Just get the hell out of here."

He turned away. I stared, angry, hurt, not wanting it to end this way. But the rebukes clouded the air between us, irrevocable. He had no right.

"Fuck you, Joe," and I let the door slam hard behind me.

DEROS

The ready-room watch touched me lightly, chanting over and over, "It's oh-five-hundred . . . it's oh-five-hundred. . . ."

I lifted up, groggy, reaching for a cigarette.

"You awake, Corporal?" he asked as my feet touched the floor. It was one of the FNGs.

"Yeah, yeah," and turned to the portion of my mind that stored flight schedules.

There was no schedule.

"You asshole. What the fuck did you get me up for?"

"Uh, gee. Gunny put you down for oh-five-hundred, right here. See?"

"Get that goddamned light out of my eyes, you piece of shit!"

I jumped from the cot, fists clenched.

"I'll kick your goddamned ass."

The watch ran away, and I lay back. Fucking Gunny. If I wasn't so short, I'd get him—put sand in his Preparation-H or something.

I lay on the canvas and pulled at the smoke. The smell of the wire washed in from the perimeter. The last morning. This was the last morning, ever. The night seemed a dream; almost as if it hadn't been at all. I smoked down to the filter, then lifted a single white rectangle from the empty shelves that had held my treasures.

The hook was at hand.

Most short-timers vowed to take no memories with them. Not me. As I walked that second-to-last walk from the hooch to the flight line, entered the mess hall for the last-ever time, borrowed a tray, and ate my final Phu Bai meal, I watched wide-eyed, took in each sound with welcome ears, knowing that each instant would be the last of its kind, forever. I wanted to remember everything about the last day.

I remember striving to be conscious of the finality of every motion. I remember wanting to absorb every clatter of aluminum on wood or metal, every slop of food and gurgle of beverage, every blur of dingy green and gray. I remember wanting to remember the final walk back to the hooch— Saint was going to drive us to the MARLOG (Marine Logistics and Supply Area), where we would hitch a ride to Da Nang, and I would not pass this way again. I would never walk through the arming area, or past the fuel pits, or past VT-12 or -11 or -25 or the others that I flew, ever again.

I remember the feeling of wanting to remember; it is so strong.

Strange. I remember so little of that last day at Phu Bai, save the need to remember.

I remember passing the flight line, heading toward the hooches. It must have been Smitty, Zeke, Tommy, and me, but I cannot see them. We would have split up and entered our hooches for the last time. I remember being conscious of each step up the splintery boards, opening the screen door. It always swelled in damp weather, and stuck and then leaped into your face when you jerked it open. A spring snapped it shut when you passed, and a thin swish of air always washed in with the echoing slam. But I don't remember that. I know I walked the thirty feet to my cubicle, that it was early and that mamasan wouldn't have come in yet, so the center aisle must have been unswept and the dry sand must have gritted under the hard rubber of my flight boots. The slow scrape and thud must have echoed the length of the hut as I neared my old home for the last time.

There was someone in it. I remember.

It was Jerome. Jerome had extended his tour like the Saint. He was certainly better than some FNG. But, I mean, I wasn't even gone yet. I looked at him, without words, sitting on my cot, unpacking his things into my desk, my nightstand, my storage box. They were all in different places. The cot—it was supposed to be on the left, parallel to the short end of the hut. Now it was at the back, under the shelves. That was stupid. He'd hit his head when he jumped for the bunker. And the nightstand, it was supposed to be there, but it was . . . and the storage unit . . .

"Nice cube, Zack," he said plainly, "wanted to grab it before someone else came along."

I just nodded. My stuff was in the aisle. At least it was piled neatly. Jerome was always neat. I gathered everything; I don't remember whether

I said anything to Jerome or not. I don't remember walking that thirty feet for the last time, nor the airy clatter of the screen door, nor that last-of-all walk to the flight line, where I had no aircraft, no bin, no helmet, and the name "Zack" meant nothing. I don't remember entering S1 to get my orders, or talking to Crisp, only that, in time, awareness seeped in, and as I looked back at the distant hooches, I began to know, really know, that it was ending, and I remember standing on the flight line, thinking . . .

Fuck you.

Fuck you, Vietnam.

Fuck you, VMO-3.

Fuck you, Marine Corps.

Fuck all of you.

Saint pulled up with the P40.

"Your limousine awaits," he waved. Tommy, Zeke, and Smitty were already aboard.

"Just get my ass out of here."

Flight ops was under way, so Saint bounced us the long way around the airstrip to the MARLOG, where a year earlier I'd had my first sight of this godforsaken place. We dumped our seabags and queued before the duty clerk to get a number for space-available to Da Nang. It was demeaning, air wing having to queue for space, but none of the squadron's ships were headed south, so we had to hitch like common grunts.

Saint stayed as long as he could, then announced that he had to get the P40 back to Ordnance. He said good-bye and shook hands all around. I looked at him and felt my face try to smile, but it wouldn't, and I felt my voice try to mouth an insult, but a lump rose high instead and choked it off. So I looked at him with tightened chest, and nodded, and was silent. We shook hands again, then he climbed into the P40 and played chicken with an Army Bronco taxiing down the ramp. Crazy fucking Saint. I watched until I could see him no longer, until he was gone. I wondered what he'd do when he found the hook in his pocket.

There were a lot of transients, so we settled in with our gear to wait. C-130s, CH-53s and -46s came and went, priority given first to supplies, then by rank. Our numbers slowly climbed to the top of the list.

Near mid-afternoon, we climbed into a cluttered CH-46. I paused on the tarmac, standing a few inches behind the helicopter's lowered ramp. I looked around, and then down, and with a sense of purpose watched my boots leave Phu Bai forever. I recall no feeling at all. I found a seat and strapped in. It was on the aircraft's port side; I knew the departing flight pattern and planned it so. I craned my neck around to peer through the dusty Plexiglas. I remember not wanting to miss a thing.

The rotors built their turns and we began to move, and then . . . we must have taxied along the ramp toward the runway. We must have, but I cannot see it. I must have stared through the portal and watched, across a year of debris, my first hooch and the bunkers and four-holers we'd built, and the old medevac hot-pad close by the old line shack. No longer VMO-3's, they belonged to another squadron. But they had been ours, first. We had built them well, and we had built them all, when we were all whole, and together. I must have seen them. The -46 had to roll past as it taxied onto the runway. I must have watched the old hooches as they passed into history, and then VMO-3's main flight line would have filled the window one last time.

I must have let the neat rows of gunships and shops make their final imprint. It is what I wanted to do. It must have been as drab as ever, under the grayness that permeated that day. We bumped along the runway, gathering speed, but I know I would have kept my attention on the shifting scene. I must have seen -12. Victor Tango 12. I know that I would have held her in my gaze longest until the racing scene stole her from my eyes. We lifted, and the expanse of the squadron's living area would have come into view. I must have looked along the distant row of huts and remembered the faces of friends who'd given them meaning. And I must have let the huts join the images in my mind as well. I would have looked out the window and back and back, until there was only the dark scar of the runway visible in the distance, until there was nothing.

It must have been that way. It is how I want to remember it.

LIMBO: DA NANG

The -46 dropped us unceremoniously at the sprawling air base. There was concrete everywhere, as far as the eye could see—more airport than combat base. We stood like country boys come to the big city.

I spotted a Marine Gunny, rice-starched and crisp in carefully faded utilities. His cover was blocked to perfection, the double-rockers of his rank pinned in mint condition over the eagle-globe-and-anchor on the forepeak. His trousers were tightly bloused and boots spit-shined.[4] God, he was pretty.

"Hey, Gunny," I yelled, "where do we catch a bus for the World?"

---

4. Blousing was a stateside practice of rolling each leg and securing it at the top of the boot with a coiled spring to give the effect of pantaloons.

He marched over, arms locked on hips, and scanned us. We were four corporals in unstarched cotton, each wearing his combat aircrewman's wings pinned over the left breast. His eyes took us in, lingered briefly on the wings, then fixed on some invisible point in the middle of us. He pointed a bony finger at his collar emblem.

"What rank is this?"

I counted on my fingers behind my back; I was E-4. Corporal, sergeant, staff, gunny. It had been a while.

"Uh, E-7?"

"This is the insignia of a Marine gunnery sergeant!"

And then he began chewing. He chewed my ass right and left for calling him "Gunny." He chewed it up and down for sounding like a wise-assed, unmilitary fuck-off (which I was) and looking as if I'd walked in off a dung heap (which I had). He chewed it backward and forward and took Zeke and Smitty in tow for not acting like or looking like his idea of what a jar-head was supposed to act or look like. He told us this wasn't the zone, and we'd better get our shit together or we'd be doing bad time. He said it all without taking a breath or smiling even once. Then he stomped off and never once gave us so much as a compass heading to the bus to the World.

"You know, we ain't shit here," Smitty said as we stood, stunned. A moment passed before it occurred to us to become pissed. Two days before, we were crewchiefs and gunners, short-timers, envied by all. Smitty was right: we weren't shit.

We asked others, but no one had the time. Eventually, a brown-bar looey wandered by. He looked as if he'd rather eat dog barf than get near us, but we saluted enough to convince him we were good at kissing ass, and he pointed the way to our destination.

The transient center was a bazaar of hooches, bunkers, and CONEX boxes. Worried Marines scurried here and there, papers flapping in their fists. Indolent Marines lounged atop the bunkers. Puzzled Marines stopped others, pointed, drew circles in the air, and everyone shook their heads. Everyone said, "Fuck" a lot—not unusual, but it seemed to mean both more and less in this place than elsewhere. We found the check-in hooch and reported to a snotty private first class dressed as prettily as the gunny. A study in insolence, and master of the situation, the private was unimpressed by the shuffling line of FNGs working up to his desk. Each moment a fresh set of balls entered his grasp. He owed us nothing, not respect nor kindness nor solicitation. He needed us not at all. Before him lay the weapons of his trade. A check-in sheet and a pencil with which he could write (or not write) my name on the sacred ticket to the World.

"Name?" he murmured without looking up.

"Zaczek, Ronald Joseph. Corporal. 2155755. 6319," I rattled back, cueing from those ahead.

"I didn't ask for the rest of that shit, Corporal," he said. I flushed, but fought the temptation to rip out his Adam's apple. I needed this turd to pass my name, else I could spend forever in this maze.

"Sorry," I choked. The private scanned the master list. His finger moved slowly, top to bottom, in spite of the alphabetical listing. I wanted to tear it from his desk and scream, "Start with 'Z,' asshole!" but I didn't. He had to find the name eventually, but how long was "eventually" when you'd passed a lifetime in waiting for this moment, and every delay seemed a lifetime in itself? I watched each page as he flipped them. Relieved, I spotted myself near the end.

"Serial number and MOS," he repeated.

"2155755, 6319," I answered tightly.

"6319? What's that?"

It was a reasonable question. Few people recognized air-wing specialities.

"Jet Helicopter Mechanic," I answered, then added with a low note of accustomed pride, "Crewchief."

I'd borne the title for a year; it had commanded respect and, I thought, a little awe. I'd sat a dozen times in a slop chute in Khe Sanh or Dong Ha, trading tales with rear-area pokes and front-line grunts alike. When I said that word, I saw heads raise, eyes flood with recognition and regard. Being crewchief had bought me one hell of a lot of drinks.

"Wing-wiper," he snorted, returning to his papers. It was the slur stateside, noncombat grunts applied to stateside, noncombat, fucking–air wing.

A flash of clerk's fear crossed his face as the blood rushed to mine. I can destroy this shit with my hands, or the Ka-Bar on my belt, I thought, then walk out and scarf ice cream at the PX for as much as it would bother me. I'd killed men who had never harmed me; how much easier it would be to snuff this rat-shit weasel clerk.

But I'd still be In Country. I could threaten him, return the insults, maybe throttle his fucking neck, but he'd simply lose my check-in or report me to the MPs, and I'd still be In Country. If I did anything to him except eat his shit, I'd stay In Country. He knew it. I knew it. His nervousness passed quickly; he knew he'd won. His eyes met mine when he gave me a hooch assignment.

"Stick tight by the hooch except for chow. There are musters in the morning and afternoon. Don't miss them or you'll miss your flight to Okinawa and you'll go to the bottom of the list."

"I have all my stuff here. I don't need to go to Okinawa," I countered.

Most Marines processed through Oki on their way to Vietnam, leaving their stateside issue to be warehoused. As far as I knew, the only purpose for returning to Oki was to retrieve uniforms. I'd arrived by carrier and had all my gear; there was no point in going to Oki. It would add days to the trip home for nothing.

"All flights go through Kadena. No choice," he explained, irritated. The line behind me was growing impatient. Too bad.

"I don't need to go to fucking Okinawa!"

"Then you can stay here and fucking rot for all I care. You want this hooch assignment or not? Next!"

Bastard! I took the chit, and my anger, and left the bastard at his desk. I knew I would hate him forever, and prayed that the motherfucker's luck would run out, that he'd be assigned to a grunt outfit, and that he'd be killed.

Tommy and I drew cots in one hut, Zeke and Smitty in another. Another PFC gave us blankets and a warning that we'd have to pay for them if they were lost. We split up, arranged to meet for chow, and entered our temporary quarters. They were Spartan, a dozen or so cots on an otherwise empty plywood floor. Green-clad bodies lay on most of them, snoozing, reading, staring blankly at the corrugated roof.

"Hey! FNGs!" one called.

"New meat!" added another.

"Welcome to my house," a third smiled warmly and swung his feet to the deck.

Defensively, Tommy and I found cots next to each other.

"Where you guys from?"

"Phu Bai," I answered, thinking that I didn't feel like adding "air wing" just at the moment. We dropped our gear between the cots.

"Better put your seabags under your cot so you can feel it if someone tries to rip them off while you're asleep. Use your ditty bag as a pillow." He pointed to my knife. "Sleep with that. The REMFs will steal you blind, and what the REMFs miss, the dinks who clean the hooches scavenge. Just the two of you?"

"Two others in another hooch," I said, thinking there wasn't much safety in numbers in this place.

"Always leave someone behind to guard your shit, and lock your blankets in your seabag during the day. The REMFs will steal them too, then make you pay before you leave. I'll watch your shit if you want me to, but

I wouldn't trust me either if I was you."

"Tough place, huh?"

"You better believe it. I'd rather be in the zone. Meet a better class of people."

"How long you been here?"

"Four days."

"Four days?"

"Four days, man. I been here since before Christmas. Freedom Bird took a little vacate, so it's all backed up. It's OK. Chow's good. I hear it's backed up four or five days on Oki, too. That's OK. I want some time on BC street and Suck-A-Hatchie Alley. You know, they say there's a broad there who gives the best head in the Pacific, but that she's swore she's going to bite off the last one she gives. How's that for Russian fucking roulette? I hear there's so much puss. . . ."

The Marine went on, but I didn't pay much attention after "four days." Four . . . fucking . . . days.

The Marine spoke truly; the REMFs of Da Nang were to be feared. Every day the curses and futile threats of the recently robbed echoed through the hooches. We exchanged our MPCs for greenbacks, which made the poorest of us attractive targets. MPCs had never seemed like real money to me; spending it had always seemed more like bartering than buying. I stored or carried it without special care; ours had been an honest outfit. Greenbacks were real money, and possessing them made me feel vulnerable. For the first time in a year, I had something that another person would steal. I so worried they'd be taken in my sleep that I secured my wallet uncomfortably under whatever part of me I placed next to the cot. I awakened every two hours, shifted a new part of my body into contact with the cot, and sandwiched the wallet under my weight once again. As the other Marine suggested, I stuffed my seabag under the cot so I'd feel it if it moved. I worried my blanket would get ripped off as I dozed and fell asleep gripping the edge in tense fingers to hold it in place.

I worried about the Freedom Bird. I worried they'd deliberately or accidentally drop me from the manifest and, after the first full day, went to the snotty clerk to see if I'd missed a flight. But it pissed him off, and he wouldn't answer my questions; then I worried he'd lose me on purpose and it would be days before I knew I'd been screwed.

I worried about the shakedowns, though I had nothing to hide. We carried our gear to a large hooch filled with rows of shallow bins. It reminded me of boot camp—the large, flat bins of Parris Island where, shaved-headed and trembling, we received our initial Marine issue. The DI would scream, "Toothbrush," and all the lowly fucking boots scrambled in the bins to find one and hold it aloft to prove he had it. In Da Nang the bins worked in reverse. We dumped our worldly possessions onto the plywood and one of the permanent assholes shoved his way through the pile, looking for drugs, weapons, or anything someone who was not you thought shouldn't be In Country or leave Country. I felt docile and lost, out of control, angry at the REMFs' manhandling the carefully culled remnants of all that had been holy or essential. I endured in silence. Here, there, a sacred item was confiscated. Silence. I just wanted to be gone, to be done. Nothing was worth staying In Country.

Days passed; one, two. Surprise shakedowns almost every day; Freedom Bird musters in the morning and afternoon. Guys leaving, others joining. Hey, did he get here after me? I'm sure he got here after me. Trips to the snotty PFC. He recognizes me now. I don't care if I piss him off. Let me out of here or I will kill you. The longer I'm here, the less I have to lose. No idea when it will end. Shit, shower, breakfast, piss, muster, lounge, lunch, lounge, muster, shit, dinner, lounge, shit, jerk off, sleep.

The third day was New Year's Eve. We found out at midnight when the assholes on the perimeter celebrated with MG. Tommy and I raced for the bunker and sat worrying that the REMFs were stealing the hooch blind. Next day, Zeke and I stopped at the USO for a burger. A Red Cross worker stuck pointy hats on our heads and party favors in our faces and told us to be happy: it was midnight in the World.

Four days. Felt like I'd been in Da Nang forever. It was getting difficult to remember being anywhere else. Lay in the cot and stared at the roof. Shit, shower, uh . . . dinner, uh. . . .

Five days. The last shakedown, so they claimed. They sealed our seabags in CONEX boxes and left us a change of clothing that we packed into puke-green ditty bags emblazoned with a puke-yellow Marine emblem and the letters "USMC." I knew I'd never see my gear again. We'd be leaving on day six, or seven, or eight, or. . . .

Day six. Same routine. Shit, shower, breakfast, piss, muster. The noncom reads; names, names, names, then . . .

"Hughes!"

"Here!" Tommy shouts.

Names, names, names!

"Smith!"

"Yo!"

Hallelujah! Me, me, me. Let it be me.

Names, names . . .

"Zaczek!"

Great God Almighty, Great God Almighty!

Zeke too.

YAHHHHHHHHHHHHHHHHHHHHHHHHHHHHHHHHHHHHHHH!

"Pick up your shit and be back in five," the noncom ordered. The trucks will take you to the airfield."

The trucks come. We jam in. Out to the runway. Bouncy, bouncy. The Freedom Bird. Gleaming, white 707. Pan Am. I look at my watch. Twelve months, twenty-eight days, and two hours, near as I can recollect. I make a mental note to mark the exact moment the Bird's gear leaves the runway. Up the stairway, dizzy. A round-eye at the top. A round-eye! A blonde round-eye! A blonde round-eye with big tits! Smiling. I smile, she smiles, everyone smiles. She touches me on the elbow and points aft, "Take any seat." She touched my elbow! God, and she talked to me! So much excitement, bursting, yet hardly a sound inside the aircraft, only the stewardess's fading voice as I move aft, "Take any seat . . . take any seat." Someone stops in the aisle to talk to a friend, halting progress. A burst of desperate anger behind them, "Fucking asshole, move it!" I take a seat on the port side, near an escape window, look for other emergency portals, and plan an evacuation route. Old habits. Smitty plops down next to me, grinning like a lunatic, not saying a word. Hardly anyone saying a word. ". . . take any seat. . . ." She stops. Closes the main hatch. Stewardesses move among us like smiling gems, staring into our laps; one stops near me: "Your seat belt. . . ."

"Huh?"

"Fasten your seat belt, please."

"Huh?"

She bends across Smitty, her hair in his face, her face close to mine. I breathe her perfume. Oh God. She takes the ends of the belt, pulls it across my pecker, "click." Smiles.

"All tucked in."

The engines turn. One. I hum with them, letting the noise fill my mind.

"We gotta get out of this place . . ."

Two. ". . . If it's the last thing we ever do."

Three. "We gotta get out of this place . . ."

Four! "Girl, there's a better life for me and you!"

The aircraft moves, rolling, creaking, roaring. To the end of the runway, brake, lurch; position, four engines, maximum thrust, going, going, going, *ba-bump*, free!

"Yahhhhhhhhhhhhhhhhhhhhhhhhh! Yaaaaaaaaaaaaahhh! God damn!
Fuck fuck fuck, we made it. God damn!" Fists pounding seats, fists pounding heads. "Yaaaaaaaaaahhhhhhhhhhhhhhh! Free free free! God damn!"

Marble Mountain passes beneath us, my last image of Vietnam: the long
line of Hueys of VMO-2. "Fuck you. You didn't get me. Fuck you!
Yaaaaaaaaaaahhhhhhhhhh!"

A soothing voice, plaintive on the loudspeaker: "Quiet, quiet please."

"Yaaaaaaaaaahhhhhhhhhhhhhhhh! Fuck fuck. Yaaaaaaaaaaaaahhh!"

"Quiet, please."

"Yaaaahhhhh!"

"Quiet!"

"Yahhhhhh!"

Dizzy, dizzy, dizzy. What's she saying?

Someone officer on the aircraft, being officer, "At ease, Marines!"

Round-eye smiling. "Yo, Baby!"

"At ease!"

Officer being officer again instead of grunt, or pilot, instead of OK guy.
Welcome to the World, motherfucker. Voice in the aft cabin: "Fuck off,
asshole."

"Who said that? Who said that?"

"Fuck you up your ass."

Asshole grabs a noncom. Grabs two noncoms: bookends. "Get control
of these people, Sergeants."

"We gotta get out of this place. . . . If it's the last thing we ever do—
screw you!"

"Quiet. Quiet!"

Somewhere out to sea I glanced at my watch. Twelve months, twenty-
eight days, and two hours. I'd forgotten to note the minutes.

LIMBO: OKINAWA

Four days on Oki. Four lousy, goddamned, miserable days. Shit, shower,
breakfast, piss, muster, Enlisted Men's Club, lunch, E-club, muster, shit,
dinner, E-club, shit, jerk off, sleep. Shakedowns salted here and there to
break the monotony. Amazing what could make it through a shakedown;
they had a huge board in the processing center wired chock full with rifles,
grenades, Claymores, RPG rounds, Bazooka, light and heavy mortars,
complete with base plate. Some of that shit weighed more than a hundred
pounds, and there was enough ammo to fight a war.

The transient center was a barracks, with real showers and hot water and beds with thick mattresses—and an asshole for NCOIC. Someone pissed him off the first night, so he restricted us to the base. No BC street. No Suck-A-Hatchie Alley. Someone else pissed him off the second and third nights, too. I learned to drink Singapore Slings in the E-Club. Made pyramids of the glasses. Smitty, Zeke, and I discovered we still had our draft cards. We burned them.

Eventually, a flight assignment, a final shakedown. Packing my ditties into my ditty bag. Next stop, California.

Reveille was at 0400. It was winter in the World, so we suited up in our dress greens although we'd sweat our asses off under the clear Oki sun. I dressed nervously.

I worked and reworked the knot in my tie. No more Da Nang, no more Oki, no more buffer zones. What would I find in the World? The newspapers had shown how terribly it had changed in the past year. Rumors were rampant about guys going home, harassed and challenged by the peace creeps, spat upon in airports. At first it was difficult to believe, but the photographs and antiwar venom proved it to be true. It was easy to believe anything; safest to believe the worst—the rumors of vets being threatened in bars, pulling frags from their tunics and blowing the bastards away.

What would I find in the World? I had pinned my hopes and dreams on DEROS, but the World. . . . The papers showed it was to be feared as much as hoped for. All those months, yearning for DEROS, and at the same time deluding myself: "Mom, don't send me any more newspapers. I don't want to read about that stuff." What would I find in the World?

The tie wasn't working. I struggled, threading it this way and that, until I realized I'd forgotten how to make a knot. It was a small thing, but frustrating. Was I so changed I couldn't remember how to tie a tie? A master sergeant, sporting ribbons from my father's war, crossed the barracks. He snugged the tie securely under my chin, a perfect Windsor with a dimple, then untied it and made me try. I did it again, and then again until I got it right—all the time not a word spoken. He stepped back, squared my tunic, tucked a loose shoulder strap under the collar, and flicked an imaginary speck of lint. "You look good, Marine." I nodded thanks and glanced in the mirror, liking what I saw. I was lean! mean! a fighting machine. I can tie a tie. I can handle the World. I can handle anything.

For six hours we cooled our heels in Kadena's terminal. Dispatchers parceled flight assignments to returning veterans, dependents, and other transients. Vets enjoyed no special status, but I don't recall that it bothered me. Someone said if we didn't get out today, there was always tomorrow. I don't recall that bothering me, either.

There was no cheering on takeoff, and I slept soon after the aircraft cleared Oki's beaches. I awakened over the Pacific to a breakfast. Sleepily, I lifted a dixie cup of viscous liquid and chugged. An acid trickle brought me erect, straining against the lap belt. Tomato juice! I hated fucking tomato juice. I'd tried it in boot camp. The DI made me swallow while standing over the garbage pail because you "took all you wanted and ate all you took" on Parris Island. It was OK to vomit, but the only way it could go into the pail was if it went through me first.

Oily beads of sweat erupted as the juice threatened to burst from my cheeks. The lowered tray blocked the barf bag, and my eyes began to tear as I searched right, left for someplace to spit. No place to put it but back in the cup, the tiniest of targets. I had to get rid of it; the stuff was going up my nose! Pink froth and slippery phlegm spewed everywhere. Into the cup, around the cup, across breakfast. The last droplets settled into my lap as I struggled hopelessly with a napkin. The stewardess rushed over, blanched at the blooded mess, and shoved a barf bag in my face. I filled it with putrid air and spittle and the rasping groans of dry heaves, but it was too late; I already looked like a medevac. I struggled to the lavatory and tried to wash the taste and stench away. The stains in my uniform yielded slowly to cold water, but the odor of vomit clouded about me as I returned to my seat.

A service brat of five or six, seated across the aisle, recoiled against his mother. She was young, blonde, and round-eyed, her eyes even rounder at the moment. I'd been staring up her skirt since Oki, trying to figure how to make conversation. She'd been the substance of months of wet dreams when I'd had the energy to make them wet, but it had been a long time since I'd spoken to an American woman. Like tying a tie, I'd forgotten how.

I smiled wanly. "Sorry for the mess—never could handle tomato juice. I'm just coming out of Country—going back to the World!" I added proudly.

She shifted uncomfortably; her skirt pulled higher, precisely to the point where you would begin sucking hickeys. She glared, then pushed the call button above her seat.

"They should have separate flights for enlisted men," she complained to the stewardess. She pulled her child closer and demanded another seat, but the plane was full. The stewardess asked if I would trade places with someone in the back, and I said I was happy right where I was. The stewardess favored the blonde with a "what can I do?" glance, and moved on.

I lay back and glanced up her skirt a last time.

"Lady, if you think this is bad, you should smell burning shitters."

DEROS!

The aircraft was silent as we flew into a purple dawn, rising to challenge the morning lights of the World. The World was big and well illuminated, and I felt vulnerable in its spaciousness. I remembered when Give-A-Fuck and I had gone on R & R in Bangkok. Everyone had departed the aircraft except us. We had sat, foolishly staring out the window as the plane had emptied. "I feel weird," I'd said to George. "Me too," he'd answered, then added, "I can't figure it out." Someone had walked back and told us to move our asses, but we hadn't, and they'd left when they saw they couldn't make us. You didn't fuck with guys coming out of Country. Finally, I'd said, "I think I know what's wrong. There's no wire and we're unarmed."

We had laughed, then joined the others, but it had taken a long time to stop feeling naked.

I stepped purposefully onto the asphalt, moved away from the flow of traffic, and savored the World beneath my feet. Occasionally, a man in uniform knelt and slowly kissed the ground, then rose eagerly and hurried into the terminal. I watched a while, wanting to witness the return. It felt both intimate and remote, like scenes from old war movies.

The dependents of husbands and fathers swarmed around the gate. I walked past their private circles of embraces. Kissing wives, tugging children. "Missed you, honey. How was your flight? Daddy, Daddy, what did you bring me?" I was home, but not entirely. It was unreasonable, I knew, to expect to feel home in an unknown airport. But the airport was in the World, and I'd sought the World for so long. It seemed unfair that there was no one to greet me. I envied the small circles. Well, I wasn't really home yet, I reasoned. Home was for tonight. It was enough that I'd survived. Tonight would be different; I'd have it all.

Four Marines gathered as we waited for our baggage, strangers drawn to one another by new ribbons on a common uniform. The journey's fatigue hung heavy on us all—so much waiting. DEROS, always one day, had proven to be so many days, indistinct and never ending. The conveyer disgorged our seabags like big green turds. We shouldered them and found a bored lance corporal directing wayward servicemen to the shuttle bus for Treasure Island.

Fatigue fell quickly as we crossed the Oakland bridge under the dawn of an awakening World. Traffic here, there, everywhere! Bright colors and clean, and the only sand was on beaches washed by the friendly Bay. No one was armed, and you could see for miles and miles, and there was no wire. We bounced in our seats, like schoolboys on field trip, waving to passersby. "Hi, World!" we chanted, "We're back!" And "Let's fuck, baby!" The military driver ignored us for a while, then warned us to can it when we began beating on the sides of the bus. We ignored him, and called him a candy-ass. He grew angrier, and shouted over his shoulder that he'd see us do bad time on the Island if we didn't shut the fuck up. We ignored him again, and began singing Petula Clark—"I Know a Place"—and the next time he threatened we told him he was a boot-camp fucking non-hacker and he could kiss our ribbons. But then he demanded our names. We grew quiet and sullen, and reined in our glee, cursing the driver under our breaths. Nothing was worth bad time.

"Orders!" The naval rating wore gray, and extended his hand without looking up from his gray counter in a gray building on Treasure Island. Outside, squids in shining white bustled to and fro, and we had even saluted a few during the short walk from the bus to the administration building.

The clerk riffled quickly through my papers.

"You're going to VMO-1 in New River, North Carolina."

"Yep. Home, and then New River."

The clerk looked up, questioning. "No, you're going directly to New River."

I didn't appreciate the humor, but decided to ignore it. "I'm going on leave."

"Your orders don't authorize leave. You're to report directly."

"Bullshit."

"Look at this guy's. His say, 'Delay en route.' Yours don't say delay en route. It means you go there from here."

"Everyone gets leave coming home."

"Maybe you get leave after you report in."

"Look, man, I just did thirteen months In Country," I wheezed desperately.

"Go over there and sit down. I'll see what I can do, but it may take a few days to figure this out."

"Days! Look, I ain't going nowhere. I want to see some fucking officer!"

"Sit down over there," he insisted, then disappeared down a gray hall.

Days! I stumbled to a row of chairs and dropped my stuff on the floor. The clerk returned. "We think we can get these recut."

"When?"

"Hard to say. Might be today, maybe tomorrow. Go back there and wait."

Two hours passed. All I needed from Treasure Island was a travel allowance for airfare. If I'd had the cash, I'd have boarded a flight in Oakland and skipped the Island and the unwelcome information in my orders. So I'd have been AWOL! So fucking what! Ignorance of the law would have been no excuse when I reported to New River, but ignorance assuredly would have been bliss. I twisted in my chair, as the minutes ticked away and the World grew cold beneath my feet. The clerk was a decent fellow, for a squid, and nodded assurances to me as he processed other returnees. I watched them jealously. Treasure Island seemed to run like a clock. Everyone came and departed swiftly; only I lingered in the anteroom to the World. The clerk suggested I grab some chow, but I vowed to rot in this spot until the orders were corrected. At last the clerk appeared with a sheaf of papers.

"Zaczek!" I was upon him in an instant, slobbering over the counter.

"You are one lucky SOB," the clerk smiled. "The duty officer called your new outfit, and they authorized the delay en route. Sign here, and here, and here; keep these papers, and these, and these too, and show them all when you get to New River."

A wad of cash accompanied the orders; I offered him a fiver, and thanks, and a French kiss on the spot, tongue and all.

"Keep it. Just promise me one thing," the clerk waved the bill away.

"What?"

"Use that to get drunk or laid."

At San Francisco International I ticketed standby on the first flight to Baltimore, then called home.

"Hello."

"Hi, Mom, I'm home," I teased mischievously.

"Who is this?"

"How many sons do you have?"

"Ronny?"

"No, Clarence."

"You're not supposed to be home yet."

I could hear my father in the background, going, "Who is it? Who is it?" He was always uncomfortable on the phone, yet invariably nagged her for every word when someone called. It was one of many foibles that drove her crazy after he retired. She shushed "Ronnies" at him in a voice of many married years, and I heard him echo, "He's not supposed to be home yet!"

My parents planned everything and were so resistant to change. Coming along so late in their lives, I must have been their biggest surprise, or thorniest decision. They could never understand—would never understand—why I joined the Marines. They were as frightened and, I thought, resentful of the change it wrought in their lives as they were of the dangers they believed I naively invited into my own.

How passionately I'd defended Kennedy and Camelot in the kitchen, to no avail. A dutiful student, quiet and shy, I'd set my sights on glory since age five. I knew they always believed I'd get over it, that it would go the way of firemen and football players. They thought I'd get over it all the way up to the day I didn't. Camelot, if anything, fanned that flame. They tried to stop me from entering the Corps. We'd never had shouting matches in our home; there was never that much passion about anything, one way or the other. It wasn't our way. I don't know who was more surprised when I'd pounded the table and shouted, "I will do this!"

They'd signed my permission for the recruiter, shocked by the unaccustomed show of anger. The sergeant, dressed proudly in the blues I coveted, smiled and nodded awkwardly when my father announced how he'd hated the Army. I was so embarrassed. My mom told the sergeant it came from watching too much John Wayne. I couldn't make them understand, and they never did accept. They were scarred by the Great Depression, and the gap between the harsh world they'd endured and the idealistic dream I pursued was as great as the difference in our ages. I was so angry at them for being them, and for the guilt that clouded my glory, understanding even as we shouted that no parents wish their only child to venture in harm's way. Still, I never knew what they expected of me, what they wanted for me—except never to grow up.

"Hey, Mom, they're not running the war on schedule. What difference does it make? I'm home!"

"Where are you?"

"San Francisco. Look, my flight is being called. It's due into Friendship at twenty-one-fifteen."

"What time?"

"Uh, lemme see—9:15—that's P.M. I'll see you at the gate."

"Call us at home when your plane arrives, and we'll come and get you."

"No, Mom, it's leaving right now. I really want you to meet me."

"Call when you get in. It might be icy. What if you can't get in? Then your father would have driven on the ice. You know he doesn't like to go out on the ice, and this way we won't have to wait if the plane is delayed."

"You can call to make sure the plane is on time." I pleaded.

"No. It'll be better if you call when you get in. That way we'll know for sure."

I replaced the phone. There'd be no one for my coming home. They'd be there within an hour or so, but it wouldn't be the same. They still didn't understand. It just wouldn't be the same.

At Friendship Airport there was a mirror on the concourse, and I paused to straighten my uniform. The world passed behind me. A quiet zone, it seemed neither friendly nor hostile. Some people glanced at the uniform and ribbons; most did not, but I remained on guard.

I studied the reflection and wondered what my parents' reaction would be. The stranger in the glass was not "normal" anymore. He was lean, a thin edge above emaciated, the thousand-yard stare occupationally seared into his eyes. The cheeks were sunken, the flesh sallow and jaundiced. No, the stranger was not "normal." The others passing in the mirror were normal. Their flesh was tan and pink. They had color, not pallor. Their lines and folds of flesh showed happiness or joy or care, untainted by the memory of the zone. I tried to see myself with my parents' eyes, and turned sideways to sneak a profile view. Christ, my head was thicker than my chest! I was all fucking head and eyes. They were going to shit! I wandered away from the mirror, thinking of the liberation photographs I'd seen of Auschwitz.

I called home, and they promised to leave immediately. Then I called Grace.

She cried. And cried. And cried. First it made me feel good, then it got frustrating. I wanted to hear her voice, not her tears.

I don't recall what we spoke about. It must have been good. I know because I don't remember many of the good things about DEROS anymore.

We were still speaking when my folks appeared, looking lost in the baggage area. I kissed the receiver good-bye and walked up behind them.

"Hi, Mom. Hi, Dad."

They turned. I saw recognition in my father's face, but my mother looked right, then left, then through me. Her mouth was open as her eyes took in the uniform and slowly recognized the frame it hung on. It must have been a shock: I was down thirty pounds.

My father shook my hand; he looked as if he wanted to cry, but that wasn't his way. I understood.

My mother hugged me and a few tears leaked. It wasn't her way, either.

I shouldered my seabag; Dad grabbed the other stuff, and we left the airport.

I understood. Really.

We drove to my cousin's house. Dolores was my favorite. Her husband, Richard, had been a Marine drill instructor and had served in Korea. He'd worn his dress blues when they married. I was the ring-bearer. Dolores had sent me ten million care packages in Vietnam. She cried when I walked in, and ran over and kissed me a lot. Richard pumped my hand, and fingered my ribbons and asked what all the fruit salad was. Dolores hung on my arm. It was pretty mushy. Kissing cousins. Then she backed away and started busting my chops about not writing as much as I should have, sort of like an older sister.

I really . . . I liked that. I'm glad I remember it.

DEROS Plus

"You should go visit the neighbors," Mom said, pushing dough.

It was as I remembered. Home, kitchen; kind of hot. I was still used to the monsoon. The radiator, snapping softly under the window, emphasized the silence of this world. It was so quiet, with solid walls that shut out the wind and defined a dwelling space. I'd been so long in the hooch, I'd forgotten what it felt like to be indoors. The cooking smell of a million meals washed in from the heating oven. Mom was always making something.

Even before I tasted Marine cooking, I knew Mom's was bad. At the age of ten I knew her spaghetti was something that only bacteria would willingly consume. But she could bake. She stood at the kitchen table, in a small-print housedress and the apron she'd sewn from the extra material. She'd been in the workhouses since the fifth grade, sewing to support her family. Pink flowers. Little, teensy ones. Mom liked pink. She worked the dough back and forth; white poofs of flour erupted like tiny smoke grenades and settled on her strong arms. Back and forth; she concentrated through bifocals in plastic frames that swooped up like wings.

"I just want to stay here. I haven't been home in fifteen months."

"They've been asking about you the whole time you've been gone. Wear your uniform."

"I don't want to wear my uniform. I'm tired of wearing a uniform."

"You can show off your medals."

"That's not what they're for."

"You don't know how many people brought me copies of *The Booster* when you saved those fellas' lives. Mrs. Wilson cut it out, and Mrs. Moxley, and Mrs. Denton. I wrapped them all up in Saran Wrap so you'd have them. Everybody was asking, 'Did you see the write-up on Ronny?' Monsignor stopped me at Church . . ."

"Monsignor gave me a zillion Hail Marys when I confessed looking at dirty books. Every time I see him I wonder if he remembers. He bawled me out so loud everyone in Church knew it."

"You never told me you looked at dirty books."

"That surprises you?"

"These people have known you since you were a baby," she pressed.

"They could visit me. I've been home two days."

"They work, and you're just sitting around, anyway. You could make the effort. It wouldn't hurt."

"Fuc . . . Sorry. Shit."

"Your language is terrible. You never heard your father use that kind of language. What did they teach you in the Marine Corps?"

"It wasn't finishing school, Mom."

I picked up the *News American*, hoping she'd drop the subject. It was good being home, but strange, like being out of phase with time. When I was In Country, life went on in the World without me. Since I'd come home, it was the war that was going on without me. It seemed that I was never where I should've been. I was always somewhere else.

The fight for Khe Sanh was in full swing; I'd missed it by days. I expected the outfit was knee deep in shit. Hue was getting hit, and it was only a few klicks up the road from Phu Bai. Poor bastards. Wonder how those stupid barracks are surviving. They must be taking incoming around the clock.

I studied the picture. There were the revetments, there was Charlie Med. A C-130 burned near the end of the runway. Another shot showed a transport making an air drop. They flew up the Ba Long and skimmed the runway at about thirty feet. The loadmaster popped a chute, and a sled packed with ammo, food, or other materials was dragged from the fuselage, hit the deck, and skidded along in a torrent of dirt. We'd watched from the revetments as the big transports roared overhead, clawing for altitude, just missing the hills off the end of the runway. I could hear them as if it were yesterday.

I pointed to the pictures eagerly. "Look at this. The hill I saved those guys on is over in this direction, about fifteen klicks. This is the Med we brought them into. Look, I think this ship is from my outfit! Hard to tell. Nah, it looks like it has a two-oh-four rotor head. We had five-forty and five-forty-A. This is probably VMO-6. Figures, those assholes always showed up for the glory."

"That's nice," she answered, her eyes fixed on the dough.

"I used to live in this hooch over here, and this is the mess hall. This used to be the showers; I think it got hit. Hard to tell; it wasn't much of a shower. You can see the edge of hills 950 and 1015 across this river. Sometimes when we walked to the showers we'd watch the grunts lob 40-millimeter all over the side of the hill. I don't think it did any good, but it was pretty. And right over in this direction are 861 and 881 North and South. I had to ferry some wounded North Vietnamese that General Walt wanted for questioning, once. His aircraft pulled them off these hills, but I had to take them back to Da Nang. I held a machete to this one gook's throat so he wouldn't get any ideas."

I looked up and saw her chew her lip slightly, still focusing on the dough, like it was all that mattered in the room.

"You didn't do the kind of things they show on television, did you?" She posed the question in a way that hoped the answer would be no.

I was still discovering how the tube portrayed us. Yeah, I'd killed, but I was seeing more of the enemy on television than I did in the zone. It seemed Americans could walk no trails, visit no villages without brutalizing something. Everything we did, in defense or in offense, seemed brutal on the screen. I didn't remember being brutal. It was all so out of proportion. The war was just a job, even though the job meant killing. They made such a big deal of the grunts' Zippo squads, torching "innocent" villages' huts. You know how often we got shot up flying over some "secure" ville? We'd get popped, report it to the grunts; they'd go in and torch the place. War's motto shouldn't be "hell"; it should be "shit happens." Sure, I'd killed. I got carried away only once. It was . . . it was . . . let's see; after Phelps. Right after? No, long time after. . . . He was killed in . . . must have been September. October? Last week, I was in Da Nang, so it was. . . .

She said no more, just kept pushing at the dough, but I hung on the tone in her voice. How can she ask me that way? It was a war. Dad was in a war. I never heard her ask him if he'd done "those kind of things." He looks at me funny, too, when I point out the places I'd been on Cronkite, and Uncle Walter shows the grunts burning hooches and the gunships attacking. I wasn't around the grunts so much in the field, just in the zones, but the gunships looked normal. Lots of action; blow 'em the fuck away. Trees, bushes, sometimes even gooks. Yeah, Dad, that's what it's like. Normal, I mean. It was a war. I saw the places he'd fought on *Victory at Sea*, and I didn't look at him the way he looked at me.

I don't like the way she asks me. I don't like it that she doesn't look at me when she speaks. It makes me feel . . . it makes me. . . .

"Nah, Mom. I was never anyplace where anyone did anything like that."

She bit her lip, and pushed the dough.

The carpet in my room was scarlet; it showed up every speck of lint. There was a teeny piece of fluff on the floor, and I bent carefully from the waist to pick it up, not wanting to crease my dress-blue trousers, nor chance crinkling the shine of my shoes. Fifteen months ago it would have remained 'til it rotted. It bothered me now. I could see it from the corner of my eye as I fastened the stiff collar, manhandling the clasps together with both hands. I picked it up, and ate it. Why mess up the trash can? Mom had looked at me funny when I'd asked her not to straighten my room anymore. I told her I was grown up. The truth was, she didn't do the job the way it needed to be done.

I checked myself in the mirror. Hours of spit and polish had raised a gloss on the peak of my barracks cover that could blind. Each emblem and bit of brass reflected star bursts on the walls as I moved. The tiny lights flickered across my collection of Orioles pennants, danced off the bureau mirror, and jittered among the soft folds of the Madonna's gown, framed over the bed. My decorations were crisp on the black tunic: Bronze Star, a bright V for Valor; Air Medal, nineteen of the motherfuckers, Vietnam service; some miscellaneous Marine crap. Combat-aircrew wings. The badge of a Marine sharpshooter. All of my pride in brass and sterling and chromed steel. God, I was pretty.

I pulled on my gloves, white, tight, buttoned at the wrist; drew my belt about me, crisp as arctic snow, and snapped the buckle with gloved hands so not to smudge the brass. Perfect. I could not breathe; it is not required that Marines breathe, but I was perfect.

Off to visit the neighbors.

I remember visiting two neighbors, but there may have been more. The first were the Moxleys. I liked the Moxleys; Rosa Lee, their daughter, had pulled me in my wagon, and Bobby was OK, even though he'd hung my tricycle on the telephone pole outside his house. It's hard to remember the visit with the Moxleys because it was a good visit. I think Mr. Moxley gave me a beer, which seemed funny since I was still underage in Baltimore, and

Mrs. Moxley smiled and told me in her deep, strong voice how proud they were. I wish I could remember more.

I visited the Wilsons last. No special reason. Well, maybe. Mary Lou had pulled me in my wagon too. I enjoyed Rosa Lee more, but I loved Mary Lou best. I always had a crush on Mary Lou. She was tall, and of course, I was small. Maybe she wasn't that tall. Things are different when you grow up. Thin as a model, with beautiful wide eyes, and long, long hair that flowed in sable rivulets across her shoulders. From the first time she pulled me in my wagon to the day I joined the Marines, I always had a crush on Mary Lou.

"John, John! Look who's here," Mrs. Wilson called to her husband, "It is so good to see you." She gave me a hug; he shook my hand. Of course, I took my gloves off first, and smiled as I wrestled with the wrist button. That's what you do with dress gloves, mostly. Take them off and then try to find some place to put them.

They brought me in, and I sat stiffly in the parlor. You do everything stiffly in dress blues. We talked, and they gave me a beer—I remember feeling pretty good by then. Maybe I did see other neighbors. They said that Mary Lou was taking a shower and would be down quickly, and they shouted, "Hurry up, Mary Lou! Ronny Zaczek is back from Vietnam!" up the stairs every now and then. And . . . and we talked, and they asked me things, I guess, and I don't remember what. Eventually, Mary Lou came down.

She stood on the steps, the second one up from the landing. Her hair was wet, and she wore a terry robe and terry slippers. Pink, I think. She looked older than I remembered, or perhaps it was just me, but oh, so beautiful. I stood and smiled, and said, "Hi" in a voice that didn't work very well. I had my own girl, now, but I always had a crush on Mary Lou.

She didn't say, "Hi" back to me. She had this funny look on her face. I'd never seen it before, at least not on anyone I knew. She looked at me like . . . like. . . .

She began to talk, real fast. Her voice was strained and bitter, each word with the taste of venom, and she kept saying over and over again, "You're a murderer. You killed babies. You killed innocent people. You should be ashamed."

Mr. and Mrs. Wilson got up really quick and told her she should be ashamed, and to go back upstairs, and she did, but she kept shouting at me over her shoulder, and I said, "I'd better go," and they said, "You sit down, Ronny. Don't you pay any attention to her," and I shook my head, and said, "No, No. . . ," and there was all this air in my voice, and I thanked them and walked out on the porch, and it was very hard to put my gloves back on, and I couldn't leave their house until I got my gloves on, and I went home and hung up my blues, and cleaned my room.

I never saw her again.

That's . . . that's all I remember.

We walked hand in hand along Charles Street, past Loyola, then Hopkins' Evergreen campus, and onward to Notre Dame—going nowhere, actually. Walking, talking, Grace and I, planning our future.

It was cold, bright. It hurt my eyes. I usually wore my smoked visor on days this bright. The bells from Mary Our Queen pierced the barren trees, clearly, as when the archbishop bestowed high school diplomas before the great altar, two years before. I remembered how the bells rang that day. We were Archbishop Curley's first graduating class, so we got the cathedral. We beat out Calvert Hall, and Loyola and St. Joe's and all those other dicks. We were special. The bells seemed to call us to the future.

Her plans were well set—another semester at Notre Dame, then on to nursing school at Columbia. Our distant future together was established—marriage after she graduated in three years. My future—that was another story.

"I don't know what to do," I said to the sidewalk. I walked with head down, avoiding the glare. It was most un-Marine-like. Gunny would have chewed me out. I didn't feel much like a Marine.

"What do you want to do?"

"I don't know. All I ever wanted in my life was to be a Marine. You know how kids switch from cowboy to fireman to ball player? Not me. God knows where I got it from. My father hated the service. Maybe I got it from being ring-bearer at Dolores's wedding. Or maybe it was too much John Wayne, like my mom said. Or maybe it was my grandfather. He fought for the Kaiser in the Franco-Prussian war. We have a picture of him in *Pickelhaube* (spiked helmet). He was a proud soldier."

I smiled, remembering the old man who was as much childhood friend as grandfather. I liked thinking I took after him. Mom had ushered me into his room the night he lay dying—my room, now. I was seven years old; it was tough to lose a friend.

"You can stay in if you want to. We've talked about that," Grace said.

"I don't know. I don't know," I puzzled, sadly. "Things aren't the same. The Crotch isn't the same now—so much chickenshit. And Vietnam . . . Vietnam . . . People over here just don't understand what's going on. It's like Vietnam has changed everything. It's hard for me to explain."

I stumbled, searching for a way to express the confusion I felt.

"You know, I was in with a great bunch of guys. Y'know, y'know. . . ." I stammered, reliving the familiar past eagerly. I understood the past.

"In July the Division planned an assault in the A Shau; it was going to be the biggest airborne assault of the war. A battalion inserted all at once—something like four full helicopter squadrons were going to be engaged in the landing. Fixed-wing up the kazoo, artillery bombardment, naval gunfire. Man, it was going to be like a real war for once. None of this snoop-and-poop, take-the-hill-and-give-it-back bullshit. We were going to take something and keep it.

"It never came off; got canceled just before zero hour. Someone said the operation plan had leaked to the enemy. Maybe that was the smart thing to do—I don't know. It would have been the biggest event in the war, and it turned into another might-have-been. The energy in the outfit was tremendous! We had twenty-two out of twenty-four aircraft up. Guys were fighting each other to get an aircraft assignment. Every senior crewchief and gunner had a slot; Jack and Mouse, Phelps and Hazel Baby, Larry and Frankie and Dutch and the Rock. Everyone but me. I was grounded—malnutrition, can you believe it? I didn't have an aircraft. You know what? At oh-two-hundred hours my friends helped me rebuild an engine for VT-12 so I could launch. Doc Kyle signed my up chit, and I got myself on the operation plan as a medevac. We were going to flight-test on the way to the goddamned A Shau! The CO called us together; he was going in with the first wave. They had radar-controlled quad 50s in that valley, and SAMs. He said whoever was left the next day could party. You know, we'd have taken that goddamned place, no matter what they threw at us. We were that good!

"We were one hell of an outfit; hell, we were one hell of a Corps! If they had turned us loose, we could have kicked ass. That's what people just don't understand. All people think we did is torch hooches and kill bab . . . and kill b. . . . Damned media; there wasn't any goddamned photographer along when Jack and I pulled those guys off that hill, or when Saint got ripped, or, or. . . .

"I don't think they want to understand. If I could just get people to understand . . . we could win, we really could."

We walked farther, the bells sounding Angelus. They tolled so slowly, too slowly to match the passion I felt. I hated the Angelus in school, when the nuns made us kneel and pray. It always seemed a dirge, holding me back from play.

I continued, hesitantly, "I've been watching the action at Khe Sanh and Hue on the news. The guys are knee deep."

I waited for Grace to answer, not knowing what her response would be. Hell, not knowing what I wanted it to be.

"You want to go back," Grace answered, half-question, half-statement. "You can do it if you need to."

I stared at the ground, thinking about the war we were not allowed to win. What would I do that could make a difference? What difference had I made? I knew two crewchiefs who had returned to Vietnam to fly, and die. It did not pay to go for seconds on DEROS. Death was not the worst thing that could happen to a man, but it should have meaning.

Yet what would I do here, in the World? All those years of wanting to be a Marine, then being a Marine, a good Marine. The Crotch was no longer home to me. The World is not home, as I knew, in those long nights on the wire, that t should be. Where is home?

"There's nothing for me in Vietnam," I answered sadly, wishing that there was.

"We're going to Golden Ring to pick up the Frog," Tom said.

"No problem,' I answered, settling into the seat of the car Tom had borrowed from one of his brothers. Tom had more brothers than I could count. He'd been borrowing something from them since we became best friends in the second grade. We were going to Golden Ring Tavern to pick up the Frog. Frog's parents owned the tavern—which always was one of the Frog's better points. We were going to meet the Crow and Tom's younger brother Joe at the game. The Clippers were playing someone or other. I mostly went for the hot dogs.

We rolled into Golden Ring. I wanted to wait in the car while Tom collected the Frog. I knew if we went in we'd have to have a beer, and then we'd be late. I didn't like being late, even though I didn't care about the game. Frog welcomed us, and insisted we follow him to get his stuff. It was a nice neighborhood bar: Big old house tar-papered here and there where the shingles were missing. Lots of small rooms, nice and gloomy, steeped with the perpetual perfume of beer and crabs. We followed Frog through a corridor of printed paneling dotted with beer ads.

Frog entered one of the public rooms, and I probably should have wondered why the hell we were down here instead of upstairs in his place, but I didn't. It was dark; he went in first, then Tom, and me last, and all of a sudden,

"Surprise!"

There was a long table, a keg and food, and a huge "WELCOME HOME RON" sign in shiny grade-school string-up lettering. There was the

Crow, and Joe, and Kenny and John, and, of course, Tom and the Frog all jumping up and down yelling, "Welcome Home." Ma Frog was there, too, and she gave me a big hug and a wet kiss on the cheek, and said she wanted me to have the best time, and everything was on the house!

She escorted me to the table. There was a white sheet cake, and in the middle a Marine emblem and two little American flags. Ma Frog pointed at it and beamed. I didn't have the heart to tell her the figure marching next to the emblem was fucking-Army. Ma Frog was good people.

She said she'd leave so we could cuss and talk dirty, and we went, "Whoa," and asked for some women, then pulled beers all around.

I'd brought my bush hat. Baltimore is damp in January, and my dome had been skinned on Oki to an eighth inch of fuzz. I had other, warmer hats, but nothing that suited me so well. I snugged it into place and grinned. The brim was snapped up, just like always, and the "Vietnam" crescent was held in place with a few less threads, but then—weren't we all? One of the guys—I don't recall who—put it on his own head, and it pissed me off a little. I laughed. We'd been doing this shit, some of us since grade school. Most of us had gone to Archbishop Curley together. It was all the usual fun, so I ignored it long enough to make it look like it didn't matter, then I took it back with controlled laughter.

Kenny or John or someone asked, "How was Vietnam?" I think all of them did, in turn. They meant well. They were all good guys.

I tried to answer. That was too bad.

"It sucked. We blistered our ass in the summer and turned to shivering prunes in the fucking monsoon. The people sucked. I preferred the enemy to the ARVN. At least you knew what to expect. What sucked was when we heard about how you guys in college were demonstrating against us. What the hell did anyone here know about the fucking war? You know, I had a friend whose parents threatened to disown him because he volunteered. What the fuck does anyone. . . ?"

I drank some more beer. I weighed 108 pounds. Shouldn't have been drinking that much.

One by one their eyes grew distant; they excused themselves to go to the tap and then found someone else to talk to as I accosted another friend and dumped on Vietnam, and the Crotch, and America, and. . . .

I ran out of steam after a while, but the party continued around me. Mostly college shoptalk. I felt like a poor dancer standing on the edge of the floor, trying to figure out how to cut in but knowing none of the movements. I thought it was stupid stuff. Professors' quirks or chasing the girls at Notre Dame, homework. ROTC bullshit. Most of the guys had done

ROTC. I thought it was really stupid. I was pulling medevacs, fixing fuck-ing aircraft. Important stuff! I looked at my watch and thought that twelve hours away the noon inserts were underway. I broke in, once in a while, with some Marine tale that easily topped whatever I'd just heard.

The guys listened, politely. After a while I stopped breaking in.

Ma Frog came in to take a picture. Someone suggested a human pyramid, like in high school, and they put me on top. I chewed on a cigar and grinned, and someone smashed the boonie hat over my ears. The guys on the bottom buckled, and the pyramid collapsed, all of us kicking and laughing.

Ma hugged me as we left. That was nice. Ma knew about the Bronze Star, told me I was a hero. I didn't like that—it made me squirm inside. She meant well. She cut the emblem out of the cake, wrapped it with the two flags and the fucking-Army figure, and gave it to me.

I have to work to remember the party. The party was a good thing. The only thing that was wrong with it, was me.

It is the last visit.

Tom and I have been seeing less and less of each other. Last time, we agreed that this would be the final visit.

I park in the familiar, pot-holed lot and approach the Center, in some ways as apprehensively as I did so long ago. Endings have never been easy for me.

The Center has grown over the years; business has been good. The staff expanded into the office space next door. Ernie painted the walls swamp green, hardly an improvement. The posters that had littered the wall had rotted, and been replaced by new ones. The fly-specked sign, "Help With-out Hassles," is faded almost beyond reading in the window. Some things never change.

Tom welcomes me. We draw a cup and take our familiar seats.

"Well?" I toast.

"Well?" he counters.

"I think I've had about all the fun I can stomach in this place."

It has always been so difficult to know, when is done?

Memories are the substance of the Center. So complicated. Would that they were bounded, well defined like an appendix, which can be excised, or a nose that can be molded under the surgeon's scalpel. When you are finished, you have an appendix, or a nose. Memories are jellyfish, some with tendrils that sting, and some, gossamer caresses upon the mind—all of them fluid, like the water that buoys them—difficult to bound.

When is done? Where is the point of diminishing returns, and when do you leave well enough alone?

I continue, "I think it's time to get on with life."

Many times, we arrived at some plateau of therapy. Each time, Tom bade me look back, into the twisted ravines from which I had emerged. He bade me look around, at the relative safety of the plateau, and then he would have me look up, at the unknown pinnacles ahead. Each time he would ask, "Do you want to go on?" The choice was always mine. It was ever as he had promised; he could make me go no further than I wished. I chose to go on, though the pinnacles challenged and frightened me. I often regretted my choice, but for five years, out of pride or curiosity, stupidity or stubbornness, I chose to go on.

We are at a plateau, and have been for some months. It is comfortable here, and life is good. I could live here for a while. A long time. A life. There are pinnacles ahead, and I have made honest forays against them. I am still crewchief, and do not shirk unknown zones. But we have found no avenues nor passes to continue the assault, and I have begun to sense that the gain is no longer worth the effort.

I still live with the dream. It comes infrequently. Perhaps two or three times a year. It is most strange. Where once it had faces of many colors, there are none. The faces vanished as I learned to look upon them. They were painful to see, and are difficult to recall, and I do not often try. There is no need. I relished the joy of resolution many times, and where I have not found resolution, I have hope.

Still, the dream hides something that I have not learned to see. When it comes there is naught but blackness, hot, choking blackness that is less than the old darkness, for darkness has the face of darkness, and blackness has none. I do not know where to look to find the last face of the dream, but sense that I have found all that may be found with Tom as my guide in the Center.

As long as the dream remains, I know there are pinnacles ahead. I would prefer that it vanish. Perhaps, someday, I will see into the blackness and excise the last remnant of those old fears, and the dream will come no more. But it is an imperfect world. In truth, who that has lived what I have lived can expect to be without dreams?

I am free of so many fears, and those that haunt me, I control. I served, and endured, and fought the war beyond its years of ending. This is the last session. I have done more than survive; I have won.

"You know, I still don't feel like I came home."

"Do you feel like you can come home?" Tom asks.

I pause, and frame the answer carefully.

"Yes, and no Sometimes I think I can come home almost anytime I want to, but that I'm just not ready. You remember what the World was like, how it rejected us. Me—us; it's still hard to differentiate. The reaction to veterans was negative across the board, but the delivery was one-on-one. I just learned to reject anything that rejected me. It was a human thing to do. Coming home seems bound to forgiving 'the World.'"

I laugh, and shake my head. "It's a large world. I'm not that big a person. No, I'm not ready to forgive. Five years of my life, coming in here, learning how to be a person again. I'm still bitter. Maybe I can't forgive. Can't, won't. It's a close call. Still, I think I won't really come home until I learn to forgive."

Tom pulls at his beard; there shouldn't be anything left of it for as much as he's done that over the years.

"Remember the coming-home party your friends from high school gave?"

"Bits and pieces. It's difficult for me to remember good things about that period. It's all lost in the clutter."

"Do you keep in touch with them?"

"Most of them. Tom went AWOL from Fort Dix to come to my wedding—Joe and the Crow smuggled him out of boot camp in the trunk of Joe's car. He's godfather to my kids. I ski his ass off one week every winter—it's the closest I get to flying crewchief, these days. The Crow was best man. He's in Denver; Tom and I visited him last year. Y'know . . . y'know . . . I just remembered something! The Crow came to my house a few days after I got home, after Frog's party. His parents wanted to give me a welcome-home dinner. He came to take my order. How about that! Mrs. Crow cooked everything I asked for, too. His dad was a fireman in Baltimore, loved Martin Denny. We sat in the living room and listened to fucking Martin Denny birdcalls all evening. God, it was boring. God, it was good. Y'know, I'd forgotten all about it. All these years! I thought the good memories had been washed away. Guess they're still inside, after all. Memories—they're funny things aren't they?" I smile.

Tom smiles in return, "What about the others?"

"Oh, Joe was an usher at my wedding. Still has all his hair—the bastard. My boys play with his daughter now and then. I saw Kenny at Curley's reunion. Same old Kenny. Lost track of the other guys."

"How long ago was that party?"

"Oh, God, twenty years."

"I seem to remember you took Vietnam out on them."

"Uh, yeah. I was always sorry about that. I guess everything was getting to me."

Tom leans back, arms pointed skyward. Oh Christ, this is the last session, and he's going to make a point.

"After twenty years, they've kept contact with you. How many people do you know who have kept up friendships for twenty years?"

He levels a stare, and continues.

"The world is only as large as you need it to be. Think about it, my friend."

He leans forward, and repeats the words, slowly. "Think about it."

We rise; Tom walks me to the door for the last time.

I have been planning for this moment; how to say good-bye, and thanks. It was easier to enter a hot zone than to say good-bye, and thanks is a pale shadow of what I feel. At the last moment, I say neither, for this is not ending, but beginning. We shake hands, and I walk into the evening.

"You know the number," he calls.

At home, I drag a chair to my closet, pull the box from its shelf, and spread the memories about me. I want to remember. The picture of that smiling pyramid is near the bottom. There is Tom—he looks like Tom, but with a lot more hair. Helluvalot more hair! Joe still looks like Joe—the bastard. Crow always did look like hell. I'm wearing my bush hat, and my teeth are clenched around a cigar. The eyes draw me: deep-set, cadaverous, a touch mad. I glance into the mirror opposite my bed. The distant echo of the zone remains.

The cake is there, its icing mostly gone, the dough spotted here and there with mold. It's like a brick; you could hammer nails with it, judging from the weight. It's held together pretty well, considering.

I bet it would take a lot before it would break.

Chapter 8

Then will he strip his sleeve and show his scars,
    And say, These wounds I had on Crispin's day.
Old men forget; yet all shall be forgot,
    But he'll remember with advantages
What feats he did that day: then shall our names
    Familiar in their mouths as household words,—
Harry the king, Bedford and Exeter,
    Warwick and Talbot, Salisbury and Gloster,—
Be in their flowing cups freshly remember'd.
    This story shall the good man teach his son;
And Crispin Crispian shall ne'er go by,
    From this day to the ending of the world,
But we in it shall be remembered,—
    We few, we happy few, we band of brothers;
For he to-day that sheds his blood with me
    Shall be my brother; be he ne'er so vile,
This day shall gentle his condition:
    And gentlemen in England now a-bed
Shall think themselves accurs'd they were not here,
    And hold their manhoods cheap, while any speaks,
That fought with us upon Saint Crispin's day.

                    —William Shakespeare, *King Henry V*, IV, iii,
                    The field before Agincourt

Crispin's Day

"You the happy camper?"

The question is a frank challenge. Although it is not hostile, there is nothing friendly in his tone, nor in his eyes, nor in any of their eyes.

Eight chairs face me, arranged along the besieging arms of a parabola. It is a large room to feel so confining, but I know how perimeters can shrink the scope of a world. Behind the men Ping-Pong and card tables neatly dot the gray linoleum, itself buffed to high gloss. Perry Point V.A. Medical Center is, after all, a federal reservation, and neatness counts.

The leader sits closer than the rest, but not too close. His hard-backed chair elevates him so that I must raise my head and eyes to meet his. We are of an age, though I deem the years have been less kind to him than to me. Of course, he might feel the same as we appraise each other for a long moment. I am seated on a tacky sofa, the kind someone would abandon furtively in the night at a Goodwill drop-off box. It is too soft, difficult to balance my ass and maintain the self-assured aura I feel the need to project. A coffee table protects me from them—the thinnest of barricades. He studies me—they all study me, with eyes that I recognize easily.

I allow a careful smile to crease my lips, one without mirth or humor, and especially without the presumption of friendship. I hope it is a smile without excessive nervousness, and utterly devoid of fear. I want it to acknowledge only shared experience. Most of all, I want to return his level gaze without flinching, in a manner that will establish equal footing with these most suspicious of men. I know how little they trust, in few but one another, and in the large man by my side.

"If the big guy here told you I was a happy camper, he's blowing smoke."

Tom laughs, pulls his beard, and reaches for the cup next to mine. His motion pulls me off balance, and I grasp the table to hold myself erect. Our cups are full, and the liquid slops dangerously near the rim as the table slides on the burnished floor. I force a laugh, but it escapes in a nervous snuffle. The leader's glance says, "What an idiot!"

I think, "This is not a good beginning."

"I want to give something back," I'd insisted.

I'd been badgering Tom, wanting to know how I could help. I needed to do something. Five years' therapy should serve a purpose larger than the reclamation of a single life. At last, he suggested the one thing I could do, that few might do as well, was to provide hope.

"Hope?" I'd questioned.

"Hope," he repeated. "You know what PTSD looks like from both sides. Remember what it was like for you? You can tell these guys what it's like, what's different when you get beyond PTSD. You can give them hope."

The notion intrigued me, and I felt the distant thrill of the medevac as I considered his proposal. But to make public what has been so private. . . .

"I don't know if I can talk about it in a group," I hesitated. "You know I never would do the rap-group thing."

Tom had hinted at group therapy during my early months in the Center. I'd refused instinctively, like pulling back from a hot stove. The violence of my refusal surprised me, and of course, he insisted we determine why. He asked me to categorize the levels of comfort, or discomfort, I felt in talking about Vietnam to different people. It was not an easy task; the aversion was instinctive, not considered. At last I realized that it was easiest to talk to interested civilians, followed by noncombatant veterans. I could handle some conversations with frontline grunts, but it was tough. Sometimes it was really tough. I could think the words; I could see the pictures so clearly. Some things, I just couldn't . . . I just . . .

"What about war protesters?" Tom asked.

"Fucking peace creeps aren't worth talking to," I dismissed.

"Who's hardest?" he'd pressed.

I couldn't meet his gaze and answered only with a shrug, hoping it would make him go away. From the corner of my eye I saw him fold his arms, then purse his lips and pull his beard. He would not go away.

"You still have friends who were crewchiefs and gunners, don't you?"

"Um, huh," I mumbled at the rhetorical question and reached for my coffee. I knew how to hide behind coffee. It takes time to pick up the cup, time to sip, time to put it down. Each stolen second gives time to quell the panic that rises with the probing questions.

"When you meet another crewchief, what do you talk about?"

"Oh, our squadrons; what great, wonderful pigs our aircraft were. R & R, and parties, and puking, and the dung heaps we lived in; the guys we knew."

I tried to make the conversation sound bright and breezy. "Go away, go away, go away," I began to think.

"Don't you talk about combat?" Tom pursued.

I worked hard at keeping my voice even.

"We never talk about the zone."

"Why?"

Finally, my throat tightened, as something inside said, "This is enough." I phrased my response carefully, in warning tones so there'd be no mistake.

"Because there's nothing we need to say."

He'd never mentioned the rap group again, and I was glad.

Hope. The offer posed a challenge, and I knew there was the need. The vets' need; my need. Still, it was frightening. I'd prepared for the evening, even jotted notes on cards, but I knew they'd be useless as soon as I opened my mouth. What would I say? What could I say? How would they respond? My thoughts churned as I sat in Tom's office, waiting for the group to assemble. "I want to help," I'd said. "Be careful of what you wish for, you may get it," I thought.

Sitting next to Tom, I feel a grim smile on my lips, and I focus on why I am there. Giving something back; that is true enough. But there are more-selfish reasons. The days of pulling medevacs—they are gone. But oh, to capture even the smallest hint of that purpose once again. To take the pain, to mold and shape it to serve a purpose, my purpose—to use it for healing. And, yes, to prove to myself that I am healed.

Tom introduces me.

"I want you to meet Ron, the Marine crewchief I told you about. He's volunteered to talk to you about his own experience with PTSD. I've known him for a long time, and he has some important things to tell you. Why don't you introduce yourselves?"

Silence. The men take their measure of me, glance from one to the other, then to the leader. No one speaks.

Tom flusters, introduces me again in different words, and again asks the men to speak.

Silence.

I enjoy the querulous glance he flashes; I've never seen him stymied before. The humor of it makes me feel more in control.

There is an awkward moment.

Fuck it. I begin speaking, very fast, very firm—at least it sounds firm. I will maintain control.

"OK! I'm going to talk. You can tell me who you are later. I'm gonna talk for half an hour. You can jump in if you want, but mostly I think you're going to watch how big an asshole I make of myself."

The energy in the announcement startles—the men, Tom, and me. The vets lower their guard a moment, look quickly one to the other, and smirk, waiting to see just how big an ass I will become. But the perimeter shifts while their guard is down, and when it rises, I am inside.

"A while back, I asked Tom how I could use my own experience in therapy to help out, and he said I could come down here to give you hope. So I'm here to tell you a bunch of shit that I think will give you hope."

The group shifts, dubiously. They do not know what to make of me, and hope is a foreign, almost threatening concept when you have lived with hopelessness for so long.

"The first thing I want you to know is that I'm not a shrink. He's a shrink. I'm a fucking victim, like you, so I don't have to dwell on how PTSD makes us feel. We all know that PTSD feels like shit. In case you're wondering, I still have PTSD. If you think that just because I'm sitting center stage, I don't have PTSD, you're wrong. I just don't have it as bad as I used to. I'm not here to hold myself up as a model—hell, I don't know whether my PTSD is, or was, better or worse than any of yours is, or was. For all I know, I'm still the worst head case in this room."

They smirk again. If I am an ass, at least I have a sense of humor.

"I'm here to talk about . . . differences, because the difference between how I was and how I am has a name, and I think the name is hope."

The mood changes at the word *hope*, an increasing focus that tells me they hear. I can only hope that they will listen.

"If you're wondering why I'm doing this, you can look at it as purely selfish motivation. All of us in this room have saved lives. Remember how good that felt? It was the best damned feeling in the world! When have any of us felt as good as we did back in the zone, when we were saving the life of another dogface, squid, airdale, or fucking jarhead? Coming here and helping just one of you motherfuckers come in from the zone is going to make me feel like a million bucks."

I gesture to Tom, sipping quietly at my side.

"And the other reason is that I want to prove to myself that I can do this. There was a time when I couldn't. I would never join a rap group like you have; Tom wanted me to, but I wouldn't. If I can talk about this stuff with you, then there's another difference, and in every difference there's a little bit of hope."

I take a fortifying sip, replace the cup carefully, then meet the leader's eye.

"The first time I walked into the Vet Center, Tom told me I could expect two things to happen, working through PTSD."

I hold up one finger; its promise captures the room's attention.

"The first was that a lot of stuff that used to bother me would simply go away."

Another finger follows.

"The second was that I'd learn to control the things that wouldn't go away, that they'd no longer control me.

"He made it clear that everything was up to me, that he couldn't make me do a damn thing that I wasn't willing to do, and that in the end I would regain something that I had lost—the ability to choose.

"The ability to choose," I emphasize. "What does that mean?"

There are raised brows, some curious, some suspicious. None answer. The leader's arms are folded pugnaciously, still challenging; another pulls his lower lip, a third teeters on the edge of his seat, as if every word is a pearl. Some face me squarely, others hold their bodies at an angle, but each gaze is fixed and intense.

"We picked up a lot of hot buttons in the war. Any of you have a 'trust' button? It's my personal favorite. I volunteered for Vietnam because I believed in America, then came home to be screwed. Do you recognize that story? Push my trust button and I'll hate you, your wife, kids, and dog for the rest of your fucking lives. For a long time I tried to make certain no one pushed it by never giving anyone the chance. Not trusting makes for a lot of loneliness. I didn't want to push people away, at least at first. Later on, I didn't not want to; being callous seemed the normal way to be. At least the safest. I used to think of it as armor. Things change, though. Armor rusts, and I guess feeling flesh wants to grow even on calluses. In the end I didn't want the life I had—I just didn't feel I had a choice.

"During therapy, you discover buttons. Some are obvious, like trust. Others affect your life in ways you may not notice. There are buttons that make you need to sit with your back to the wall in restaurants, and buttons that steal the joy from your work, that make you think everything you do is worthless. Buttons that make you feel like you can't feel, that rob you of a night's sleep. Some go away almost as soon as you recognize them, and it feels really good when that happens. Learning to deal with the ones that don't go away is tough, but once you do, you've regained control, and with control the ability to choose.

"Look at it this way," I emphasize. "Any button you don't control, controls you, and as long as it does, it robs you of the ability to make choices."

The vets look as puzzled as I was, years before. I smile inwardly and glance at Tom, watching the men who watch me, and wonder if he deems me an apt student. In the years since I left the Center, I have not forgotten the need to understand each fragment of our burden: why it is there, how to be rid of it, how to be certain it is gone. These are men, like me, who in surviving war learned to leave nothing to chance, to whom control is synonymous with survival, and who despise and fear the doubt they cannot shed.

"This sounds abstract as hell, doesn't it? Well, I learned it from him, and he's a shrink, so what do you expect?"

Tom winces at the word, mutters something like, "No respect," and returns to his cup.

"I'm going to make this real for you. I'm going to talk about some things that used to bother me, that don't bother me anymore, and some other things that still get to me, but that I've learned to control. In other words, differences.

"After Vietnam, I hated to be outside in the full moon. When I drove home late at night, I'd run in from the car like the bogeyman was chasing me. I got cranky and irritable; I used to say it was my time of the month. My wife said it was my werewolf gene.

"What do you think of that?" I toss, rhetorically. "Here I am, a fucking Marine crewchief, and I'm afraid of the goddamned moon!"

I force a laugh at the absurdity, but in their faces I see reflections of their own "absurd" fears, entanglements of memories that prevent them from doing what they wish to do, or becoming what they wish to be.

"We covered a lot of ground during therapy: incoming, the gunner's moon, the nights on the perimeter. Remember what it felt like, being under the full moon on the perimeter?"

Nods. They remember the fear of the wire.

"All of those things—those obvious things—I worked through them in therapy. And after I was done, what do you think happened?"

Silence. Though attentive, they will not be drawn in by casual questions. I do not blame them; I was not.

"I still had problems with that fucking moon. Tom kept pushing me, making me look at things in different ways. That's what you have to do, look at the same thing different ways until you see something new, or in a way that makes it fit into place.

"One day, I remembered the flight to Dong Ha."

Dong Ha needed blood, and the moon was bright, and VT-12 sparkled like a jewel above Highway One. Oh, the tracers that flew up at us. Deadly fire-flies, reaching, reaching. Tracers were pretty when you weren't the target. Head-on, they were tiny pinpoints of angry orange, closing, brighter, brighter, flashing by in Morse dots and dashes of crazy lightening. Miles and miles and miles of them.

We were quite high, at the limit of small-arms range. It made no differ-ence. It was the idea of them—those burning, reaching fingers. I was scared. I never told anyone how scared I was. I didn't tell myself.

I was crewchief, you know. Scared wasn't allowed. I remember think-ing that Dong Ha had better damn well need the blood. I remember think-ing that some bastard better be on the goddamned table. Don't recall who the pilots were. They were scared too. Could tell it in their voices.

We doused our running lights—even the instruments were down low. Didn't matter; the moon washed the panel as if it were day. That fucking moon! Got shot at on final at Dong Ha. Got shot at taking off too. Shit, we got shot at everywhere. About a half hour flight between Phu Bai and Dong Ha.

Damn.

We delivered that blood. Must have been important; a corpsman was waiting on the medevac pad. He looked desperate. Raced off with it before our skids even settled. We took off right away. Didn't even drop rotor turns; no sense waiting around. Then we flew the gauntlet back home. God. Know what? Dong Ha called Roseanne, and Roseanne contacted us en route. We'd delivered the wrong blood type! Ha! Those fucking assholes at the fucking Med, they gave us the wrong fucking type!

Back up the road, beneath the full moon, and those deadly fireflies. Didn't fly so high that time. Couldn't afford to. Saw how desperate they were. Too long to achieve safe altitude and then descend. Minutes up, minutes down. Too long. Red-lined that mother. Jesus. Highway One. Fucking point blank, all the goddamned way. Oh, man.

I reach for the coffee; the cup is empty. When did it empty? Someone rises without speaking, takes the cup, fills and returns it. I nod thanks, and drink, and let the warming liquid still the quiver in my voice. It is hard to wrap the urgency of that night in words, but words bind, and the men now look at me with different eyes, understanding. These men were frontline grunts and riverine; each knows the zone. I am crewchief, combat aircrew—they know we would have done anything for them. Anything. The parabola is no longer the wire, but embraces.

"Know what I found when we finally shut down? The whole damned tailcone was slicked with oil. A round had come up the hell hole and unseated a plug on the transmission. Most of the oil drained; don't think more than a quart bled out when I replaced the plug. If we'd flown another half hour, quarter hour, minutes—who knows? I fixed it that night. Had to get it fixed; I was going on R & R the next day. The Jew took my ship up in the morning. Guess what? He got shot down. 12.7 grazed his flight helmet, shock wave blew out his ear bone.

"Differences!" I announce suddenly, brightly. "The moon doesn't bother me anymore! Tom and I had talked about the nights on the wire, and incoming, and a whole bunch of things, but it wasn't until I remembered that mission that it went away, and when it was gone, it was gone for good. It sounds like such a simple thing, doesn't it? I can go outside in the full moon. I don't get bitchy, and I don't gripe at my wife and kids, and I feel normal. That fucking button is disabled!"

I feel a child's relief, like when you turn the light on and find out the bogeyman really isn't in the room; it's only the shadow of Teddy Bear cast on the wall.

I watch them watching me. Some seem to look inward to their own fear, and I wonder if they wonder what it will take to make it go away. I hope they understand that it can go away.

I chew a finger and plan the next thing to say. Still, no one speaks. I hadn't expected it to be like this. I'd known they'd be mistrusting, but I'd hoped our shared experience would pave the way for discussion. I should have known better. Tom told me this group has been together for many months. I am as much an outsider to them as any FNG In Country.

In preparing for this night, I had asked Tom what I should say, or should avoid. His only advice was to be myself. "Being myself might scare them," I'd joked.

"They accept honesty. They'll accept and find value in whatever you say because you're one of them, and you've been where they are going. Remember how things were for you in the beginning—what could anyone have said to make things worse? Be yourself. They need to see someone who made it to the other side. But remember, this isn't only about them. This is something you need to do, to complete your own journey."

Facing them, I remember the pain I'd felt in opening the old memories, and have no desire to raise it in others. I have told them, truly, what Tom told me—some things would go away. What will they think when they see the pain that never leaves? Will they believe that I control?

I tell them of the hill, beginning at the beginning, for it is the only way I can bear the tale. I tell almost all of it. Almost all. I tell them of the pleading operator on Fox-Mike, the dead wall, and the Marine in my arms, but I do not say the words. I never say the words. For me it is part of staying in control.

The zone fills the room; it settles in dust and smoke on the polished floor. A Huey pilot sits opposite. His left hand betrays him, gripping an imaginary collective at his side, pushing air where the throttle governor once rested. He rocks forward in his chair, as I rock in mine, mouthing, "Go, go, go," willing the laden aircraft into the sky, as all around us the earth screams.

"We were too fucking heavy! I tell people how we were too fucking heavy, and they listen and nod, and tell me that I shouldn't feel bad, that it wasn't my fault, that they were dead anyway.

"Anyway. . . . Anyway! There isn't any 'anyway' about being dead. I was the last one into the aircraft. I was the fucking crewchief! And crewchiefs aren't allowed. . . . They aren't allowed. . . .

"I didn't check the bodies. Maybe I should have. No excuse for not checking. So what, that the operator told me they were dead? What the hell did he know? It's not logical they were alive. I saw the wounds on some, the faces of others, and the living wouldn't have piled the living on the dead. But so what? I should've got the bodies out. I knew exactly how much weight my ship could carry. I never thought to check the fuel. Damn. Three weeks earlier—the grunts were hit bad the far side of 1015. Chopped meat. We went ammo minus and returned to arm and refuel. Copilot didn't watch the gauge—we took on too much. Couldn't lift. I crawled under the ship and opened the petcocks right in the fucking fuel pits. You should've seen the pilots' faces! Six hundred fucking pounds of JP into the dirt. Rotor wash threw it everywhere. God, we were a fucking bomb! Crawled though the muck to cut them off. We took the top strand of concertina with us on takeoff. Christ. I was soaked. No-smoking flight. I should have known."

The mission fast-forwards, and for the thousandth time I analyze each frame, strip out every second, demanding, *What could I have done better?*

"I could've stripped the ship in the zone. Jack and I dumped shit at Khe Sanh, but we could have lost more. You can always do more! Ditched our flak vests; the tool box weighed twenty-five pounds. Fucking useless rifles eight apiece. Me and Jack, we could've got . . . could've . . . oh, man. One more, two more. Oh, man . . . So what, someone was pouring his . . . pouring . . . uh. . . . So what if we were being rushed, and our cover was evaporating. Fucking excuses. I was crewchief!"

I bury my face in my hands and breathe, deeply. Inside, the voice I learned to hear in the Center whispers. It tells me that I am wrong. It tells me that the cruel equations of weight and temperature and humidity were more powerful than a nineteen-year-old's image of self and Corps. It tells me with logic, cool and clear, that the memory that will never go away is undeserved punishment, and that in that hour of that day I did the best a man could do, and should be proud. I listen to the voice that gives control, and try to believe, then lower my hands and turn to the leader.

"Still think I'm a happy fucking camper?"

I rub my forehead, wondering if they can see that this is control. I think of how to make my point and force the tremor of the hill from my voice.

"I'm a perfectionist. I was a perfectionist before I went into the Marine Corps, but not nearly as bad as I became. It probably had something to do with choosing the Crotch in the first place, else I'd have been a squid, or even a fucking doggy. Any Air Force here?" I ask quickly.

One of the men motions slightly.

"I wouldn't have been Air Force. There's a limit."

The slurs draw smiles from the jarheads and scowls from the others.

"I can be a real asshole about being a perfectionist. Big things, little things, anything that goes wrong drives me up a wall. There's a saying that if your only tool is a hammer, every problem looks like a nail. Nickel problems, dime problems, thousand-dollar problems—I solved them all with the same hammer. A twenty-pound sledge.

"That was me. It's still me, probably always will be. There are things about being a perfectionist that appeal to me; I choose to be that way. Sometimes it wants to get out of hand, though. And, well, I'm not perfect—sometimes it does get out of hand, so I've learned to control it.

"I have a name for being a perfectionist. I call it *crewchief*."

I pick a man. "What was the price of a mistake, In Country?" I ask suddenly, then turn to another.

"What were the consequences when you did your best, and your best just wasn't good enough?

"What do you say when you're the one who has survived, and you've left four bodies to bu . . . burn. . . ?"

I knit my fingers, blow into them, then breathe deeply, until the ache of remembering passes and calm returns to my voice. Each eye meets mine as I look around the room.

"You're thinking, 'This guy is cured? This is control? Bullshit!'

"Yeah," I continue quietly, "this is control. And what I'm showing you right now, is what you're working toward.

"I'm in control because I can sit here, and discuss this mission without going completely apeshit, like I did years ago. Each of you has some memory you can't bear to talk about. This was mine. I've developed enough control that I can force my mind and voice box to get most of these memories into the open where I can deal with them. I still can't say some things, but I think that someday I will. I understand the feelings I have around the stuff that's still trapped inside. When I torture myself for leaving those dead Marines behind, I stop and think that this is being crewchief, and I did the absolute best I could.

"When I run into a nickel problem, or a thousand-dollar problem at work or at home, and I want to beat it into the ground with a twenty-pound sledge because the consequences of failure are unthinkable— because my training says that failure is unthinkable—I ask myself if I'm just being crewchief, and if I'm still trying to get those guys off that damned hill. Then I try to put down the sledge and let the nickels and dimes sort themselves out. Remember how we used to say, 'Don't sweat the small stuff' In Country, but had to sweat everything in the zone? I try to remem-

ber that I'm no longer in the zone, and that I can control being crewchief
if I try.

"Differences!" I remind.

"I no longer let job stress or family stress put me into attack mode
because of what happened on that hill. When I feel survival mode coming
on, I question why, and look for the causes that are trying to take away my
choices. I try not to take things out on my loved ones like I used to. That's
one of the real differences that happens working through PTSD. You
become a hell of a lot nicer person to be around. I function, in the World. I
have a job, and I can keep a job, and sometimes I stop being crewchief long
enough to let myself be proud that I did save three of those bastards. I have
a life, and people I love, who love me despite all the guilt I still carry about
what happened on that hill, and the other hills and valleys where I fucked
up, or didn't fuck up, or just had the bad luck or the good luck to survive.

"What I want you to understand is this. Being in control, working
through PTSD doesn't mean things stop hurting. It doesn't mean the memories go away. It doesn't mean you don't question your actions, including
your own survival. You always wish you could have done more. It means
you can live with all those things, and control the way you react to them
so they don't fuck up your life, so that you relearn how to function in the
World.

"I function, and every fucking day I think I function just a little bit better. That sounds like what a recuperating stroke victim does, doesn't it?
Well, no shit. For people who have forgotten how to build relationships
that aren't based on surviving the zone, relearning how to function in the
World is pretty accurate. Some of us are so dys*functional* that guys who cap-
tained aircraft, and led squads and companies of men, can't hold a job at a
fast-food joint.

"If you think *functioning* doesn't sound like much of an existence, I've
got some real good news for you. Do you know what happens to something that you control for a long, long time? The need diminishes. Very quietly. One day, you wake up to realize that something you had to keep at
the forefront of your mind just doesn't occur to you anymore, like going
outside in the full moon without even thinking that there was ever a problem. You just do it because it feels normal. When that happens, you don't
feel like you're merely functioning.

"You feel alive. That, my friends, is the biggest difference of all."

The vets save one gaze inward as I finish, and I think that they have listened, and felt, and heard. I have touched all but the leader. Only he main-
tains a defensive posture; I cannot touch him, nor influence, nor share even

the hope of hope unless he chooses to accept it. I am tired, and no shrink, and can offer only what I have learned. The leader's eyes still challenge, and I think, sadly, that I have given full measure in this zone and cannot reach him.

No. There is one last measure to give, for them. For me.

"I had a friend, Ron Phelps. He was the best, and he died," I begin gently.

Twenty years and more past his death. The word is no longer so hard; I have found the tears to soften it.

"He was a crewchief!" I exclaim, and the tears want to come, but as much in pride as in sadness.

"I used to think about him every day. I felt like I owed it to him, as if scavenging every detail of our friendship kept him alive. Trouble was, I always felt empty when I thought of him. It didn't seem right; the more I thought of him, the emptier I felt. Sometimes I felt guilty, sometimes angry. Sometimes I felt cheated that it wasn't me who died. Sometimes I envied him his place on The Wall.

"The sense of loss wouldn't go away. They say people who lose a limb experience phantom pain. I couldn't see how I could be whole again unless I could grow that limb back. I knew that was impossible, but the only alternative seemed to be admitting that it was gone. I couldn't do that. You know, almost every combat veteran I've met feels he lost his best friend In Country. I wonder . . . if we lost more of ourselves, the best in ourselves, when we lost these friends, than we knew.

"Differences," I say, almost in a whisper, "the last one.

"I miss him differently now. Do you know which feeling is gone? Emptiness. Anything that touched the emptiness is gone as well: the anger, the despair, the loneliness. It all faded away. All the important feelings are still there, but the emptiness is gone. I've brought the memories forward. They define a piece of who I am, not all of what I was. When I think of him, and all the men I knew, I feel full, and I feel proud of them, and me. I feel proud of us all.

"I visit Ron at the Wall; it doesn't make me sad anymore. I say, 'Hi, you old bastard,' and I tell him about my family. I think he hears me. I leave little things to pay him back for all the cigarettes I bummed, and a 'present' I made him give me. It was a Filipino butterfly knife. I still have it. For a while I considered leaving it at the Wall—that returning it would be the true test of letting him go.

"Know what? There's no hurry. It makes me feel close to him, and I don't think that's bad. I'll give it to him the next time I see him. I believe in God, and I like the idea of friends coming to lead you to the light when you die. Phelps was a crewchief! He'll cover me in that zone, like he always did. I don't have to miss him anymore."

I wipe my eyes with both hands, hard, and let myself enjoy the tiny sparkles that linger in the darkness. Flip me the finger, you bastard! They dance a long time.

I think, it must be time to go, but there is a shifting of chairs, then an uncertain whisper. I lower my hands, but cannot tell who has chosen to speak, so low are his words. I follow the questioning eyes of his fellows to find a spare man in plaid shirt and Levis with the look of sun and stream.

"Vietnam was the biggest thing in my life. The guys I knew—they were the best. Everything since then has been a bummer." He shakes his head sadly.

I nod, understanding his dismay, but feel relieved at the sound of another voice. I try to draw them into conversation.

"That feeling is hard to shake. For years I felt I'd peaked out at the ripe age of nineteen. I used to wonder how I could ever top that emergency extract, or how I could have friends as close as those I knew In Country.

"A chance conversation helped me put things into perspective. This guy had nothing to do with it!" I jab Tom, and the group snickers.

"I was hospitalized; thought my ticker was on the fritz. I shared a room with a vet, a REMF who went through some rear-area mortar and rocket attacks but saw no real combat. We couldn't sleep; I had chest pains, and he had a concussion. We told each other war stories late into the night. We must have made a pair, two middle-aged invalids with our heinies hanging out, living the old days."

Tom interrupts, "You never told me that!"

"Who says I solved all my problems in the Vet Center?" I retort, enjoying his surprise.

"I was talking about medevacs, and that as rough as those days were, everything since Vietnam had been a bummer. Know what? He said that he would give his eyeteeth to save a life, that I'd accomplished many times what thousands of people would only ever dream about. He said it in a way that made me think the war experience should be fulfilling, and not diminish the life that followed.

"A large part of working through PTSD is putting things into perspective. Remember when we came home and people told us to forget the war, and get on with our lives? They didn't understand that it couldn't be over so quickly for us. What the hell were we supposed to replace the war with? Demonstrations and a 'bad war–bad soldier' label? Were we supposed to replace the friends who enriched us with people who screamed,

'Babykiller'? With family or friends who were frightened of what they thought we'd become?"

I face the vet squarely, letting my body show the passion and energy I feel as the fatigue drops from me. It is good to be passionate; it is good to feel.

"We mustn't forget the past, but we can't live there. We have to bring it forward, so that it sustains instead of diminishes. We have to let the pride of who we were give us hope of who we can become. Never forget—most people can only fantasize about accomplishing the deeds that we remember.

"Besides, these are the '90s," I joke. "Make the most of being 'in.' Politicians wear Vietnam on their sleeves to get elected. Don't you know that was the place to be? I could make a mint if I could sell my flashbacks to the Vietnam wannabees."

The leader rises on the word. "You have flashbacks?"

"Well, I never thought of them as flashbacks." I laugh, still enjoying the joke. "I thought of them as oozebacks."

He dismisses the humor, and I see I've tripped upon his shadow.

"What are they like for you?" he demands.

"Well, I could be walking along, anywhere—work, home, church—and I'd feel a creeping sense of being in two places at the same time. I'd be 'here,' wherever that was, but 'back there' as well. It was like looking at a photograph, and instead of seeing it in your hand, being inside it, so that you can feel the space and sound of the picture all around you.

"It was very pleasant," I conclude. "Oozebacks were one of the things that went away. I'm sorry about that. I miss them."

"You enjoyed them," he leans forward, wide-eyed.

"Yeah," I answer sheepishly, "it was the ultimate in escapism. I couldn't control when they came on. That was probably for the best, otherwise I'd have lived in them all the time."

"God. I enjoy mine. I thought I was fucking crazy."

"Hey, maybe you are fucking crazy. Who says that what I think is sane? I think it makes sense that we enjoyed them, though.

"Do you remember what we felt like, back then? I sat on top of four thousand rounds and fourteen rockets, screaming out of the sky at 140 knots. Man, I felt like God."

The helicopter pilot sitting opposite nods vigorously. "We really did feel like God, didn't we? I guess I never thought about it like that; sometimes I felt invincible."

"You don't have them anymore?" the leader continues.

"No. They faded away. They got fewer and farther between, then stopped altogether. Same as the dreams. I took it as a sign that I was getting close to the end."

"How can you tell you're done?" he presses. "That's what I want to know. I am so fucking sick of coming here. I mean—there is no goddamned way I want this to go on. I don't even know why the hell I'm here anymore. All I want to know is, when is done?"

So many times, as I journeyed to the Center, long into therapy, I wondered, "When will this be over?" How many times did I harden my mind to Tom's questions, thinking, "Go to hell away"? How often did I probe this way and that, wanting to give up in anger because the barrier seemed too great?

"Pretend you're an onion," I offer simply. "Working through this stuff is like peeling an onion to get to an ache inside. Since you're not a surgeon, you find one to help. He probes the dried outer skin, and gets you to take off a few layers. It's thin, right? Removing one or two layers doesn't shrink the onion very much. Maybe you figure, nothing's happening, and want to give up, but you can't 'cause the ache is pretty deep and draws you on. You peel and peel. At times the ache will seem to go away on its own, and peeling onions isn't pleasant, so you think, 'Maybe I can stop.' Trouble is, the surgeon knows it's still there, so he pushes you. It's easy to get pissed at surgeons when they want to cut and you don't.

"Eventually, you touch wounded flesh, and it hurts a lot! You'd like anesthesia, but there isn't any. You'd like it to scab over—get all dry-skinned again. You think, as bad as the ache was, the peeling's made it worse."

I shake my head, "You can't put a layer back even if you try. Raw onions don't skin over—the flesh is too thick. Oh, they dry out a little on the outside, and maybe the inside will stop throbbing once in a while. You can pretend you're done if you want, but you're still a half-peeled onion that's going to be hurting as soon as the outer part shrivels and cracks. Nah, you're better off going on. Eventually, you hit the rotten parts and cut them out. Some of them hurt a lot, but some of them have no feeling within themselves; they just make everything around them feel lousy. It's like having an achy tooth pulled. Your mouth feels better, and the tooth doesn't feel at all.

"Eventually, you end up with a really small onion. That's when you begin to think you might be done. 'Done' with PTSD is one of these things that you can't define, but you know when you've reached it."

"What do you do with the center of the onion?" the leader asks, humor in his tone but also the need to know.

I pause; I'm good at analogies, but this one seems exhausted. I think of onions. A gardener, I carefully cull through bins at the feed store each spring, selecting well-formed sets of Stuttgart and Yellow, bulb by bulb. I never use the scoop, and my search for perfect sets pisses off the other customers. Suddenly the end is obvious.

"You plant it, and let it grow."

The leader smiles, and nods, and I think he, too, understands.

"No one can tell you you're done. Tom isn't going to tell you. He could guess, but he won't because he doesn't know what's done for you. You have to make that decision. I spoke about functioning. From one perspective, you can be done when you function in society: hold a job, be self-dependent, and control yourself well enough to do all of the above. I think that's the minimum definition of done, and it's probably not such a bad place to stop. I had to take it a little further. Tom gave me many chances to stop, but I turned them down—jarheads are pretty stubborn. Remember that bit about some things going away, and the rest you learn to control? I don't like being wrapped as tightly as I still am. For me, done is when most things have gone the way of that moon over Dong Ha. My trust button is still hot; placing confidence in my fellow man is very difficult for me. Overcoming the bitterness I feel about Vietnam isn't easy, either, and I know it's the one thing that keeps me from truly coming home. But I figure this; if I came to terms with Phelps, I can come to terms with anything."

A vet in plaid shirt and jeans captures my attention. "I get confused trying to sort out what problems are due to Vietnam and what aren't. I know I'm a real asshole about some things, but Vietnam isn't responsible for every problem I have. How the hell can you figure out when you're done if you don't even know what you're supposed to work on?"

"Man, you don't know how much aggravation you're going to avoid if you follow this bit of advice," I answer wryly. "Don't try to figure out 'what's due to Vietnam, and what isn't.' It's not important. Work on making your life right. That's what's important.

"I can't tell you how much time I wasted trying to compartmentalize Vietnam. I had the idea that everything I discussed in the Center had to have happened between 1966 and 1968. Sometimes Tom or I would touch on something from before the war, or after, and I'd say, 'This doesn't have anything to do with Vietnam.'

"Bullshit. There are two things wrong with that approach. One is that if I could clearly distinguish the things that were caused by the war from those that weren't, I wouldn't have fucking PTSD, and I wouldn't be here. The other is that it's a little stupid to become aware of something in your life that might need to be fixed, and refuse to fix it because you don't think it has anything to do with Vietnam.

"Even after I figured out that I had to work on things that weren't obviously tied to Vietnam, I wondered if I wasn't blaming every ill in my life on the war. For a long time, PTSD seemed like a cop-out for not being able to handle life. Sometimes I didn't want to talk about things because I didn't want to feel like I was whining."

I shake my head vigorously, flinging the thoughts from my head.

"You can double-think yourself to death, worrying about what to worry about.

"Look at it this way. We were people before we went to Vietnam, and we're people now. Don't try to separate who you were, who you are, and Vietnam. You are the sum of your life's experience. Trying to separate Vietnam from your life is like trying to strip colors from a rainbow. If you were ever successful, whatever you had left certainly wouldn't be a rainbow. If whatever is bothering you happened in Vietnam, work on it. If it happened before the war, work on it. If it happened today, work on it. Don't think that you're necessarily going to like what you discover about yourself. Don't think that working through PTSD is going to let you off any hooks, or make you into a wonderful person. We're talking about differences, not exoneration."

I strike a bargaining pose, as if putting forth a deal.

"Look, PTSD sucks, but I believe in turning a disadvantage into an advantage. We did it all the time in the zone, didn't we? As long as Uncle Sam is paying this guy big bucks to shrink your head, you may as well have it dry-cleaned and fucking pressed so it comes out looking better than when it went in.

"You're going to come out of this knowing more about yourself than most of the people who walk this planet ever know about themselves. Don't blow the chance to improve yourself. If you were a worthless asshole when you went to Vietnam, and you got all fucked up In Country for all the reasons we've been talking about tonight, and all you do is walk in one end of a Vet Center, concentrate on polishing up Vietnam, and then walk out the other end, do you know what you're going to be?

"You're going to be a right shiny asshole, and not worth dick as a human being.

"The last thing I have to say is this. Our wartime baggage doesn't excuse our responsibility to manage our lives. Vietnam dealt us a rotten hand. Sure, we wanted the homecoming our fathers got. I would've settled for a simple 'You done good.' Know what? That's too fucking bad. You want fair? Go to Disney World. We have to play the hand we were dealt. We have lives to rebuild, that can be rebuilt if you work at it. Remember the zones you survived. PTSD is the last one. Don't fuck it up."

The room falls silent, the last echo of passion absorbed by the air, transmuted into the flesh around me.

I have given them much to consider, said much to complicate their lives. That is good, as well. These are not stupid nor cowardly men, nor men who shrink from challenge. Such did not survive the zone. They are only lost. I have passed the thread that was passed to me, and tried to impart the will to follow it. It matters little how Gordian it weaves through their lives. They have intellect and courage; these are never lost. Pain and doubt can erode the will, but its roots, which are founded in courage and intellect, can flourish with tending.

Tom breaks the silence. "It is our custom, when a friend visits, to tell our names," he says firmly.

The formality in his voice surprises me; I have not heard it before.

I look, with the others, to the leader.

"I'm Frank Walker."

The arc of men continues, "John Petersen, Bob Kenney, Lou Yeatman. . . ."

They gather around; there are handshakes, pats on the back. A memory stirs—the flight line at Marble Mountain, bright satin in the presence of friends. I never thought I'd feel that way again.

Tom walks me to the door. I know he will spend private time with them, discussing, probing, posing quiet questions.

"Well?" I ask.

He glances at the vets. Here and there a hand waves, a smile. Tom turns to me, recognizing the old need. There are differences, but the things that make you who you are do not change.

"You done good."

To Ron,

Now we're old and grey, Fernando,
and since many years I haven't seen
   a rifle in your hand.

Can you hear the drums, Fernando,
Do you still recall the frightful night
   we crossed the Rio Grande?

I can see it in your eyes
   how proud you were to fight for
   freedom in this land.

There was something in the air, that night,
  the stars were bright, Fernando.
They were shining there for you and me,
  for liberty, Fernando.
Though we never thought that we could lose,
  there's no regret.
If I had to do the same again,
I would my friend, Fernando.

Yes, if I had to do the same again,
I would my friend, . . .

Ronny Joe.

Gotcha, you old bastard.

Glossary

The definitions in this glossary come from:

The Guidebook for Marines, circa 1965–69, oral-history tapes, exhibits, maps, and archival material in the U.S. Marine Corps Historical Center, History and Museums Division, Navy Yard, Washington, D.C.

The Vietnam collection of U.S. and enemy ordnance in the U.S. Army Ordnance Museum, Aberdeen Proving Ground, Maryland.

The glossary section of *U.S. Marines in Vietnam: Fighting the North Vietnamese—1967* by Maj. Gary L. Telfer, USMC, Lt. Col. Lane Rogers, USMC, and V. Keith Fleming, Jr.; History and Museums Division; Headquarters U.S. Marine Corps, Washington, D.C.; 1984.

Wayne Mutza, crewchief, 240th AHC, U.S. Army.

The NATOPS (Naval Air Training and Operating Procedures Standardization) Crewchief Pocket Checklist for Bell Helicopter's UH-1E (types 540 and 540-A rotor aircraft).

The author's records and recollections.

The military jargon of Vietnam varied dramatically between service organizations, theaters of combat, and time of service. Marston matting, for example, was known as perforated steel plate (PSP) in the southern Corps.

The terms in this glossary were in popular use by Marine ground and air units in northern I Corps during the author's tour (1966–68).

12.7-mm: Soviet DShK M1938 dual-purpose heavy machine gun. When mounted on a tripod in antiaircraft configuration, firing at 600 rounds per minute, it could reach an aircraft flying at 3,000 feet. 12.7-mm is the same size round as .50-caliber.

A-1E Douglas Skyraider: A propeller-driven, single-engine attack aircraft.

A-4 Douglas Skyhawk: A single-seat jet, carrier- or land-based attack aircraft.

A-6A Grumman Intruder: A twin-seat, twin-jet, all-weather, night-attack aircraft.

AAR: After-Action Report; filed by the pilot after each mission, logging the mission coordinates, crew, expended ordnance, with a terse description of the mission. Sometimes used to vent feelings against uncooperative or incompetent units or aircraft participating in the mission. (Note: Some AARs make interesting reading in the Marine Archives.)

AK-47: Russian-made Kalashnikov, gas-operated, 7.62-mm automatic assault rifle; 30-round banana-shaped magazine, range 400 meters. Standard weapon of the North Vietnamese and Viet Cong.

Ammo-minus: Out of ammunition.

Arc Light: Code name for B-52 bombing missions.

ARVN: Army of the Republic of (South) Vietnam; a South Vietnamese soldier.

Autorotate: The procedure of landing a helicopter without engine power. If a helicopter is high enough when it loses engine power, the pilot drops the collective (flattens the blade pitch, thereby decreasing wind resistance on the blades) and allows the aircraft to free-fall. The wind rushing up through the blades keeps them turning, developing enough inertia so that when the pilot raises the collective (increases the blade pitch) as he approaches the ground he can slow the aircraft and bring it to a heavy, but safe, landing. A low-flying helicopter does not have the altitude to autorotate, and instead smears itself and its crew across the earth.

BDA: Battle Damage Assessment; an estimate of the damage an aircraft has inflicted upon a zone, often in terms of percentage of blast coverage.

Blooper: M-79 grenade launcher—a 40-mm, breech-loaded, shoulder-fired weapon firing small, spin-armed grenades. Range 375 meters. Also called the "thumper."

Boonie/Boonie Hat: The boonies were anyplace in the field where you shit directly onto the ground; boonie hats were soft, floppy hats worn in the boonies. Grunts favored short-brimmed models. Air wing often wore Aussie hats with the brim snapped up.

Brown Bar: Slang for a second lieutenant.

C & C: Command and control missions, usually flown from a Huey slick.

C-4: Plastic explosive.

C-rats: C-rations, provided to the field in cartons of twelve, containing such delicacies as ham and chopped eggs, beans and franks, chicken, cookies, cocoa powder, cigarettes, toilet paper, matches, etc.

Call Signs: Each squadron had a unique radio call sign, which was changed on an irregular basis. An eavesdropping enemy would learn to associate the call sign with the type of aircraft. VMO-3 used "Scarface," "Oakgate," and "Cyclone" (SOG only). VMO-2 used "Deadlock," and VMO-6 "Klondike." The author elected to use "Scarface" for all missions flown in Vietnamese territory, although it was in fact used only for the latter half of 1967, to avoid confusion, and because the author (as well as the other aircrew of VMO-3) always thought "Oakgate" was a pussy call sign.

CH-46 Boeing Vertol Sea Knight: The tandem-rotor, jet-powered aircraft that replaced the UH-34D in Marine resupply and medical evacuation. The CH-46 was the carrier version of the Army's CH-47 "Chinook" (Shit-hook). Crewed by four, carrying up to seventeen combat troops, the -46 was armed with two .50-caliber machine guns.

Cheap Charlie: Catcall flung at Americans by kids who didn't get an adequate amount of candy, cigarettes, etc., or by a hooker unhappy with her compensation.

Collective: The control linkages on a helicopter that adjust the pitch on all blades equally, or "collectively," thereby enabling the aircraft to go up or down.

Concertina: Barbed wire, shipped wound in coils; it was easily stretched like an old fashioned concertina. Concertina wire was liberally employed on the outer perimeters. It was self-supporting and could be piled several diameters high. Automatic fire could pass through it with little damage.

CONEX: Large steel box with a single door used to transport supplies, bodies, etc. A CONEX was large enough to serve as a shelter.

Copy: Radio jargon meaning, "I understand."

CP: Command Post.

Crewchief: Enlisted man assigned to maintain a helicopter. All members of the crew except the crewchief rotated among the aircraft in the squadron. Crewchiefs carried the MOS (Military Occupational Specialty) of Mechanic, and were responsible for the overall health of the aircraft. In flight, the crewchief was in charge of the passengers, and acted as a loadmaster, machine gunner, and medic. Although not trained as pilots, most crewchiefs learned to fly while sitting "left seat" on missions in slicks.

CS and CN Gas: Aircrew carried several irritating and harassing agents in canister-grenade form. Developed during the First World War, CN gas (chloro acetophenone) is tear gas used in training and as a riot-control agent. CS gas (2-chlorobenzal malononitrile) was developed in England in the 1950s. The effects are almost immediate and range from a slight prickling sensation in the eyes and nose to severe chest pain, tears, running nose, retching, and vomiting. (Source: Edward M. Spiers. *Chemical Warfare.* 1986)

Cyclic: The control linkages on a helicopter that cause the pitch on a blade to change as the blade cycles through its arc of rotation, causing the aircraft to move in a forward, backward, or lateral direction.

DEROS: Date of Estimated Return from Overseas; thirteen months from the day of entry In Country for Marines, twelve months for all other services. However, most viewed it as the date of *expected* return.

DMZ: Demilitarized Zone; the no-man's land between the Vietnams. Both sides were supposed to remain clear of this area.

ETA: Estimated Time of Arrival.

F-4B McDonnell Phantom II: A twin-seat, twin-engine, all-weather interceptor and bomber.

FAC(A): Forward Air Controller (Airborne).

Five-by-five: Radio jargon meaning, "I read you loud and clear."

FNG: Fucking New Guy; anyone who was going home after you.

Fox-Mike: FM radio; used primarily for air-to-ground communications with infantry units or ground controllers.

Frag: Fragmentation grenade. Also "to frag": to kill someone with a grenade.

Freedom Bird: The aircraft that would take you home to the World.

Gook: Derogatory slang for any Oriental; also dink, slope, gooner, slant, zip.

Grease Gun: A .45-caliber machine gun; short-barreled, easily fired from a sling suspended around the neck, noisy, scary, and inaccurate.

Grunt: Any ground-pounding infantry man.

Gunship: An armed UH-1E used to provide fire support for ground troops. VMO-3's gunships carried four M-60s and two 7- or 19-shot, 2.75-inch folding-fin rocket pods fired by the pilot; two M-60s in a chin turret controlled by the copilot; and one M-60 for each crewchief and gunner to fire from inside the aircraft. Typical ordnance load was 4,000 rounds of 7.62-mm ball and tracer, and fourteen to thirty-eight 10- or 17-pound High Explosive Anti-Tank (HEAT) or white phosphorous (Willy Pete) rockets (used to mark targets for fixed-wing).

H & Is: Harassment and Interdiction fire; sporadic artillery fire employed to keep the enemy off balance and generally piss them off.

HEAT: High Explosive Anti-Tank 10- or 17-pound warheads carried on 2.75-inch, electrically fired air-to-ground rockets.

Heavy Shit: Any enemy presence or level of firing that exceeds "normal" combat experience.

Hill xxx: Except in rare instances, hills were identified by their elevations; therefore, Hill 918 was nine hundred and eighteen feet tall.

HMM: Helicopter, Marine Medium. The UH-34D and CH-46 were considered "medium"-sized aircraft.

Hooch: A dwelling, whether the plywood or canvas huts we lived in, or the thatched buildings in a village.

Hook, The: Two meanings, both relevant to the last day In Country: (1) "230 and a hook" was synonymous with "230 and a wake up"; (2) "The Hook" was the skyhook (Freedom Bird) that took you out of country. Sometimes pronounced *hoo* by incredibly salty short-timers.

I & E: Insertion and/or extraction of combat troops in the field. Insertions usually called for a "zone prep" by helicopters or fixed-wing to ensure there were no enemy nearby, and that the landing zone was not mined. Emergency extracts were any extractions carried out under enemy fire, and differentiated from emergency medevacs in that all friendly forces were removed because of the probability of being overrun.

I Corps: The military subdivision containing the five northern provinces of South Vietnam; from top to bottom—Quang Tri, Thua Thien, Quang Nam, Quang Tin, Quang Ngai. I Corps ranged from 50 to 100 kilometers wide and was about 300 kilometers long.

ICS: Inter-Communication System. The aircraft's internal communication system, activated by the pilots by a push button on the cyclic stick or a floor switch, and by the crewchief and gunner with a push button on a hand-held switch box on a spiral cord connected to their flight helmets. The crewchief and gunner normally held the switch in their hand during flight, but it was easily misplaced in combat.

John Wayne: To demonstrate motion-picture bravery during combat in a manner devoid of common sense, exhibiting death-wish tendencies; also a C-rat can opener.

Ka-Bar: Marine Corps combat knife issued during WWII. Though no longer issued, many men purchased their own knives before going In Country.

KIA: Killed In Action.

Klick: Kilometer.

LAW: Light Antitank Weapon; a 66-mm, one-shot rocket fired from the shoulder. Intended as a last-ditch weapon in the perimeter bunkers.

LPH: Landing Platform, Helicopter; a carrier and assault ship designed to land troops by helicopter.

LSA: Logistics Support Area.

LZ: Landing Zone; "the zone."

M-14: The standard U.S. infantry weapon during the early days in Vietnam, firing a 7.62-mm NATO cartridge from a 20-round magazine. Sustained rate of fire of 30 rounds per minute; range 460 meters.

M-16: Began replacing the M-14 as the standard U.S. infantry weapon in 1967, firing a 5.56-mm cartridge from a 20-round magazine with a sustained rate of fire of up to 15 rounds per minute; range 460 meters.

M-60: The standard U.S. light machine gun, firing a 7.62-mm NATO round. Sustained rate of fire of 100 rounds per minute; range 1,450 meters.

M-79: See "Blooper."

Main Force: Refers to organized VC or NVA battalions and regiments, rather than guerrilla groups. See "Heavy Shit."

Marston Matting: Sheets of corrugated, perforated, interlocking steel laid on the earth to serve as landing pads, air bases, etc. Also called PSP (perforated steel plate).

Med: "Medical Battalion." Medevac helicopters usually transported wounded to intermediate medical facilities staffed by U.S. Navy personnel assigned to the 1st and 3rd Medical Battalions and the 1st Hospital Company at Chu Lai, Da Nang, Phu Bai, Dong Ha, and Khe Sanh. The "meds" were usually the closest field hospitals to the zone. Charlie Co. of the 3rd Med at Khe Sanh performed emergency surgery and stabilized patients for transport to hospitals at larger bases or to the hospital ships *Repose* and *Sanctuary*.

Medevac: Medical evacuation. Routine medevacs were walking wounded or dead. Priority medevacs were seriously wounded. Emergency medevacs were men near death.

Mike-mike: Millimeter.

Mortar, 82-mm: Communist, smooth-bore mortar, single shot, 25 rounds per minute, range 3,040 meters.

Mortar, 120-mm: Communist, smooth-bore mortar, 15 rounds per minute, range 5,700 meters.

MOS: Military Occupational Specialty.

MPC: Military Payment Certificates; paper "funny money" in denominations from 5 cents to 10 dollars.

Napalm: Canisters containing 125 gallons of jellied gasoline. Developed during World War II, dropped from fighter-bombers, napalm ignited on impact and spread a torrent of flame through hundreds of yards of dense jungle that a fragmentation bomb could not penetrate.

NATOPS: Naval Air Training and Operating Procedures Standardization. The set of procedures required to operate an aircraft in flight, as well as the training regimen required of aircrew. "NATOPS" qualification was required of crewchiefs. The senior crewchief in the squadron was the "NATOPS crewchief," who trained others.

No joy: Variably, "I can't read you," "I can't make this thing work," "(Whatever) isn't going to work."

Numbah 1: Vietnamese slang for "very good."

Numbah 10: Vietnamese slang for "very bad." "Numbah Ten Thou" was the pits.

NVA: North Vietnamese Army.

O-1B: Cessna, single-engine observation aircraft; often carried seven WP rockets for marking targets.

OP: Outpost, or Observation Post.

Pickle: Gunship slang—fire all rockets as fast as possible. Also "ripple fire."

Pucker Factor: A dimensionless quantity; the ratio of the opening of an unstressed asshole to that of a stress-constricted asshole.

REMF: Rear-Echelon Motherfucker; also Pogue, Saigon Warrior.

Rock In: Gunship and fixed-wing slang—to attack with guns and rockets. Also "rolling in hot."

Rocket, 122-mm: Communist rocket with booster, battery-fired from a launch rail. The original Soviet weapon had a single motor; the VC added a second in tandem for greater range. An adapter between the motors contained a decelerometer that separated the rear motor from the second stage. So much for guerrilla warfare! The 122 was 10 feet long and weighed 170 pounds, with a range of 16–18 kilometers. Elevation and range were adjusted by moving the launching bipod; there was no optical sight. See "Heavy Shit."

RPG: Rocket-Propelled Grenade; a shoulder-fired rocket launcher used by the NVA and VC against tanks, personnel, and very low flying helicopters. After the grenade was expelled from the launcher by propellant, a rocket motor cut in to give greater range. The RPG could penetrate 10-inch armor.

S-*: Designations for functions within a unit: S-1, Personnel; S-2, Intelligence; S-3, Operations; S-4, Supply; S-5, Civil Affairs.

SAM: Soviet surface-to-air missile. Rarely used against helicopters.

Sav-A-Planes: Slang for radio reports on firing artillery missions. Aircraft entering a tactical area of responsibility requested "Sav-A-Planes" to avoid flying through them.

Sit-rep: Situation report.

Six: Code for the unit commander. "Get your six on the line."

Slick: Unarmed UH-1E used for observation, ferrying VIPs, and some emergency medevacs. Marine slicks were unarmed (unlike Army "slicks," which mounted M-60s in the doors). A Marine slick stored a single M-60 with a single can of ammo under the seat for use in aircraft defense if shot down. Two pilots and a crewchief crewed a slick, although the crewchief frequently flew in the copilot's position because of the squadron's shortage of officers.

Snakeye bomb: When providing close air support, an attacking fighter-bomber could be directly over its bomb as it exploded. "Snakes" were high-drag bombs with fins that popped open on release, acting as air brakes so that the bomb fell well behind the attacking aircraft.

SOG: Special Operations Group—clandestine operations in Laos. Marine gunships occasionally provided support for Air America Hueys and ARVN and Green Beret troops conducting operations across the border. We were required to leave identifying U.S material (dog tags, wallets, etc.) behind, and flew in aircraft from which U.S. and Marine insignia had been removed.

SOP: Standard Operating Procedure.

Sortie: An operational flight by one aircraft.

Sparrowhawk: A standby grunt reaction force ready for rapid insertion into hot zones by helicopter. Frequently identified themselves by small red patches sewn into their covers and trousers legs.

Strobe light: A hand-held, brilliantly flashing signal used to mark zones at night, or to signal rescuers.

TAC(A): Tactical Air Controller (Airborne)—an airborne observer who controlled close air support, frequently from an O-1, or a Huey slick.

TACAN, ADF: Tactical Aircraft Navigation system and the Automatic Direction Finder were two means of locating an aircraft's position relative to a broadcast radio signal. TACAN indicated direction and distance, ADF direction only.

TAOR: Tactical Area of Responsibility—the area for which a particular base, unit, or squadron was responsible. For example, the Phu Bai TAOR was an irregular region within a two to eight mile radius of the airstrip, but VMO-3's TAOR included Hue/Phu Bai, Dong Ha, Khe Sanh, and the DMZ—the top two provinces of I Corps.

TAT 101-E: Emerson Electric Tactical Armament Turret mounted two M-60 machine guns beneath the chin of a Huey, which tracked with an optical aiming device held by the copilot.

UH-1E Bell Helicopter "Huey": Although its primary mission was intended to be observation, the addition of guns and the willingness of aircrew to work closely with ground forces quickly extended its role to fire support. See Gunship and Slick for armament.

UH-34D Sikorsky Sea Horse: The Marines' primary troop-transport, resupply, and medevac helicopter until 1969. Originally designated the "HUS," the -34 was so fondly regarded by Marines that "cut me a Huss" (do a favor for me) was common slang. Crewed by four, carrying up to twelve combat soldiers, the -34 was armed with two M-60 machine guns.

UHF: Ultra High Frequency radio—used primarily for air-to-air communications with fixed-wing or other helicopters. The "Guard" UHF band was monitored by all UHF receivers, and was for emergency use by aircraft broadcasting in the blind, usually to request help. Spotters in North Vietnam would often broadcast MiG sightings on Guard to alert U.S. fighters.

VC: Viet Cong—a contraction of Vietnamese Communists. Also Charles, Charlie, Mr. Charles, Chas., Chuck.

Ville: Short for a village.

VMA: Literally, Heavier than Air, Marine Attack Squadron.

VMO: Literally, Heavier than Air, Marine Observation Squadron.

VNAF: Vietnamese Air Force.

WIA: Wounded In Action.

Willie Peter: White Phosphorous, available in 2.75-inch rocket warheads or grenades. WP rockets were frequently loaded as the first two shots in the aircraft's pods, and were used to mark targets for other aircraft to attack. The remaining shots in the pod would be loaded with HEATs.

World: The U.S., or any place that was not In Country.

About the Author

Ron Zaczek enlisted in the Marines in his hometown of Baltimore at the age of seventeen. From December 1966 to January 1968 he flew 393 combat missions over Hue, Dong Ha, and Khe Sanh, Republic of Vietnam, as a UH-1E helicopter-gunship and medevac crewchief with Marine Observation Squadron 3 (VMO-3). He was awarded nineteen Air Medals and the Bronze Star for valor.

Ron was diagnosed with Post-Traumatic Stress Disorder in 1981 and received therapy through the Vet Center Outreach Program until 1987. Begun in 1982 on the day of the dedication of "The Wall," written during and after therapy for PTSD, *Farewell, Darkness* is his first book.

Ron resides in Elkton, Maryland, with his wife, Grace, and two children, Christopher and Matthew.